Numerical Methods for Seakeeping Problems

Bettar Ould el Moctar · Thomas E. Schellin ·
Heinrich Söding

Numerical Methods for Seakeeping Problems

Springer

Bettar Ould el Moctar
Institute of Ship Technology
and Ocean Engineering
University of Duisburg-Essen
Duisburg, Germany

Thomas E. Schellin
Institute of Ship Technology
and Ocean Engineering
University of Duisburg-Essen
Duisburg, Germany

Heinrich Söding
Lüneburg, Germany

ISBN 978-3-030-62560-3 ISBN 978-3-030-62561-0 (eBook)
https://doi.org/10.1007/978-3-030-62561-0

This Springer imprint is published by the registered company Springer Nature Switzerland AG
The registered company address is: Gewerbestrasse 11, 6330 Cham, Switzerland

Preface

Predicting wave-induced ship motions and ensuing loads is often part of the design of ships, their structure, and possibly their operation. When, instead of predictions, rules are used, these are based on numerical predictions, which in turn have been verified by model experiments.

Knowledge of a ship's seakeeping is required to assess the safety of the people on board, the feasibility and efficiency of the ship for its intended purpose, and the integrity of its cargo. Today, numerical methods play a crucial role in predicting the seakeeping of ships.

Ship motions cannot be determined without accurate knowledge of the water flow around the hull. To determine this flow, most methods are based on potential theory or solving the Reynolds-averaged Navier-Stokes equations for incompressible flow. In each of these two groups of methods, the procedures differ in many details of discretization, simplifying assumptions, and implementation.

The book presents a range of numerical methods of increasing complexity. In Chap. 5, a linear (with respect to wave amplitude) strip method is presented: an efficient and widely used technique accurate enough for many practical problems. In Chaps. 6 and 7, this is followed by linear three-dimensional boundary element methods dealing with ships in regular waves. The first group of methods is based on Green functions without forward speed and includes also a nonlinear correction; the second uses Rankine sources. The latter group considers also the interaction of the stationary flow due to the ship's speed with the periodical flow in regular waves.

To further improve the accuracy in steep regular or irregular waves, effects depending nonlinearly on wave amplitude must be taken into account. The relevant published literature includes several approaches based on potential flow, which include certain, but not all nonlinear effects. For example, some include only the nonlinear Froude-Krylov force, i.e., the force produced by the wave pressure, unchanged by the body, acting up to the actual waterline. Fully nonlinear boundary element methods that account for the influence of forward speed and all substantial nonlinear effects of the waves are hardly found. In Chap. 8, the authors present such a method.

Then an overview of common methods is provided that solve the mass conservation equation and the Navier-Stokes equations for the fluid surrounding the ship (Chap. 9). The chapter includes a guidance as to the choice of approximations appropriate for predicting ship motions and loads. In addition, the associated numerical errors are elaborated, and common procedures are presented to estimate the discretization errors. Also, different methods for generating waves and for avoiding wave reflections at the outer boundary of the computational grid are discussed and illustrated.

Several additional chapters deal with empirical formulas for the effect of fins, bilge keels, sails, etc. (Chaps. 11 and 12), and, besides others, with determining hull pressure and loads in transverse cross sections (Chap. 13), and with second-order drift forces (Chap. 14). Of the latter, especially the added resistance in waves is of practical interest.

An additional Chap. 15 compares ship motions in regular waves and linear and nonlinear loads in transverse sections, determined in model experiments and computed by various numerical methods. Another Chap. 10 deals with vibrations of the elastic ship structure generated by waves, excited either continually (springing) or induced by slams of the water surface against the hull (whipping).

The book also contains a description of the basic equations for incompressible flows of Newtonian fluids and of the motion of rigid bodies (Chap. 2). In addition, there are chapters on methods for computing incompressible potential flows (Chap. 3), on wind waves and the seaway (Chap. 4), and on statistical methods for dealing with Gaussian and non-Gaussian stochastic processes (Chap. 16), which model ship responses in a natural seaway. A final Chap. 17 sketches special topics: fast roll simulation methods for intact and damaged displacement ships and for planing boats; a method aimed to simplify second-order response calculations, and a comparison of motions between common and staggered-hull catamarans (Weinblums).

The book's material is based on many years of experience in the development and application of numerical methods for computing wave-induced motions of and loads acting on ships. Part of these methods, developed by one of us (Söding), resulted in codes PDStrip, GLRANKINE, SIS, and ROLLS. These codes are just examples of several currently applied programs used to deal with practical design-related problems. In addition, the text is based on lectures at the Technical University Hamburg and the University of Duisburg-Essen given by the authors. Of same importance are the many years of professional experience of the authors' affiliation with a classification society, and in providing services to the maritime industry.

The book is aimed at master's and doctoral students and engineers in practice. The authors thank Dr. Udo Lantermann, Guillermo Chillcce, and Dirk Hünniger for their support in generating figures.

Duisburg, Germany — Bettar Ould el Moctar
Duisburg, Germany — Thomas E. Schellin
Lüneburg, Germany — Heinrich Söding

Contents

Chapter 1
Introduction

Abstract This chapter shows that total losses of seagoing ships decrease worldwide at an impressive rate: from about 150 to roughly 50 per year within the last 15 years. However, the share of losses caused by adverse weather conditions rises slowly and amounts now to about 50%, thus showing the relevance of this book's topic.

The primary purpose of computing motions and loads of ships in a seaway is to assure the safety of persons on board, the integrity of the ship and, if present, its cargo, and to improve its performance and efficiency. Excessive motions may not only shift cargo or cause damage from loosened deck containers or equipment, but may even cause dangerously large heel angles and capsizing. Furthermore, ship motions affect the comfort of persons on board, leading to sea sickness or, in extreme cases, to render it impossible for the crew to accomplish the ship's mission. Knowledge of wave-induced loads is necessary to assess the integrity of the ship's structure. Most important for this are vertical and horizontal bending moments, torsional moments, and sometimes shear forces in transverse sections of the hull girder. Wave-induced local pressure acting on the hull determines the necessary strength of plates, stiffeners, and web frames. Furthermore, steady wave- and wind-induced forces and moments should not prevent the ship from arbitrary course changes and from making some speed ahead.

Due to continuous effort for improving ship safety, the number of total losses of ships per year is decreasing (Fig. 1.1) in spite of the increase in the number of ships worldwide. The decreasing trend is present for centuries. However, according to the International Union of Marine Insurance [1], the percentage of ship losses due to severe weather conditions is increasing and is, at present, approximately 50% (Fig. 1.2). In this figure, also part of the losses due to hull damage and perhaps grounding are caused by heavy seaway.

The marine sector is unique in organizing the safety of ships and marine structures on a private basis by classification societies. They edit and supervise rules ships and marine structures have to fulfill to be 'classed'. Traditionally, the rules are the result of experiences with casualties as well as theoretical and experimental studies including full-scale measurements. These rules are adequate for most ships. Since about 60 years, computational methods have been used to improve and extend the rules related

B. O. el Moctar et al., *Numerical Methods for Seakeeping Problems*,
https://doi.org/10.1007/978-3-030-62561-0_1

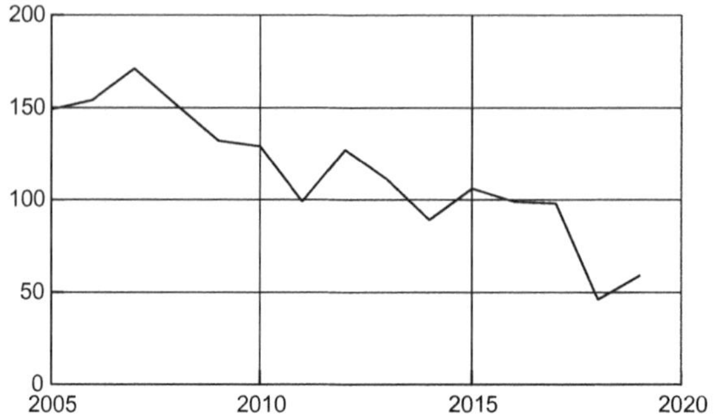

Fig. 1.1 Worldwide total losses of seagoing ships over 100 gross tons per year. *Source* Statista

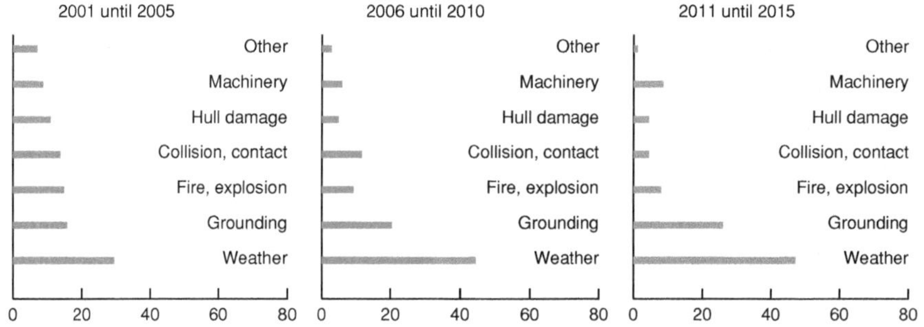

Fig. 1.2 Percentage of total losses of ships exceeding 500 GT by different causes between 2001 and 2015. *Source* [1]

to wave loads and seakeeping, and to investigate wave responses for newbuilds that differ substantially from those for which the rules were prepared.

Reliable load predictions enable reducing the allowances specified to account for uncertainties of loads and structural strength. It may be advantageous to apply advanced, possibly costly computations to reduce a ship's scantlings or the probability of structural failures. Regarding ship motions, numerical simulations may help to estimate the probability of excessive motions and accelerations. This may help to extend the safe limits of metacentric height.

In principle, wave-induced motions and loads can be assessed by using full-scale measurements, model tests, and numerical methods. Full-scale measurements are possible only if the ship, or a similar one, has been built already. They are expensive; the wave conditions cannot be controlled; and assessing the wave conditions during the measurements with the required accuracy is usually impossible. Model test results can be converted to full-scale data except for the influence of viscosity, which is small in most cases. More important is the limited size of the model basin, the degree of sophistication of the equipment of the test facility, and cost and time to perform such experiments. In irregular seaways, long test runs are required to obtain representative results. Thus, for seakeeping model, experiments are used today mostly to validate numerical methods. An exception is the sloshing of fluids in tanks, where small-scale effects like wave breaking and the collapse of bubbles may be important for practical questions, but are difficult or impossible to simulate accurately.

Currently, in designing and approving new ships, often a linear strip or panel method is used to determine motions and structural hull girder loads for a ship advancing at constant forward speed in small amplitude regular waves under various combinations of wave frequency and heading. For any seaway described by a wave spectrum, the results are combined to obtain root mean square values of loads extrapolated linearly over wave amplitude. Results for different seaway conditions are then combined to a long-term probability distribution of loads. For suitably selected design conditions, nonlinear corrections to the linear loads can be applied. If more accuracy is required, solvers for Navier-Stokes or Euler equations may be applied, which take into account the water/air interface. Today, using such a code is the obvious choice to compute free-surface waves around the ship including breaking waves, sprays, and air trapping: phenomena that should be considered to predict slamming loads in severe seas accurately.

A full understanding and an accurate prediction of hydrodynamic wave body interactions is challenging. The associated nonlinear effects become critical when large-amplitude body motions and/or high surface waves are involved. Indeed, no numerical method is yet capable to perform accurate, fully nonlinear seakeeping calculations during the required long-time simulations. The objective of this book is to present the current computational methods for seakeeping problems suitable for numerical predictions of acceptable accuracy. Among these methods is a new, fully nonlinear Rankine panel method as well as an advanced technique for avoiding wave reflections for arbitrary wave encounter angles when using a Navier-Stokes equations solver.

Reference

1. IUMI. International Union of Marine Insurance (2016), http://www.iumi.com/

Chapter 2
Fundamental Governing Equations

Abstract In this chapter, the conservation equations for mass and momentum are derived to describe the flow of Newtonian fluids. Subsequently, the basic equations for incompressible potential flows, namely, the continuity equation (in the form of the Laplace equation) and the Bernoulli equation, are described. Finally, the equations of motion for rigid bodies are derived.

2.1 Governing Equations of Fluid Flow

Fluids are substances that do not resist external shear forces. This definition includes not only liquids, but also gases. Fluids are treated here as continua, neglecting the motions of their molecules. Furthermore, we confine ourselves to Newtonian fluids, i.e., we consider only mechanical forces and assume that the shear stress in the fluid is proportional to the shear rate.

Flows of Newtonian fluids can be described by four fields (physical quantities depending on space): density ρ, velocity $\mathbf{v}$, pressure p, and temperature T. Three equations that govern the flow are obtained from conservation principles:

- Continuity equation (mass conservation)
- Navier-Stokes equation (momentum conservation)
- Heat capacity equation (energy conservation)

Furthermore, there is an equation of state, which relates pressure, density, and temperature to each other and which is specific for the fluid material.

The heat capacity equation and the equation of state are only required when computing the flow of compressible fluids. Because these occur only exceptionally in ship technical flows, they will not be discussed here.

2.1.1 Conservation Principles

Extensive quantities for which a conservation equation exists, here mass, momentum, and energy, depend on the amount of matter, whereas the intensive quantities density

B. O. el Moctar et al., *Numerical Methods for Seakeeping Problems*,
https://doi.org/10.1007/978-3-030-62561-0_2

(ρ), velocity (v), and temperature (T) are the corresponding extensive quantities per volume or mass. If Φ is an extensive quantity, and ϕ is the corresponding intensive quantity per mass, we have

$$\Phi(t) = \int_{V_{\text{CM}}} \rho(\mathbf{r}, t)\, \phi(\mathbf{r}, t) dV, \tag{2.1}$$

where $\mathbf{r}$ is the position vector to a point in the fluid, and V_{CM} is the volume occupied by the 'control mass' CM, i.e., an arbitrarily selected subset of the fluid particles. The boundary of V_{CM} is moving with the local velocity $\mathbf{v}(\mathbf{r}, t)$ of the fluid.

For numerical calculations of fluid flow, it is preferred to have 'control volumes' CV, the boundaries of which can move with arbitrary velocity $\mathbf{v}_b$ (including $\mathbf{v}_b = 0$ for stationary volumes). In the following, it is assumed that, for each CM, we have a corresponding CV, and that at time t both volumes are the same. Then, for any conserved quantity Φ, the following relation holds between integrals over CM and CV:

$$\frac{d\Phi}{dt} = \underbrace{\frac{d}{dt}\int_{V_{\text{CM}}} \rho\phi dV}_{\text{Rate of Change of } \Phi \text{ in CM}} = \underbrace{\frac{d}{dt}\int_{V_{\text{CV}}} \rho\phi dV}_{\text{Rate of Change in CV}} + \underbrace{\int_{S_{\text{CV}}} \rho\phi\, (\mathbf{v} - \mathbf{v}_b) \cdot \mathbf{n}\, dS}_{\text{Net flux of } \Phi \text{ through } S_{\text{CV}}}, \tag{2.2}$$

where
$V_{\text{CV}} = V =$ volume of the CV,
$S_{\text{CV}} = S =$ surface enclosing the CV, and
$\mathbf{n}$ = unit vector $\perp$ to S directed outwards.
The index CV will be omitted in the following.

2.1.2 Mass Conservation (Continuity Equation)

Using $\phi \equiv 1$ and $\mathbf{v}_b \equiv 0$ in (2.1) and (2.2) gives an equation for conservation of fluid mass M within V_{CM}:

$$\frac{dM}{dt} = \frac{d}{dt}\int_{V_{\text{CM}}} \rho\, dV = \frac{\partial}{\partial t}\int_V \rho\, dV + \int_S \rho\mathbf{v} \cdot \mathbf{n}\, dS = 0. \tag{2.3}$$

We apply the Gauss theorem

$$\int_S f(\mathbf{r})\mathbf{n}\; dS = \int_V \nabla f(\mathbf{r}) dV, \tag{2.4}$$

which holds for continuously differentiable scalar functions f, to the surface integral in (2.3), using each component of $\mathbf{v}$ as f and adding the result. This transforms the mass conservation equation into

$$\frac{\partial}{\partial t}\int_V \rho\, dV + \int_V \nabla \cdot (\rho \mathbf{v})\, dV = 0. \tag{2.5}$$

Because V is fixed in space, this can be written as

$$\int_V \left(\frac{\partial \rho}{\partial t} + \nabla \cdot (\rho \mathbf{v}) \right) dV = 0.$$

In the limit of V shrinking to a point, this results in the continuity equation (mass conservation) as partial differential equation:

$$\frac{\partial \rho}{\partial t} + \nabla \cdot (\rho \mathbf{v}) = 0. \tag{2.6}$$

2.1.3 Momentum Conservation (Navier-Stokes Equation)

Using for ϕ the components of the velocity $\mathbf{v}$, we obtain from (2.1) the definition of the momentum

$$\mathbf{P} = \int_{V_{\text{CM}}} \rho \mathbf{v}\, dV. \tag{2.7}$$

Newton's second law states

$$\frac{d}{dt}\mathbf{P} = \sum_i \mathbf{F}_i,$$

where the $\mathbf{F}_i$ are all external forces acting on the CM. Applying (2.2) with $\mathbf{v}_b \equiv 0$ for each component v_j of $\mathbf{v}$ gives

$$\frac{\partial}{\partial t}\int_V \rho \mathbf{v}\, dV + \int_S \rho \underbrace{(\mathbf{v}\mathbf{v}) \cdot \mathbf{n}}_{v_j v_k n_k}\, dS = \sum_i \mathbf{F}_i. \tag{2.8}$$

The term $v_j v_k = \mathbf{v}\mathbf{v}$ is the outer product of the velocity $\mathbf{v} = (v_1, v_2, v_3)^T$ with itself:

$$v_j v_k = \mathbf{v}\mathbf{v} = \begin{pmatrix} v_1\, v_1 & v_1\, v_2 & v_1\, v_3 \\ v_2\, v_1 & v_2\, v_2 & v_2\, v_3 \\ v_3\, v_1 & v_3\, v_2 & v_3\, v_3 \end{pmatrix}. \tag{2.9}$$

The external force $\sum_i \mathbf{F}_i$ contains surface forces, which act on S, and mass forces acting on the fluid inside the CV.

Surface forces are forces, which must be applied at the imagined dissection of the fluid inside the CV from that outside of it, to take account of the internal stress tensor $\mathbf{T}$:

$$\mathbf{F}_S = \int_S \mathbf{T} \cdot \mathbf{n} \, dS.$$

For Newtonian fluids, the stress tensor depends on the pressure p and on the velocity field $\mathbf{v}$:

$$\mathbf{T} = -\mathbf{E}\, p + 2\,\mu\, \mathbf{D} + \frac{2}{3}\,\mu\,(\nabla \cdot \mathbf{v})\mathbf{E} \quad \text{with} \quad \mathbf{D} = \frac{1}{2}\left[\nabla \mathbf{v} + (\nabla \mathbf{v})^{\mathsf{T}}\right], \tag{2.10}$$

Where $\mathbf{D}$ denotes the rate of the strain tensor, $\mathbf{E}$ the unit tensor, p the pressure, and μ the dynamic viscosity.

In index notation using Cartesian coordinates, this is

$$T_{ij} = -p\delta_{ij} + 2\mu D_{ij} + \frac{2}{3}\mu\delta_{ij}\frac{\partial v_k}{\partial x_k} \quad \text{with} \quad D_{ij} = \frac{1}{2}\left(\frac{\partial v_i}{\partial x_j} + \frac{\partial v_j}{\partial x_i}\right),$$

where δ_{ij} is the Kronecker tensor: $\delta_{ij} = 1$ for $i = j$, otherwise 0.

The weight of the fluid causes the mass force

$$\mathbf{F}_g = \int_V \rho\,\mathbf{g}\, dV,$$

where $\mathbf{g}$ is the gravity acceleration vector. In rotating or accelerating reference systems, further mass forces must be added. In the following, an inertial reference system is presumed.

Taking together the above momentum equations give

$$\frac{\partial}{\partial t}\int_V \rho\mathbf{v}\, dV + \underbrace{\int_S \rho\mathbf{v}\mathbf{v} \cdot \mathbf{n}\, dS}_{\text{convective flux of momentum}} = \underbrace{\int_S \mathbf{T} \cdot \mathbf{n}\, dS}_{\text{diffusive flux of momentum}} + \underbrace{\int_V \rho\mathbf{g}\, dV}_{\text{source term}}, \tag{2.11}$$

The surface integrals can be transformed to volume integrals using the Gauss theorem (2.4) for each tensor component:

$$\frac{\partial}{\partial t}\int_V \rho\mathbf{v}\, dV + \underbrace{\int_V \nabla \cdot (\rho\mathbf{v}\mathbf{v})\, dV}_{\text{convective flux of momentum}} = \underbrace{\int_V \nabla \cdot \mathbf{T}\, dV}_{\text{diffusive flux of momentum}} + \underbrace{\int_V \rho\mathbf{g}\, dV}_{\text{source term}}. \tag{2.12}$$

Because V does not move, the differential operator and the integral may be interchanged in the first term. Taking again the limit of V shrinking into a point results in the Navier-Stokes equation in inertial coordinate systems:

$$\frac{\partial}{\partial t}(\rho\mathbf{v}) + \nabla \cdot (\rho\mathbf{v}\mathbf{v}) = \nabla \cdot \mathbf{T} + \rho\mathbf{g}. \tag{2.13}$$

In some cases, the viscosity of the fluid can be neglected. Then the stress tensor **T** in (2.10) simplifies to

$$\mathbf{T} = -\mathbf{E}p, \tag{2.14}$$

and

$$\nabla \cdot \mathbf{T} = -\nabla p. \tag{2.15}$$

If this simplified **T** is used in (2.13), the equation is called Euler equation.

2.1.4 Flow of Incompressible Fluids

Fluids can be approximated as incompressible if the maximum flow velocity is less than about 0.3 times the velocity of sound waves. That condition is usually satisfied in the flow of air and water around ships; an exception may be water-air mixtures treated as a single substance.

If the fluid density ρ is constant over time and space, the continuity Eq. (5.1) becomes

$$\nabla \cdot \mathbf{v} = 0, \tag{2.16}$$

and the Navier-Stokes equation (2.13) simplifies to

$$\frac{\partial \mathbf{v}}{\partial t} + \nabla \cdot (\mathbf{v}\mathbf{v}) = \nabla \cdot \mathbf{T}/\rho + \mathbf{g}. \tag{2.17}$$

2.1.5 Potential Flow

The rotation of the velocity field is defined as

$$\text{rot } \mathbf{v} = \nabla \times \mathbf{v} = (v_{3y} - v_{2z},\ v_{1z} - v_{3x},\ v_{2x} - v_{1y}), \tag{2.18}$$

where indices x, y, z designate partial derivatives with respect to the Cartesian space coordinates x, y, z. If an inviscid fluid motion starts from rest, rot **v** will remain zero. Therefore, in many real flows rot $\mathbf{v} = 0$ is a reasonable approximation. Flows with rot $\mathbf{v} = 0$ everywhere within the flow field are called potential flows because their velocity field $\mathbf{v}(x, y, z, t)$ can be expressed as the gradient of a scalar function $\phi(x, y, z, t)$ called flow potential:

$$\mathbf{v} = \nabla\phi. \tag{2.19}$$

If ϕ is twofold continuously differentiable, the condition rot $\mathbf{v} \equiv 0$ is satisfied identically:

$$\text{rot}\,\mathbf{v} = (\phi_{zy} - \phi_{yz},\ \phi_{xz} - \phi_{zx},\ \phi_{yx} - \phi_{xy}) = (0, 0, 0). \tag{2.20}$$

For a potential flow of an incompressible fluid, the continuity equation follows from (2.16) as

$$\nabla \cdot (\nabla\phi) = (\nabla \cdot \nabla)\phi = \Delta\phi = \phi_{xx} + \phi_{yy} + \phi_{zz} = 0. \tag{2.21}$$

This second-order linear differential equation for ϕ is called the Laplace equation.

For a potential flow of an incompressible fluid, also the momentum conservation (2.17) with (2.15),

$$\frac{\partial \mathbf{v}}{\partial t} + \nabla \cdot (\mathbf{v}\mathbf{v}) = -\nabla p/\rho + \mathbf{g}, \tag{2.22}$$

can be simplified: If $\dot{\phi}$ is the partial time derivative of the potential, g is the scalar gravity acceleration, and the z coordinate is directed downward, one obtains

$$\nabla\dot{\phi} + \nabla \cdot (\mathbf{v}\mathbf{v}) = -(\nabla p)/\rho + \nabla(gz). \tag{2.23}$$

The second term on the left-hand side is

$$\nabla \cdot (\mathbf{v}\mathbf{v}) = \nabla \cdot \begin{pmatrix} \phi_x\phi_x & \phi_x\phi_y & \phi_x\phi_z \\ \phi_y\phi_x & \phi_y\phi_y & \phi_y\phi_z \\ \phi_z\phi_x & \phi_z\phi_y & \phi_z\phi_z \end{pmatrix} \tag{2.24}$$

$$\begin{aligned} = \ & (\phi_{xx}\phi_x + \phi_{yy}\phi_x + \phi_{zz}\phi_x,\ \phi_{xx}\phi_y + \phi_{yy}\phi_y + \phi_{zz}\phi_y,\ \phi_{xx}\phi_z + \phi_{yy}\phi_z + \phi_{zz}\phi_z) \\ & +(\phi_x\phi_{xx} + \phi_y\phi_{xy} + \phi_z\phi_{xz},\ \phi_x\phi_{yx} + \phi_y\phi_{yy} + \phi_z\phi_{yz},\ \phi_x\phi_{xz} + \phi_y\phi_{yz} + \phi_z\phi_{zz}) \end{aligned} \tag{2.25}$$

The upper line of (2.25) is zero because of the continuity Eq. (2.21), while the lower line is one half of the expression

$$\begin{aligned} \nabla(\mathbf{v} \cdot \mathbf{v}) &= \nabla(\phi_x^2 + \phi_y^2 + \phi_z^2) \\ &= 2\phi_x\phi_{xx} + 2\phi_y\phi_{xy} + 2\phi_z\phi_{xz},\ 2\phi_x\phi_{xy} + 2\phi_y\phi_{yy} + 2\phi_z\phi_{yz}, \\ &\quad 2\phi_x\phi_{xz} + 2\phi_y\phi_{yz} + 2\phi_z\phi_{zz}. \end{aligned} \tag{2.26}$$

Thus

$$\nabla \cdot (\mathbf{v}\mathbf{v}) = \frac{1}{2}\nabla(\mathbf{v} \cdot \mathbf{v}) = \frac{1}{2}\nabla|\mathbf{v}|^2. \tag{2.27}$$

Inserting this into (2.23) gives

$$\nabla\left(\dot{\phi} + \frac{1}{2}|\mathbf{v}|^2 + \frac{p}{\rho} - gz\right) = 0. \tag{2.28}$$

From this follows the Bernoulli equation for incompressible potential flows:

$$\dot{\phi} + \frac{1}{2}|\mathbf{v}|^2 + \frac{p}{\rho} - gz = C(t), \tag{2.29}$$

in which C may depend on time, but not on position within the fluid. In potential flows, this scalar equation substitutes the vector equation for momentum conservation in more general flows.

2.2 Rigid Body Motions

Ships are elastic bodies. Their flexibility has a considerable influence on the stresses caused by wave loads. To compute these stresses, ship motions are superimposed from rigid body motions and elastic deformations. The latter are discussed in Chap. 10; here, the equations for rigid body motions are elaborated.

2.2.1 Coordinate Systems

We introduce an inertial Cartesian right-handed coordinate system $\mathbf{O}(x, y, z)$ to describe the motions of the ship and the fluid surrounding it. Its z-axis is directed downward. Position vectors expressed in inertial coordinates are $\mathbf{x}$; especially the center of gravity G of the ship is located at $\mathbf{x}_G = (x_G, y_G, z_G)^T$.

Sometimes, especially for ships sailing in regular waves, an inertial coordinate system is chosen that follows the mean forward speed of the ship. Then $\mathbf{x}_G$ specifies the translational motion of the ship (at its center of gravity), excluding the steady forward speed. The longitudinal motion $x_G(t)$ is then called surge, the transverse motion $y_G(t)$ is called sway, and the vertical motion $z_G(t)$ is called heave; however, sometimes the ship-fixed reference point for these motions is chosen differently from G. On the other hand, to describe maneuvering motions, the inertial system should be either fixed to the earth, or it should follow the constant stream of water not disturbed by the ship.

A second Cartesian coordinate system $\mathbf{o}(\underline{x}, \underline{y}, \underline{z})$ is fixed to the ship's body. Its $\underline{x}$ axis is directed forward, its $\underline{y}$-axis to starboard, and its $\underline{z}$-axis normal to the plane decks downward. Its origin is at the center of gravity G (see Fig. 2.1). It is used to describe the hull shape and other quantities (for instance, the mass distribution) which are constant in ship-fixed, but not in inertial coordinates.

Position vectors of the same point, expressed (a) in ship-fixed coordinates as $\underline{\mathbf{x}}$ and (b) in inertial coordinates as $\mathbf{x}$, are related by

$$\mathbf{x} = \mathbf{x}_G + \mathbf{T}\underline{\mathbf{x}}. \tag{2.30}$$

Here, the 3 by 3 matrix $\mathbf{T}$ takes account of the different directions of the inertial and the ship-fixed coordinates.

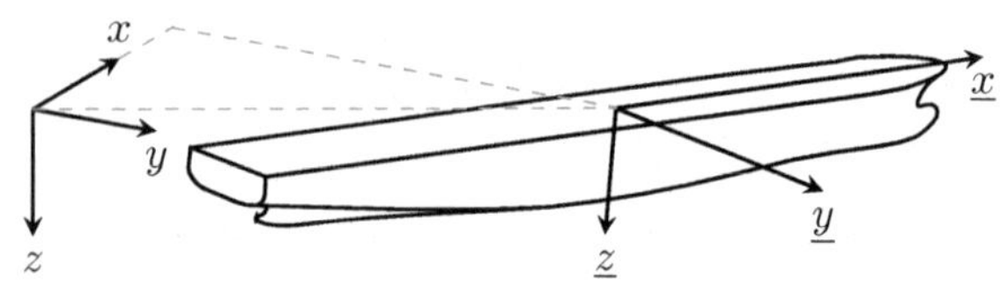

Fig. 2.1 Coordinate system used to describe ship motions

The ship's inclination (compared to the upright floating condition) about the longitudinal axis is the heel or roll angle φ; it is defined positive for a right-hand rotation around $\underline{x}$. Correspondingly, the trim or pitch angle θ and the course or yaw angle ψ describe the rotations about the transverse and vertical axes, respectively. In case of a combined rotation around all three axes, the ship is imagined to be starting from the initial orientation $\varphi = \theta = \psi = 0$, then rotating about the x-axis by φ, then about the inertial y-axis by θ, and at last about the inertial z-axis by ψ. This sequence of rotations allows determine the matrix $\mathbf{T}$ in (2.30): The heel rotation corresponds to a left multiplication of $\underline{\mathbf{x}}$ by the matrix

$$\mathbf{T}_\varphi = \begin{pmatrix} 1 & 0 & 0 \\ 0 & \cos\varphi & -\sin\varphi \\ 0 & \sin\varphi & \cos\varphi \end{pmatrix}, \tag{2.31}$$

and the trim and course rotations, correspondingly, by left multiplications by

$$\mathbf{T}_\theta = \begin{pmatrix} \cos\theta & 0 & \sin\theta \\ 0 & 1 & 0 \\ -\sin\theta & 0 & \cos\theta \end{pmatrix} \quad \text{and} \quad \mathbf{T}_\psi = \begin{pmatrix} \cos\psi & -\sin\psi & 0 \\ \sin\psi & \cos\psi & 0 \\ 0 & 0 & 1 \end{pmatrix}, \tag{2.32}$$

respectively. Thus, the matrix $\mathbf{T}$ combining the three rotations is

$$\begin{aligned} \mathbf{T} &= \mathbf{T}_\psi \mathbf{T}_\theta \mathbf{T}_\varphi \\ &= \begin{pmatrix} \cos\psi\cos\theta & \cos\psi\sin\theta\sin\varphi - \sin\psi\cos\varphi & \cos\psi\sin\theta\cos\varphi + \sin\psi\sin\varphi \\ \sin\psi\cos\theta & \sin\psi\sin\theta\sin\varphi + \cos\psi\cos\varphi & \sin\psi\sin\theta\cos\varphi - \cos\psi\sin\varphi \\ -\sin\theta & \cos\theta\sin\varphi & \cos\theta\cos\varphi \end{pmatrix} \end{aligned} \tag{2.33}$$

The rows of matrix $\mathbf{T}$ are pairwise orthogonal; thus, the inverse of $\mathbf{T}$ is its transpose:

$$\mathbf{T}^{-1} = \mathbf{T}^T. \tag{2.34}$$

Table 2.1 summarizes the motion components of ships.

Table 2.1 Definition of ship motions

DOF	Name	Description	Velocity	Position, Euler angle
1	Surge	Translation in x direction	u	x
2	Sway	Translation in y direction	v	y
3	Heave	Translation in z direction	w	z
4	Roll	Rotation around x-axis	p	φ
5	Pitch	Rotation around y-axis	q	θ
6	Yaw	Rotation around z-axis	r	ψ

2.2.2 *Kinematics*

To simplify the following, the so-called Euler angles φ, θ, and ψ are combined to a pseudo-vector $\boldsymbol{\alpha} = (\varphi, \theta, \psi)^T$. Its time derivative $\dot{\boldsymbol{\alpha}}$ is related to the angular velocity $\boldsymbol{\omega}$ expressed in inertial coordinates:

$$\boldsymbol{\omega} = \mathbf{R}\dot{\boldsymbol{\alpha}}. \tag{2.35}$$

The 3 by 3 matrix $\mathbf{R}$ follows from the relation (2.30) applied to a body rotating about G with angular velocity $\boldsymbol{\omega}$. Ship-fixed points move then with velocity

$$\dot{\mathbf{x}} = \boldsymbol{\omega} \times \underbrace{(\mathbf{x} - \mathbf{x}_G)}_{\mathbf{T}\underline{\mathbf{x}}} = \dot{\mathbf{T}}\underline{\mathbf{x}}. \tag{2.36}$$

With (2.35) follows

$$\mathbf{R}\dot{\boldsymbol{\alpha}} \times \mathbf{T}\underline{\mathbf{x}} = \dot{\mathbf{T}}\underline{\mathbf{x}} \tag{2.37}$$

for arbitrary vectors $\underline{\mathbf{x}}$. Using (2.33), it is easily verified for three linearly independent vectors $\underline{\mathbf{x}}$, e.g., for $(1, 0, 0)^T$, $(0, 1, 0)^T$, and $(0, 0, 1)^T$, that (2.37) is satisfied if

$$\mathbf{R} = \begin{pmatrix} \cos\psi\cos\theta & -\sin\psi & 0 \\ \sin\psi\cos\theta & \cos\psi & 0 \\ -\sin\theta & 0 & 1 \end{pmatrix}. \tag{2.38}$$

To determine $\dot{\boldsymbol{\alpha}}$ from $\boldsymbol{\omega}$, we need the inverse of $\mathbf{R}$:

$$\mathbf{R}^{-1} = \begin{pmatrix} \cos\psi/\cos\theta & \sin\psi/\cos\theta & 0 \\ -\sin\psi & \cos\psi & 0 \\ \tan\theta\cos\psi & \tan\theta\sin\psi & 1 \end{pmatrix}. \tag{2.39}$$

It exists because, in practice, $|\theta|$ will be less than 90°.

2.2.3 Motion Equations

Rigid body translations are governed by Newton's second law

$$m\ddot{\mathbf{x}}_G = \mathbf{F}, \tag{2.40}$$

where m is ship mass, and $\mathbf{F}$ is the total force acting on the ship. Further terms would occur if the body motion were defined not as that of the center of gravity $\mathbf{x}_G$, but for any other point. Here, we prefer to use G as a reference point; after translations at G and rotations have been determined, the motion of any other point P is easily determined from (2.30).

Rigid body rotations are governed by the equation

$$\frac{d}{dt}(\mathbf{I}\boldsymbol{\omega}) = \mathbf{I}\dot{\boldsymbol{\omega}} + \boldsymbol{\omega} \times \mathbf{I}\boldsymbol{\omega} = \mathbf{M}, \tag{2.41}$$

where $\mathbf{I}\boldsymbol{\omega}$ is the angular momentum of the body. Here, $\mathbf{M}$ is the external moment about G acting on the body and $\mathbf{I}$ is its inertia matrix referring to G, both expressed in inertial coordinates. The inertia matrix is constant in ship-fixed coordinates if all mass items m_i belonging to the ship and its load do not move relative to the ship:

$$\underline{\mathbf{I}} = \begin{pmatrix} \underline{I}_{xx} & -\underline{I}_{xy} & -\underline{I}_{xz} \\ -\underline{I}_{yx} & \underline{I}_{yy} & -\underline{I}_{yz} \\ -\underline{I}_{zx} & -\underline{I}_{zy} & \underline{I}_{zz} \end{pmatrix} = \begin{pmatrix} \sum_i m_i(\underline{y}_i^2 + \underline{z}_i^2) & -\sum_i m_i \underline{x}_i \underline{y}_i & -\sum_i m_i \underline{x}_i \underline{z}_i \\ -\sum_i m_i \underline{y}_i \underline{x}_i & \sum_i m_i(\underline{x}_i^2 + \underline{z}_i^2) & -\sum_i m_i \underline{y}_i \underline{z}_i \\ -\sum_i m_i \underline{z}_i \underline{x}_i & -\sum_i m_i \underline{z}_i \underline{y}_i & \sum_i m_i(\underline{x}_i^2 + \underline{y}_i^2) \end{pmatrix}. \tag{2.42}$$

To apply this ship-fixed inertia matrix in (2.41), we change its right-hand factors in (2.41) to ship-fixed coordinates by left multiplication with $\mathbf{T}^{-1} = \mathbf{T}^T$, then multiply with $\underline{\mathbf{I}}$, and then change the result back to inertial coordinates by left multiplication with $\mathbf{T}$. From the associative law for matrix products follows that this procedure is equivalent to using the relation

$$\mathbf{I} = \mathbf{T}\underline{\mathbf{I}}\mathbf{T}^T. \tag{2.43}$$

Equations (2.40) and (2.41) are used to determine the linear and angular accelerations:

$$\ddot{\mathbf{x}}_G = \mathbf{F}/m \quad \text{and} \quad \dot{\boldsymbol{\omega}} = \mathbf{I}^{-1}(\mathbf{M} - \boldsymbol{\omega} \times \mathbf{I}\boldsymbol{\omega}). \tag{2.44}$$

To compute the body motions, the linear and angular accelerations $\ddot{\mathbf{x}}_G$ and $\dot{\boldsymbol{\omega}}$ are determined from $\mathbf{F}$ and $\mathbf{M}$ using (2.44). Any method of integrating ordinary differential equations may then be used to determine $\dot{\mathbf{x}}_G$ and $\boldsymbol{\omega}$. The latter quantity is transformed to $\dot{\boldsymbol{\alpha}} = \mathbf{R}^{-1}\boldsymbol{\omega}$ using (2.39). $\dot{\mathbf{x}}_G$ and $\dot{\boldsymbol{\alpha}}$ are again numerically integrated to obtain $\mathbf{x}_G$ and $\boldsymbol{\alpha}$.

In bodies which are partly or fully immersed in a fluid, the fluid force depends on the initially unknown linear and angular acceleration of the body. For one-dimensional motion, the derivative $-dF/d\ddot{x}$ is called added mass; for six-dimensional motion (three components each of translation and rotation), there is, instead, a 6 by 6 added mass matrix **A** that depends on the actual immersion and on the body orientation:

$$\begin{pmatrix} \mathbf{F} \\ \mathbf{M} \end{pmatrix} = -\mathbf{A} \begin{pmatrix} \ddot{\mathbf{x}}_G \\ \dot{\boldsymbol{\omega}} \end{pmatrix} + \begin{pmatrix} \mathbf{F}_r \\ \mathbf{M}_r \end{pmatrix}, \tag{2.45}$$

where $\mathbf{F}_r$ and $\mathbf{M}_r$ are the rest force and rest moment, respectively, after subtracting the acceleration-dependent fluid force and moment contributions.

Various methods may be applied to resolve the difficulty of **F** and **M** depending on the initially unknown acceleration:

- Equation (2.44) is solved by iteration using under-relaxation, starting with the accelerations found in the previous time step. This is often done if an iteration is required anyway to determine **F** and **M**; both iterations should be included in the same loop. However, erratic acceleration and force histories may result from this procedure, and for light bodies in heavy fluids (for instance, airplanes ditching in water in case of emergency), small under-relaxation factors are required to enforce convergence. That may increase the required number of iterations.
- If the added mass matrix **A** is known, (2.40) and (2.41) should be combined to the form

$$\left[\begin{pmatrix} m\mathbf{E} & \mathbf{0} \\ \mathbf{0} & \mathbf{I} \end{pmatrix} + \mathbf{A}\right] \begin{pmatrix} \ddot{\mathbf{x}}_G \\ \dot{\boldsymbol{\omega}} \end{pmatrix} = \begin{pmatrix} \mathbf{F}_r \\ \mathbf{M}_r - \boldsymbol{\omega} \times \mathbf{I}\boldsymbol{\omega} \end{pmatrix}, \tag{2.46}$$

  where **E** is the 3 by 3 unit matrix. The accelerations follow from solving this 6 by 6 linear equation system.
- Söding [1] describes a method to approximate the added mass matrix **A** from the accelerations found in previous iterations and previous time steps; this method may then be used in (2.46).

2.2.4 *Linearized Equations of Motion*

If the translation amplitudes are small compared to the ship's draft d (say, $<d/10$), and if the body's rotation amplitudes are small compared to unity (say, <0.1 rad), nonlinear terms of the motion equations can be neglected. Then cosine functions of rotations are approximated as unity and sine functions are substituted by their argument. Thus,

$$\mathbf{T} = \begin{pmatrix} 1 & -\psi & \theta \\ \psi & 1 & -\varphi \\ -\theta & \varphi & 1 \end{pmatrix} \tag{2.47}$$

and

$$\boldsymbol{\omega} = \dot{\boldsymbol{\alpha}}. \tag{2.48}$$

In the motion equations, the gyroscopic term $\boldsymbol{\omega} \times I\boldsymbol{\omega}$ is omitted because it is proportional to $|\boldsymbol{\omega}|^2$.

Reference

1. H. Söding, How to integrate free motions of solids in fluids, in *Numerical Towing Tank Symposium NuTTS, Hamburg* (2001)

Chapter 3
Numerical Methods to Compute Incompressible Potential Flows

Abstract This chapter illustrates the principles of computing potential flows in an unbounded fluid (without a free surface). It applies a strongly simplified source-sink method to the non-lifting flow around a body of arbitrary shape, and the steady lifting flow around an airfoil, both in two space dimensions. In the latter example, the patch method is explained: a variant of the usual panel method, if uses aims simpler formulas and yields better accuracy of the pressure force.

Potential flows are flows in which the rotation of the velocity vector rot$\mathbf{u} = \nabla \times \mathbf{u}$ is zero in the whole fluid domain Ω. As shear stresses, but not the pressure p, change rot$\mathbf{u}$ (especially near walls), potential flows occur only in inviscid fluids. In many cases, especially for highly accelerated flows (ship in waves), the influence of viscosity is small. Then potential flow is a suitable approximation for $\mathbf{u}$ outside the boundary layer and, for pressure and the resulting normal forces, often also on the body surface.

Viscous flow simulations require a higher effort in grid generation and computing time than potential flows. Therefore, potential flow simulations are still widely used and the preferred choice in industry except for ship resistance.

In the following, an inertial (not accelerated, not rotating) reference system is assumed, where x, y, z are Cartesian space coordinates; z is pointing downward. t designates time. For potential flow, the gradient of the potential $\phi(x, y, z, t)$ gives the velocity:

$$\mathbf{u} = \nabla\phi. \tag{3.1}$$

This assures that rot$\mathbf{u} = \nabla \times \nabla\phi = 0$. The pressure follows from the Bernoulli equation

$$p + \frac{1}{2}\rho\mathbf{u}^2 + \rho\phi_t - \rho g z = \text{ constant}, \tag{3.2}$$

where g is the gravity acceleration, ρ the fluid density. In our applications, the pressure is meant to be the difference to air pressure, which is approximated as constant. Then the 'Bernoulli constant' on the right-hand side is constant in space and time.

Furthermore, we presuppose that the flow speed is always much less than the speed of sound, which is about 1480 m/s in bubble-free saltwater; thus, the water

B. O. el Moctar et al., *Numerical Methods for Seakeeping Problems*,
https://doi.org/10.1007/978-3-030-62561-0_3

can be assumed incompressible. An incompressible inviscid fluid is called an ideal fluid. In chapters dealing with potential flow, we always treat the water as an ideal fluid.

In an incompressible fluid, conservation of mass is expressed by the Laplace equation:

$$u_x + v_y + w_z = \phi_{xx} + \phi_{yy} + \phi_{zz} = \Delta\phi = 0. \tag{3.3}$$

We do not need equations for conservation of momentum, as the flow is determined solely by the Laplace equation and the boundary conditions.

3.1 Source-Sink Method

The Laplace equation is linear in ϕ. Thus, it is possible to approximate an unknown potential $\phi(x, y, z, t)$ as superposition of elementary solutions $\varphi_i(x, y, z)$ which all fulfill the Laplace equation in Ω:

$$\phi(x, y, z, t) = \sum_{i=1}^{n} a_i(t)\varphi_i(x, y, z). \tag{3.4}$$

Then ϕ also fulfills the Laplace equation for arbitrary a_i. The functions $a_i(t)$ are chosen such that the boundary conditions are fulfilled, at least approximately. The most important elementary solution functions φ for this purpose are:
Parallel flow: e.g.,

$$\varphi = -Ux \tag{3.5}$$

for a parallel flow in negative x-direction with speed U.
Three-dimensional sink located at (ξ, η, ζ):

$$\varphi = 1/\sqrt{(x-\xi)^2 + (y-\eta)^2 + (z-\zeta)^2}. \tag{3.6}$$

The streamlines of this flow are straight lines of arbitrary direction through (ξ, η, ζ). The sink takes in a volume per time of 4π.
Two-dimensional source located at (ξ, η):

$$\varphi = \log[(x-\xi)^2 + (y-\eta)^2]. \tag{3.7}$$

The source ejects an area per time of 4π (if log is the natural logarithm, as is meant also hereafter).

These functions φ fulfill the (2D and 3D) Laplace equation everywhere except at the source or sink point. Therefore, sources or sinks should be arranged only outside the fluid domain Ω. At the location of the source or sink, φ becomes singular. Therefore, methods based on (3.4) with singular elementary solutions φ (source, dipole, vortex) are called singularity methods.

3.2 Example: Two-Dimensional Flow Around Smooth Body Without Lift

We consider the example of a two-dimensional flow of an ideal unbounded fluid around a rigid body with a smooth contour. Far away from the body, the fluid is at rest. The body moves with speed U in x-direction. U may be time-dependent; this plays no role as neither differential equation nor boundary conditions contain time derivatives. The flow is assumed free of circulation; this excludes flows around wings and other lifting surfaces for now.

The body contour is described by

$$r(x, y, t) = r(x - Ut, y, 0) = 0. \tag{3.8}$$

The condition for $r = 0$ is that a fluid particle on the body contour at time t (i.e., at $r = 0$) is also on the contour at time $t + dt$. Let D/Dt denote the substantial derivative (i.e., the time derivative at a point moving as a fluid particle with speed (u, v)). Then we have the following boundary condition at the body contour:

$$\frac{Dr}{Dt} = r_t + ur_x + vr_y = 0. \tag{3.9}$$

With $u = \phi_x$, $v = \phi_y$ and (3.8), we have:

$$\phi_x r_x + \phi_y r_y = Ur_x. \tag{3.10}$$

Far away from the body, the fluid is at rest:

$$\lim_{x^2+y^2\to\infty} \nabla\phi = 0. \tag{3.11}$$

This condition is automatically fulfilled if we use for φ_i only source potentials (at locations ξ_i, η_i inside the body).

The initially unknown coefficients a_i are chosen such that the body boundary condition (3.10) is fulfilled at n points on the body contour, the so-called collocation or control points (x_k, y_k), $k = 1$ to n. Such a method is called collocation method.

The function $r(x, y, 0)$ describing the body contour (at time $t = 0$) is approximated here in the vicinity of each collocation point by the equation of a straight line through the point and parallel to the straight line through the neighboring collocation points:

$$y = y_k + \frac{x - x_k}{x_{k+1} - x_{k-1}}(y_{k+1} - y_{k-1}). \tag{3.12}$$

We assume here that the collocation points 1–n circumscribe the body contour (almost) equidistantly. Then (3.12) converges for $n \to \infty$ to the tangent at the collocation point, provided the body has no sharp corners. Sharp corners require a special treatment. Here and in the following, we calculate the indices modulo n; i.e., for $k-1=0$ we take n, and for $k+1=n+1$ we take 1. To avoid a zero denominator, we rewrite (3.12):

$$r(x, y, 0) = (y - y_k)(x_{k+1} - x_{k-1}) - (x - x_k)(y_{k+1} - y_{k-1}). \tag{3.13}$$

Inserting this in (3.10) yields:

$$-(y_{k+1} - y_{k-1})\phi_x + (x_{k+1} - x_{k-1})\phi_y = -(y_{k+1} - y_{k-1})U \quad \text{for } k = 1 \text{ to } n. \tag{3.14}$$

Inserting here expression (3.4) for ϕ yields a linear system of equations for the unknown coefficients a_i:

$$\sum_{i=1}^{n} a_i \left[-(y_{k+1} - y_{k-1})\varphi_{i,x}(x_k, y_k) + (x_{k+1} - x_{k-1})\varphi_{i,y}(x_k, y_k)\right]$$
$$= -(y_{k+1} - y_{k-1})U \text{ for } k = 1 \text{ to } n. \tag{3.15}$$

Here $\varphi_{i,x}$ and $\varphi_{i,y}$ denote the partial derivatives of φ_i; they follow from (3.7) as

$$\varphi_{i,x} = \frac{2(x - \xi_i)}{(x - \xi_i)^2 + (y - \eta_i)^2} \quad \text{and} \quad \varphi_{i,y} = \frac{2(y - \eta_i)}{(x - \xi_i)^2 + (y - \eta_i)^2}. \tag{3.16}$$

3.3 Demonstration Program

The following Fortran90 program demonstrates this method. It reads the coordinates of the collocation points and the source points, calculates the coefficient matrix of the system of linear equations, solves the system for the unknown source strengths a_i, and calculates the velocity (u, v) at all collocation points.

```
module simq90;
!Linear equation system solved by Gauss elimination; line pivoting
implicit none; private; public:: simq; contains
subroutine simq(a,n,nr,i,s)
integer:: i,n,nr,j,k,imax,ir
real:: s,biga,save
real, dimension(:,:):: a  !coefficient matrix, r.h. sides (in) and solution (out)
Outer_loop: do i=1,n
 biga=a(i,i); imax=i                                          !Pivoting
 do j=i+1,n
  if(abs(a(j,i))>abs(biga))then; biga=a(j,i); imax=j; endif;
  if(abs(biga)<s)return
 enddo
 do k=i,n+nr; save=a(imax,k); a(imax,k)=a(i,k); a(i,k)=save/biga; enddo
 Middle_loop: do k=i+1,n+nr
  Inner_loop: do j=i+1,n; a(j,k)=a(j,k)-a(j,i)*a(i,k); enddo Inner_loop
 enddo Middle_loop
enddo Outer_loop
Backsubstitution: do i=n-1,1,-1;
```

```
  do j=i+1,n; do ir=1,nr; a(i,n+ir)=a(i,n+ir)-a(i,j)*a(j,n+ir); enddo; enddo
 enddo Backsubstitution; i=0
 end subroutine simq
 end module simq90

 program example1                   ! 2d flow around smooth body; collocation method
 use simq90; implicit none
 integer:: i,k,n,ks,kp1,km1
 real, dimension (50):: x,y,xi,eta          !coordinates of contour points, of sources
 real:: c(50,51),u,uk,vk            !influence matrix, inflow, fluid vel. components
 read *,n,(x(k),y(k),k=1,n),(xi(i),eta(i),i=1,n),u      !input contour points, sources
 print *,n,(x(k),y(k),k=1,n),(xi(I),eta(I),I=1,n),u
 ForAllCollocP: do k=1,n
  kp1=mod(k,n)+1; km1=mod(k-2+n,n)+1                          !k+1, k-1 projected to [1,n]
  ForAllSources: do i=1,n
   c(k,i)=2*(-(y(kp1)-y(km1))*(x(k)-xi(i))+(x(kp1)-x(km1))*(y(k)-eta(i))) &
          /((x(k)-xi(i))**2+(y(k)-eta(i))**2)                        !influence matrix
  enddo ForAllSources
  c(k,n+1)=-(y(kp1)-y(km1))*u                               !column n+1 (inhomogeneous)
 enddo ForAllCollocP
 call simq(c,n,1,ks,1.e-5)                                      ! Solves equation system
 if(ks/=0) stop ' Equation system singular'
 ForAllColl: do k=1,n                                       !computes panel mean velocities
  uk=u; vk=0.
  ForAllSs: do i=1,n
   uk=uk+c(i,n+1)*2*(x(k)- xi(i))/((x(k)-xi(i))**2+(y(k)-eta(i))**2)
   vk=vk+c(i,n+1)*2*(y(k)-eta(i))/((x(k)-xi(i))**2+(y(k)-eta(i))**2)
  enddo ForAllSs
  print *,k,x(k),y(k),uk,vk        !output coordinates and fluid velocity components
 enddo ForAllColl
 end program example1
```

The subroutine `simq` solves the system of linear equations. Its parameters are

- coefficient matrix, containing as last column(s) the negative right-hand side(s) of (3.15);
- number of unknowns;
- number of right-hand sides (here 1);
- flag `ks`, indicating whether the system of equations could be solved (then `ks` = 0) or whether it is (almost) singular. In the latter case, `ks` gives the rank (number of independent lines) of the system of equations. Due to the finite computational accuracy, the rank cannot be determined exactly. Lines are then seen as dependent if, for solving the system of equations, a division by a quantity with absolute value below s would be required;
- above-mentioned threshold s.

For equation systems containing more than, say, 1000 equations, `simq` should be substituted by a program which exploits the computer architecture to minimize the required computer time; we recommend the LAPACK (Linear Algebra Package) routines.

3.4 Program Test

Before working with any such program it should be tested as follows:

- Are differential equation and boundary conditions fulfilled at test points? For such tests, derivatives may be determined both by exact formulas and, approximately, as numerical differences.
- Is the solution independent of the measuring units? To test this, input and output of programs should be given using the same measuring units (including powers and products of these units) for length, time, and mass. If physical results (numbers times measuring units) change when a unit is changed (e.g. km instead of meter, hour instead of second, kg instead of ton), that indicates an error.
- For symmetric bodies: Does the solution fulfill the expected symmetry conditions?
- Does the solution converge with increasing n to expected values?
- Which n gives approximations of acceptable accuracy compared to the solution for large n?
- Do numerical solutions agree well with analytical solutions (e.g., for a circular cylinder)?
- How does the solution change for different arrangements of sources and collocation points?

For the program given above, applied to a circle of radius 1 with $U = 5$, the results given in Table 3.1 were obtained.

Plots may help to check for plausibility and accuracy. This is the purpose of subroutine `phiiso.f90` (see next page). It generates a Postscript file isolines.eps showing isolines of the potential, here of the 2D flow around a unit circle at $U = 1$. $n = 10$, with sources located at radius 0.3 (Fig. 3.1). Note that the body moves, not the water far away from the body. The potential for a body at rest in parallel inflow

Table 3.1 Exact and numerically computed velocities for flow around the unit circle

	Point 1	Point 2
x	0.8090173	0.3090180
y	0.5877848	0.9510562
ϕ_x analytic $= -U(y^2 - x^2)/r^2$	1.545090	−4.045079
ϕ_y analytic $= 2Uxy/r^2$	4.755281	2.938935
ϕ_x with program for $n = 10$; ϕ_y coll. points equidistant over circumference; sources at r = 0.3	1.545329 4.754978	−4.044495 2.938745
ϕ_x with program for $n = 20$; ϕ_y coll. points equidistant over circumference; sources at r = 0.3	1.545087 4.755285	−4.045080 2.938936

is obtained by adding the potential for parallel uniform flow (Fig. 3.2). The lines of constant potential are orthogonal to the body contour because the flow is tangential to the body contour there. Exceptions are the stagnation points (grad$\phi = 0$) on the x axis, where the flow direction is undefined.

```
program phiiso
! Plots isolines of function phi in grid dx,dy within rectangle xmin,xmax,ymin,ymax
! Calls function phi(x,y). Grid lines may not pass through singular points of phi!
! x to the right, y downward
implicit none
integer:: i,j                       !cell indices
integer:: line,c,c1,is,ks           !index of line,corner,intersection
real:: p(4)                         !phi values at cell corners (clockwise)
real:: pline                        !potential of an isoline
real:: dphi                         !phi difference between contour lines
real:: pmin,pmax                    !minimum, maximum of phi in cell
real:: dx,dy,xmin,xmax,ymin,ymax    !grid data
real:: xs(4),ys(4)                  !intersection isoline with cell boundaries
real:: xc(4),yc(4)                  !x, y of cell corners
print *,'Enter dx,dy,xmin,xmax,ymin,ymax,Delta phi'
read *,dx,dy,xmin,xmax,ymin,ymax,dphi                                   !reads grid data
open(10,file='isolines.eps')                             !generates plot file isolines.eps
write(10,'(a)')'%!PS-Adobe-2.0'
write(10,'(a,4i9)')'%%BoundingBox:',nint((/xmin,-ymax,xmax,-ymin/)/0.03528)     !in points
write(10,'(a)')' 72 2.54 div dup scale 0.015 setlinewidth'                 !all lengths in cm
Columns: do j=0,ceiling((xmax-xmin)/dx)-1
 xc(1)=xmin+j*dx
 Rows: do i=0,ceiling((ymax-ymin)/dy)-1
  yc(1)=ymin+i*dy
  xc(2:4)=xc(1)+(/dx,dx,0./)
  yc(2:4)=yc(1)+(/0.,dy,dy/)
  p=(/(phi(xc(c),yc(c)),c=1,4)/)
  pmin=minval(p); pmax=maxval(p)
  Isolines: do line=max(-50,ceiling(pmin/dphi)),min(50,floor(pmax/dphi))   !<=101 Isolines
   if(line==0)write(10,'(a,2f10.3)')'gsave [',0.4*dx,0.6*dx,'] 0 setdash' !broken line p=0
   pline=line*dphi; if(any(p==pline))pline=(line+1e-5)*dphi         !to avoid coincidence
   is=0                                      !Index of intersection isoline / cell boundary
   Corners: do c=1,4
    c1=mod(c,4)+1                                              !Index next point clockwise
    Intersection: if((pline-p(c1))*(pline-p(c))<0)then
     is=is+1
     xs(is)=xc(c)+(pline-p(c))/(p(c1)-p(c))*(xc(c1)-xc(c))
     ys(is)=yc(c)+(pline-p(c))/(p(c1)-p(c))*(yc(c1)-yc(c))
    endif Intersection
   enddo Corners
   OneOrTwo: do ks=1,is,2
    write(10,'(2f10.3,a)')xs(ks),-ys(ks),' moveto'
    write(10,'(2f10.3,a)')xs(ks+1),-ys(ks+1),' lineto stroke'
   enddo OneOrTwo
   if(line==0)write(10,'(a)')'grestore'                                !next line continuous
  enddo Isolines
 enddo Rows
enddo Columns
write(10,'(a)')'showpage'
close(10)

contains
real function phi(x,y)
real:: x,y                !arguments
!Here the function determining phi is to be inserted.
end function phi

end program phiiso
```

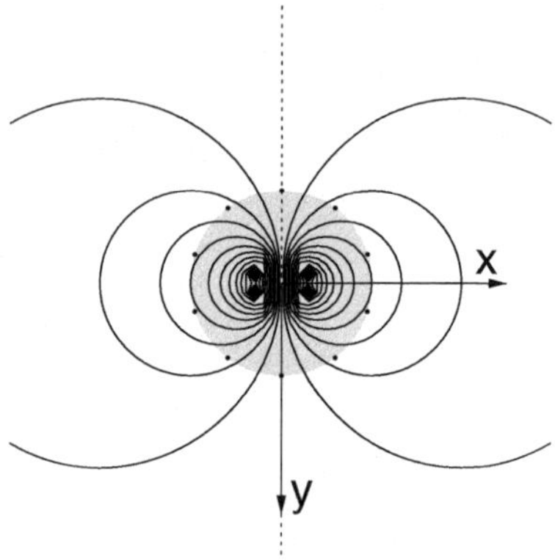

Fig. 3.1 Isolines ($\Delta\phi = 0.25$) of the potential around a moving ($U = 1$) unit circle (gray)

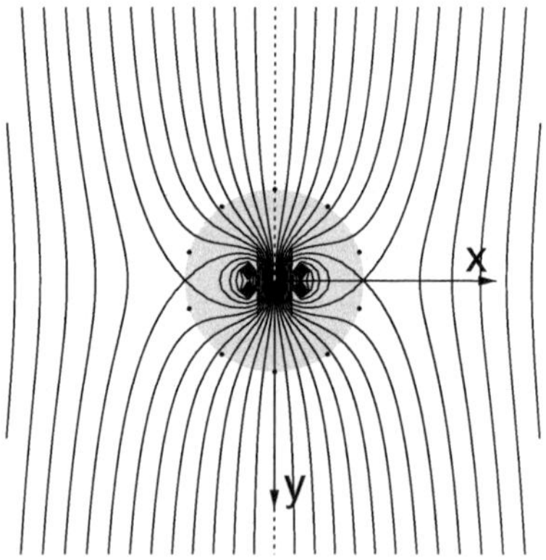

Fig. 3.2 Isolines of the potential around a stationary unit circle in parallel flow ($U = 1$)

3.5 Arrangement of Source Points

The sources must not be too close to each other to avoid nearly singular systems of equations, because in such systems, discretization errors will cause large inaccuracies of the computed flow. (If two sources coincide, not all conditions can be satisfied; if two collocation points coincide, the flow is not uniquely specified; in both cases, the solution of the system of linear equations fails.) To avoid nearly singular systems of equations, sources, and collocation points should face each other (i.e. the line between a collocation point and its closest source are approximately orthogonal to the body contour), or (better) staggered by 1/2 distance between collocation points. The results for staggered arrangement and non-staggered arrangement should be similar. For this end, sources should not be closer to the body contour than approximately 1.5 times the distance between adjacent collocation points. Otherwise, the body boundary condition is strongly violated between the collocation points. Figure 3.3 shows such a case, where $n = 10$ and a source distance of $r = 0.8$ was chosen. Here, the sources are too close to the body contour; thus, the isolines of the potential are not perpendicular to the circle.

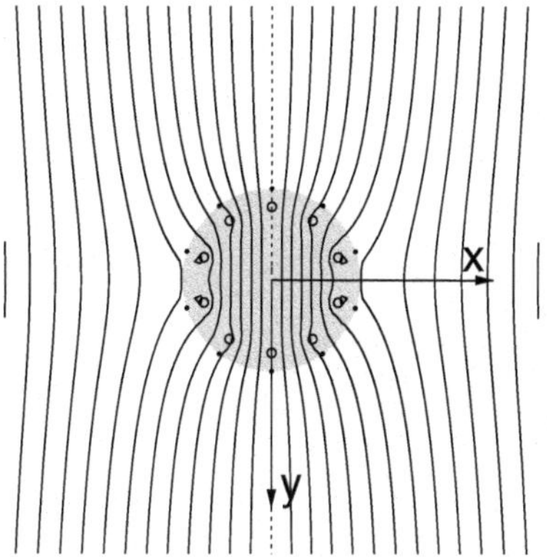

Fig. 3.3 Flow like in Fig. 3.2; however, with sources located at radius 0.8 instead of 0.3

On the other hand, to avoid a nearly singular equation system, the distance of the sources from the body contour should be not much more than half of the curvature radius and 2.5 times the distance between collocation points.

This results in a recommendation that the body normal should not change direction by much more than 15° within a single panel, and sources should be shifted inside by about twice the panel length. It is obvious that sharp corners on bodies must be rounded off.

Slender bodies pose problems: Here, the sources may be too close to the opposite side of the body if one follows the above rules. Increasing n can help, but a better solution is to avoid the use of point sources combined with collocation points. This most simple kind of a singularity method is described here as an introduction; it is not recommended for accurate flow computations around other than circular bodies.

3.6 Alternative Singularities

Would it be possible to express ϕ by dipole elements rather than sources, e.g., with dipoles in x-direction

$$\varphi = \frac{\partial}{\partial x} \log[(x-\xi_i)^2 + (y-\eta_i)^2] = \frac{2(x-\xi_i)}{(x-\xi_i)^2 + (y-\eta_i)^2} \tag{3.17}$$

or by dipoles perpendicular to the body contour? The dipole potential fulfills also the Laplace equation and the condition of vanishing velocity far away from the body. The body boundary condition can be fulfilled as in the source-sink method numerically. However, a pure 'dipole method' performs poorly. For small n acceptable results may be attained, but with increasing n the performance gets worse. This is due to the nearly singular coefficient matrix and can be explained as follows. The coefficient matrix remains unchanged if we change the right-hand side of (3.10). Imagine the right-hand side modified such that the boundary condition corresponds to an expanding body. The singularities must then be able to describe the volume increase inside the body. This requires at least one source because dipoles alone cannot change the volume.

The velocity decay with large distance from the body should also be appropriately reflected by the chosen singularities. The velocity decays as 1/distance for a source and the expanding body, but as 1/distance^2 for a dipole and for several sources with an added source strength of zero. Another aspect is that dipoles parallel to the body contour affect mainly the tangential velocity and less the normal velocity; however, the normal velocity is decisive for the boundary condition. Thus, one source and $n-1$ dipoles perpendicular to the body contour constitute a sound numerical method.

Could we use one source and $n-1$ vortices with potential

$$\varphi = \arctan \frac{y-\eta_i}{x-\xi_i} \; ? \tag{3.18}$$

The circulation strength (= sum of the strengths of all vortices) around a body does not follow from the boundary conditions formulated so far; thus, potential flows with arbitrary circulation may satisfy the boundary condition. This is illustrated by the Magnus effect for a circular cylinder. The flow of a real (viscous) fluid around a rotating circular cylinder depends on sign and magnitude of the cylinder's rotation rate about the cylinder axis. But the rotation has no influence on the boundary condition (no flux through the boundary) for potential flow. Thus, the approximation of the real flow by a potential flow can have different circulation. Therefore, the circulation strength must be determined from an additional condition; for instance:

- The circulation is 0 (as in the above program).
- The tangential velocity along the contour vanishes at a given point, e.g., at the trailing edge of a foil (Kutta condition).

Thus, one source, n vortices, no penetration at n collocation points, and one such additional condition may form a possible numerical method for two-dimensional potential flow around a body.

3.7 Symmetry; Mirror Principle

We can exploit symmetry to reduce the number n of unknowns and thus the complexity (computing effort), which—for large n—is nearly proportional to n^3. For a symmetric flow, we can arrange pairs of symmetrically located sources of same strength (with same or opposite sign, depending on the type of symmetry) for φ_i. For the flow in x-direction around a circle with center at the coordinate origin, we have flow symmetry about the x-axis and anti-symmetry about the y-axis. We can then write

$$\varphi_i = \log[(x-\xi_i)^2+(y-\eta_i)^2]+\log[(x-\xi_i)^2+(y+\eta_i)^2]$$
$$-\log[(x+\xi_i)^2+(y-\eta_i)^2]-\log[(x+\xi_i)^2+(y+\eta_i)^2].$$

Then we need collocation points only on a quarter of the circle; on the other parts of the circle, the boundary condition is then automatically fulfilled.

Also for an asymmetrical flow around a symmetrical body, the mirror principle can be applied to reduce the computing time. An example is a ship being symmetrical to the plane $y = 0$ in an oblique wave. Here, the flow potential ϕ is split into a symmetric and an anti-symmetric part:

$$\hat{\phi}^{sym}(x,y,z) = \frac{1}{2}\left(\hat{\phi}(x,y,z)+\hat{\phi}(x,-y,z)\right);$$
$$\hat{\phi}^{ant}(x,y,z) = \frac{1}{2}\left(\hat{\phi}(x,y,z)-\hat{\phi}(x,-y,z)\right). \quad (3.19)$$

To compute the symmetric potential, pairs of sources of unit strength are arranged on both sides of the plane $y = 0$, while for the anti-symmetric part pairs of a unit source and a unit sink (source of strength -1) are used as Green functions. Collocation points are arranged only on the starboard side ($y > 0$). Thus, instead of one equation system for n unknowns, two systems each of size $n/2$ must be solved. Compared to a direct solution, the complexity is reduced to be about 1/4 for large n. From the symmetrical and the anti-symmetrical source strengths $\hat{q}_j^{sym}$ and $\hat{q}_j^{ant}$, the source strengths on starboard ($y > 0$) and port ($y < 0$) side are easily determined:

$$\hat{q}_j^{stb} = \hat{q}_j^{sym} + \hat{q}_j^{ant}; \quad \hat{q}_j^{prt} = \hat{q}_j^{sym} - \hat{q}_j^{ant}. \quad (3.20)$$

3.8 Steady Two-Dimensional Flow Around a Foil; Patch Method

Foils are slender; thus, a method combining point sources with collocation points is not suitable here because the sources, especially in thin foils, must be arranged near to the contour; otherwise, they would be near to or even outside of the contour on the other side of the foil, except in case of very small panels. Therefore, we must use either a panel or a patch method. Instead of point sources, panel methods use distributed (smeared) sources (and sometimes normal dipoles) on the panel, or along a line parallel to the panel within the body.

Here, however, we use the foil flow to demonstrate the 'patch method' [1], a variant of the panel method intended to increase the accuracy of computed pressure forces, and to simplify the formulas for determining the potential and its patch-averaged gradient. Contrary to [1], it is applied here to a two-dimensional flow, and it is simplified with respect to computing the average pressure on a patch.

The patch method differs from the panel method in the following details:

1. The disturbance potential ϕ is a superposition of the potentials of point sources somewhat outside of the fluid domain (within the body).
2. Boundary conditions are not satisfied at collocation points, but in the average over each patch.
3. The average tangential flow speed over a patch is determined from the difference of the potential at the patch endpoints. (In [1], also the gradient of the potential at the patch endpoints is used to increase the accuracy of the computed average pressure; however, this is avoided here to simplify the method, and because it did not always result in greater accuracy.)
4. The average normal flow speed follows from the angle (in three-dimensional flows, the solid angle) under which the patch appears seen from the source.

These details depend on each other. The increased accuracy of computed forces follows essentially from details 3 and 4. It requires detail 1; in typical panel methods, the potential is infinite at the patch endpoints. Detail 1 is responsible for the relatively simple formulas of the patch method. It requires detail 2 for correct source strengths and details 3 and 4 for correct velocity. For a sufficiently smooth body surface, the average velocity over a patch is also a correct approximation of the velocity at the patch center. Here, correct means that the computed flow speed vector converges to the actual vector in the limit of vanishing panel size. That would not be the case if the fluid velocity at points on or near the boundary would be determined directly from the sum of source point potentials.

Validation studies have shown that the patch method improves the accuracy of computed fluid forces significantly not only compared to the above-described point-source/collocation method, but also compared to the usual panel method. In the patch method, the point sources may be located relatively close (e.g. 1/10 patch length or less) to the body surface, provided each source has nearly the same distance from the corners of the associated patches.

The flow around a foil may create a lift force (perpendicular to the inflow direction), which is often the main purpose of the foil and the flow quantity of major interest. The lift is proportional to the circulation around the foil. To model it, at least one vortex is required. A single vortex, however, produces flow irregularities at nearby parts of the foil contour. In principle, these can be canceled by the sources, but better accuracy is attained if many vortices are distributed on the mid-line of the foil profile. Here, the vortex strengths are chosen proportional to the profile thickness, because this results in small vortex strengths near the trailing edge. The vortices generate most of the difference in flow speed between pressure and suction side. Because this difference approaches zero at the trailing edge (Kutta condition), the vortex strength should vanish at the trailing edge. The proportionality constant between profile thickness and vortex strength is the additional unknown in the equation system, which is required to satisfy also the Kutta condition.

For demonstration, we determine the potential flow around a non-moving foil in uniform, stationary inflow. The foil profile is assumed symmetrical to the x-axis; the symmetry is not used here to decrease the computing time. The flow has to satisfy the following boundary conditions:

1. There is no flow through the profile contour:

$$\phi_n = \mathbf{n} \cdot \nabla\phi = 0, \tag{3.21}$$

where $\mathbf{n}$ is the unit normal vector on the contour of the profile directed into the profile.
2. At the trailing edge, there is no transverse velocity (Kutta condition):

$$\phi_y = 0. \tag{3.22}$$

3. Far away from the foil, there is a parallel flow of velocity U with angle α relative to the negative x direction:

$$\phi \to U(-x\cos\alpha + y\sin\alpha). \tag{3.23}$$

The profile is defined by n different points $x_i, y_i,\ i = 0$ to n, where point 0 and point n coincide at the trailing edge. The points surround the profile clockwise and n must be even. The $n/2$ points on the pressure side and the corresponding points on the suction side should have equal x coordinates, while the y-coordinates have the same absolute value and opposite sign.

The flow is approximated as a superposition of the parallel flow, of n sources located at (ξ_i, η_i), and of $n/2 - 1$ vortices located on the profile mid-line at $(x_i, 0)$ for $i = 1$ to $n/2 - 1$:

$$\phi(x, y) = U(-x\cos\alpha + y\sin\alpha) + \sum_{i=1}^{n} a_i \log[(x - \xi_i)^2 + (y - \eta_i)^2]$$
$$+\, a_{n+1} \sum_{i=1}^{n/2-1} (y_{n-i} - y_i) \arctan\frac{y}{x - x_i}. \tag{3.24}$$

Here, $y_{n-i} - y_i$ is the profile thickness at $x(i)$.

The function arctan in (3.24) is not applicable if $x = x_i$. To include also this case, it should be substituted by the function arctan2(dy, dx) which is equal to arctan(dy/dx) if $dx \neq 0$; otherwise, it is $\pi/2$ for $dy > 0$ and $-\pi/2$ for $dy < 0$. (For $dx = dy = 0$, both arctan and arctan2 are undefined.) This substitution is always implied where, in the following, the function arctan occurs.

As the potential of sources and vortices decays at large distance, condition 3 is automatically fulfilled; conditions 1 and 2 need to be fulfilled numerically by choosing appropriate a_i, $i = 1$ to $n + 1$.

The no-flux condition, integrated over the panel length coordinate s, yields for panel k ($k = 1$ to n):

$$\int \mathbf{n}_k \cdot \nabla\phi\, ds = 0 = U\mathbf{n}_k \cdot (-\cos\alpha, \sin\alpha)\Delta s + \sum_{i=1}^{n} a_i \int \frac{\partial}{\partial n} \log[(x - \xi_i)^2 + (y - \eta_i)^2] ds$$

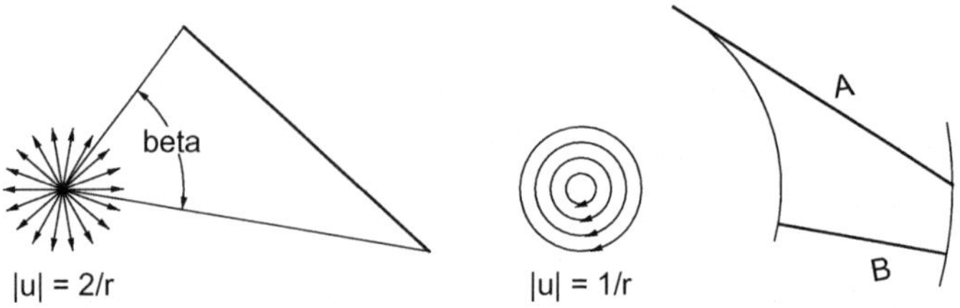

Fig. 3.4 Source (left) and vortex flow (right) through a panel. The vortex flux through panels A and B is the same

$$+ a_{n+1} \sum_{i=1}^{n/2-1} (y_{n/2+i} - y_{n/2-i}) \frac{\partial}{\partial n} \int \arctan \frac{y}{x - x_i} ds. \tag{3.25}$$

Here, Δs is the panel length and $\mathbf{n}\Delta s = (-y_i + y_{i-1},\ x_i - x_{i-1})$. $\partial/\partial n$ is the derivative in direction $\mathbf{n}$.

The integrals in (3.25) are easily evaluated:

$$\int \frac{\partial}{\partial n} \log[(x - \xi_i)^2 + (y - \eta_i)^2] ds = 2\beta. \tag{3.26}$$

This is the volume flux through a panel. The potential $\log[(x - \xi_i)^2 + (y - \eta_i)^2]$ corresponds to a source emitting a flux of 4π over an angle of 2π. Thus, we have a flux through the panel of 2β, where β is the angle under which the panel is seen from the source (Fig. 3.4). To avoid special cases, β is computed as $\arctan[\sin(\beta)/\cos(\beta)]$, where the vector product of the vectors from the source point to the panel ends is used for $\sin(\beta)$, and the scalar product for $\cos(\beta)$. This gives

$$\beta = -\arctan[(dx_1\, dy_2 - dy_1\, dx_2)/(dx_1\, dx_2 + dy_1\, dy_2)], \tag{3.27}$$

where $dx_1 = x_{k-1} - \xi_i$, $dy_1 = y_{k-1} - \eta_i$, and dx_2, dy_2 are defined correspondingly for k instead of $k - 1$.

The integral over the derivative of the vortex potentials in (3.25) can be found analytically or by physical insight, see Fig. 3.4:

$$\frac{\partial}{\partial n} \int \arctan \frac{y}{x - x_i} ds = \log r_k - \log r_{k-1}, \tag{3.28}$$

where r_{k-1} and r_k are the distances of the panel endpoints to the vortex point.

The Kutta condition (3.22) cannot be accurately satisfied directly by a patch method, because the patch method determines flow speed averages over panels, not the flow speed at points on the contour. As the Kutta condition has great influence on the most interesting quantity of the flow, the lift, we should satisfy it as accurately as possible. We notice that any flow around the trailing edge is caused by the pressure difference between pressure side and suction side of the profile at the trailing edge. As the normal velocity is zero along with the profile, vanishing pressure difference between pressure and suction side at the trailing edge is attained if the tangential velocity there is the same on both sides. To approximate the tangential velocity at the trailing edge on either side, we determine the panel-averaged tangential velocity $\bar{u}_1$ for the panel immediately in front of the trailing edge, and $\bar{u}_2$ for the next upstream panel; we interpret them as the velocity at the panel midpoints; and we use them to extrapolate u to the trailing edge. If the two panels have the same length Δs, the extrapolated tangential velocity at the trailing edge is

$$u_{TE} = \bar{u}_1 - \frac{1}{2}(\bar{u}_2 - \bar{u}_1). \tag{3.29}$$

The panel average of the tangential velocity is simply the difference between ϕ values at both panel ends divided by the panel length Δs. This results in

$$u_{TE} = (-1.5\phi_0 + 2\phi_1 - 0.5\phi_2)/\Delta s, \tag{3.30}$$

where ϕ_0, ϕ_1 and ϕ_2 are the potentials at the trailing edge and the next two panel endpoints further upstream. Thus, a suitable form of the Kutta condition is

$$(-3\phi_0 + 4\phi_1 - \phi_2)_{PS} - (-3\phi_0 + 4\phi_1 - \phi_2)_{SS} = 0, \tag{3.31}$$

where PS and SS designate pressure side and suction side, respectively. The ϕ values are determined from (3.24). ϕ_0 on the pressure side is $\neq \phi_0$ on the suction side, although both refer to the same point. That is due to the arctan2 function which can assume, besides the principal value between $-\pi$ and π, also values differing from the principal value by an integer number of 2π. Thus, when using the arctan2 function in a program, one has to assure that its value must possibly be modified to fit the neighboring points.

After solving the system of equations from (3.25) (n equations) and (3.31) (1 equation) for a_i, $i = 1$ to $n + 1$, the potential at the panel endpoints can be computed using (3.24). The Bernoulli equation (3.2) (without the term $\rho g z$) then gives the dynamic pressure (i.e., the difference to the pressure far away from the foil) on each panel:

$$p = \frac{1}{2}\rho(U^2 - \bar{u}^2) = \frac{1}{2}\rho\left(U^2 - \frac{(\phi_j - \phi_{j-1})^2}{(x_j - x_{j-1})^2 + (y_j - y_{j-1})^2}\right). \tag{3.32}$$

The following program illustrates this. It determines the dynamic pressure force components in x- and y-direction, the moment about the trailing edge, the lift L (normal to the inflow) and the drag D (in inflow direction), for the case $U = 1$ and $\rho = 1$.

```
!Module simq90 of example 1 must be included.
program example2                    !Lift and moment symm. foil. U=rho=1.
use simq90; implicit none
integer:: i,j                       !index of singularity, boundary condition
integer:: n                         !number of panels
integer:: ks                        !flag for solution
integer:: fact(0:2)=(/-3,4,-1/),fact1!factor for Kutta condition
real, dimension (0:50):: x,y        !coord. of panel endpoints
real, dimension(50):: xi,eta        !coord. of source points
real:: mid(2),norm(2)               !panel midpoints; normal vect.times Delta s
real:: c(51,52)                     !coefficient matrix incl. right-hand sides
real:: alpha,cosa,sina,phi,p,dx1,dy1,dx2,dy2,fx,fy,mz,phiprev  !p=pressure/rho
! Input: n (even), (x_i,y_i,i=0 to n), alpha in deg (see. Fig.~4)
! Points clockwise, point 0 and n at trailing, point n/2 at leading edge.
read  *,n,(x(i),y(i),i=0,n),alpha
print *,n,(x(i),y(i),i=0,n),alpha; if(n>50)stop'n too large'
y(0)=-1.e-5; y(n)=+1.e-5            !required because of discontinuous atan2(0,
!neg)
cosa=cos(alpha/57.296); sina=sin(alpha/57.296)
ComputeSourcePoints: do i=1,n
  mid=(/x(i)+x(i-1),y(i)+y(i-1)/)/2
  norm=(/-y(i)+y(i-1),x(i)-x(i-1)/)
  if(abs(norm(2))>0.67*abs(mid(2)))norm=norm*abs(mid(2))*0.67/abs(norm(2))
  xi (i)=mid(1)+norm(1); eta(i)=mid(2)+norm(2)
enddo ComputeSourcePoints
! Compute coefficient matrix c
ContourPoints: do j=1,n
  c(j,n+2)=(-y(j)+y(j-1))*cosa-(x(j)-x(j-1))*sina
  c(j,n+1)=0
  SourcePoints: do i=1,n
    dx1=x(j-1)-xi(i); dy1=y(j-1)-eta(i); dx2=x(j)-xi(i); dy2=y(j)-eta(i)
    c(j,i)=-2*atan2(dx1*dy2-dy1*dx2,dx1*dx2+dy1*dy2)
    if(i<n/2)c(j,n+1)=c(j,n+1)-y(i)* &      !for vortex at (x_i,0) ~ -2y_i
     (log((x(i)-x(j))**2+y(j)**2)-log((x(i)-x(j-1))**2+y(j-1)**2))
  enddo SourcePoints
enddo Contourpoints
c(n+1,1:n+2)=0.
Kutta:do j=0,n
 if(j>2.and.j<n-2)cycle
 if(j>n/2)then; fact1=-fact(n-j); else; fact1=fact(j); endif
 c(n+1,n+2)=c(n+1,n+2)-(-x(j)*cosa+y(j)*sina)*fact1
 do i=1,n
  c(n+1,i)=c(n+1,i)+log((x(j)-xi(i))**2+(y(j)-eta(i))**2)*fact1
 enddo
 c(n+1,n+1)=c(n+1,n+1)+sum(-2*y(1:n/2-1)*atan2(y(j),x(j)-x(1:n/2-1)))*fact1
enddo Kutta
call simq(c,n+1,1,ks,1.e-5)                              !Solve equation system
if(ks.ne.0)stop'System of equations singular'
!Compute force
fx=0; fy=0; mz=0
SurfPoints: do j=0,n
  phi=-x(j)*cosa+y(j)*sina &
    +sum(c(1:n,n+2)*log((x(j)-xi(1:n))**2+(y(j)-eta(1:n))**2)) &
    +c(n+1,n+2)*sum(-2*y(1:n/2-1)*atan2(y(j),x(j)-x(1:n/2-1)))
  if(j>0)then
    p=0.5*(1-(phi-phiprev)**2/((x(j)-x(j-1))**2+(y(j)-y(j-1))**2))
    fx=fx-p*(y(j)-y(j-1)); fy=fy+p*(x(j)-x(j-1))
    mz=mz+p*(y(j)-y(j-1))*(y(j)+y(j-1))/2+p*(x(j)-x(j-1))*(x(j)+x(j-1))/2
  endif
  phiprev=phi
enddo SurfPoints
print *,'fx=',fx,'  fy=',fy,'  mz=',mz
print *,'L= ', fx*sina+fy*cosa,', D= ',-fx*cosa+fy*sina
end program example2
```

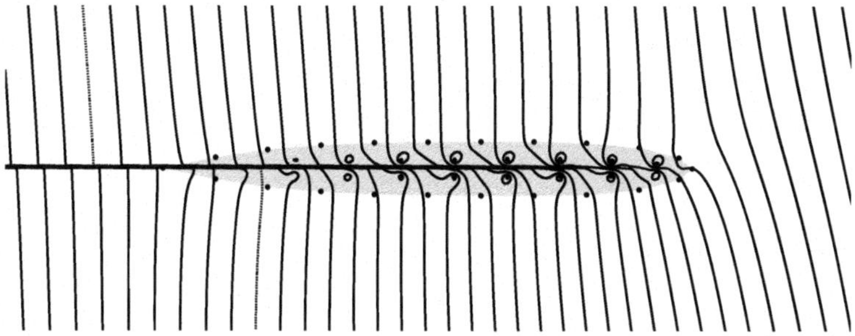

Fig. 3.5 Isolines of the potential of the flow around a symmetrical profile at 5.73° angle of attack

For a symmetrical foil profile of 10% thickness at $\alpha = 0.1$ rad angle of attack, described by $n = 22$ points (on both sides), the program gave a lift coefficient $c_L = L/(\frac{1}{2}\rho U^2 c)$ of 0.311 instead of the theoretical value $0.1\pi = 0.314$, and a drag coefficient $c_D = 0.018$ instead of 0. The potential of the flow is illustrated by Fig. 3.5.

Reference

1. H. Söding, A method for accurate force calculations in potential flow. Ship Technol. Res. **40** (1993)

Chapter 4
Water Waves

Abstract This chapter starts with the potential of linear regular waves in deep water, proceeds to nonlinear Stokes waves and continues with periodic long-crested steep waves of arbitrary order for deep and shallow water. Then methods to simulate a natural (irregular) seaway are discussed, using both linear and second-order superposition of regular waves; a higher order method is only cited. Typical wind wave spectra are described, and diagrams and formulas are given to approximate their parameters from wind speed, fetch, and duration. Finally, an example of a scatter table (probability distribution of significant wave period and height) taken from the modern 'wave atlas' is given.

Natural wind-generated water waves are described as a stochastic process depending on space and time. In most numerical methods, it is superimposed from regular waves, i.e., periodic progressive waves constituting a two-dimensional flow. Thus, this chapter starts with regular waves.

4.1 Regular Waves

4.1.1 Airy Potential

A regular wave propagating into direction μ is, by definition, periodic in time t and in the horizontal coordinate

$$\xi = x\cos\mu - y\sin\mu. \tag{4.1}$$

If the coordinate axes x, y, z are, relative to a ship, directed forward, to starboard and downward, respectively, then $\mu = 0$ holds for following waves, 90° for waves from starboard, and 180° for head waves.

Formulas for the motion of water particles in a regular wave, and of the water surface height at a fixed point may be developed into a Taylor series over wave amplitude $A = \frac{1}{2}$ wave height. Linear regular waves are waves the height of which

B. O. el Moctar et al., *Numerical Methods for Seakeeping Problems*,
https://doi.org/10.1007/978-3-030-62561-0_4

is so small that terms proportional to higher than first powers of A can be neglected because they are small compared to terms proportional to A. The detailed description of linear regular waves was given by Airy [1].

If a water body is initially at rest, and then a wind wave is generated only by air pressure fluctuations at the water surface, the motion of the water will be (almost) free of rotation: rot$\mathbf{u} = 0$, where $\mathbf{u} = (u, v, w)^T$ is the velocity vector of water particles. Then, the velocity $\mathbf{u}$ can be described by a potential $\phi(x, y, z, t)$ with

$$\nabla\phi = \mathbf{u} \tag{4.2}$$

at arbitrary time and for every location within the fluid. Actually, the wind generates also a shear stress at the water surface, causing rot$\mathbf{u} \neq 0$. This is neglected to allow the simpler description of the linear wave motion by the potential ϕ. A better approximation would be to add the potential and the shear flow field.

The periodicity of ϕ over t and ξ suggests for ϕ, using (4.1):

$$\phi(x, y, z, t) = \mathrm{Re}(f(z)e^{i(\omega t - k\xi)}) = \mathrm{Re}(f(z)e^{i\chi}). \tag{4.3}$$

with

$$\chi = \omega t - kx \cos\mu + ky \sin\mu. \tag{4.4}$$

Here, ω is the angular frequency $2\pi/T$ of the wave with period T, and k is the wave number $2\pi/\lambda$, where λ is wave length. The negative sign before $k\xi$ is chosen because, if t is increasing, a constant phase angle $\omega t - k\xi$ requires that ξ is increasing with time; that means, the wave is running into positive ξ direction.

The function $f(z)$ describes the dependence of ϕ on depth coordinate z. Because the speed of fluid particles is far below the speed of sound in water, the fluid behaves incompressible. This is described by the Laplace equation

$$\Delta\phi = \phi_{xx} + \phi_{yy} + \phi_{zz} = u_x + v_y + w_z = 0, \tag{4.5}$$

where indices x, y, z designate partial derivatives. Inserting (4.3) into (4.5) results in

$$\mathrm{Re}\left(\{f(z)\underbrace{[(-ik\cos\mu)^2 + (ik\sin\mu)^2]}_{-k^2} + f''(z)\}e^{i\chi}\right) = 0. \tag{4.6}$$

This equation is satisfied for all $\mathbf{x}$ and t if

$$f''(z) = k^2 f(z). \tag{4.7}$$

For a real wave number k, i.e., positive k^2, the general solution is

$$f(z) = c_1 e^{kz} + c_2 e^{-kz}. \tag{4.8}$$

For shallow water, where a boundary condition must be satisfied at the water bottom, we need both terms of $f(z)$. Here, however, we deal only with deep water. Then, for positive k the parameter c_1 must be zero to avoid infinite ϕ in the limit $z \to \infty$. Thus, we have

$$\phi(x, y, z, t) = \text{Re}(c_2 e^{-kz+i\chi}). \tag{4.9}$$

This deep-water approximation and its corollaries are held applicable if the water depth exceeds about $\lambda/2$.

From the above expression, the vertical component of the fluid speed results as

$$w = \phi_z = \text{Re}(-kc_2 e^{-kz+i\chi}). \tag{4.10}$$

Integrating w over time gives the vertical position z_w of a water particle having the average depth coordinate z_0:

$$z_w = z_0 + \int_0^t w(\tau)d\tau = z_0 + \text{Re}\left(\frac{k}{\omega} i c_2 e^{-kz+i\chi}\right). \tag{4.11}$$

Here,

$$\frac{k}{\omega} = \frac{2\pi}{\lambda} \cdot \frac{T}{2\pi} = 1/c, \tag{4.12}$$

where c is the phase velocity of the wave (i.e., the propagation speed of, e.g., wave crests). For $\mathbf{x} = (x, y, z)^T = \mathbf{0}$, Eq. (4.11) results in

$$z_w(t) = z_0 + \text{Re}\left(\frac{c_2}{c} i e^{i\omega t}\right). \tag{4.13}$$

In the following, the coordinate origin is assumed to be on the average water surface. Thus, for $z_0 = 0$, Eq. (4.13) describes the sinusoidal motion of a water particle at the water surface, which is equal to the vertical motion of the water surface itself. (Due to the horizontal motion of the water particle, the surface height at a fixed point differs from the vertical motion of a water particle on the surface; this difference is neglected here because it nonlinearly depends on wave amplitude.)

In (4.13), ic_2/c is the complex amplitude of the wave. Thus, we define

$$ic_2/c = A. \tag{4.14}$$

Depending on the argument of c_2, A may be real or complex; its modulus is the real wave amplitude.

Inserting (4.14) into (4.9) gives the linear wave potential in its final form

$$\phi(x, y, z, t) = \text{Re}\left(-icAe^{-kz+i\chi}\right). \tag{4.15}$$

The vertical coordinate of the water surface z_s follows from (4.11), (4.12) and (4.13):

$$z_s(x, y, t) = \mathrm{Re}\left(Ae^{i\chi}\right). \tag{4.16}$$

4.1.2 Dispersion Relation

Omitting terms depending nonlinearly on the amplitude A, Bernoulli's equation for a point on the water surface gives

$$p = p_0 + \rho g z_s - \rho\dot{\phi}. \tag{4.17}$$

Here, p_0 is the air pressure. Because of the small density of air compared to that of water, p_0 is assumed as constant over time and space. The equation neglects the difference between air and water pressure at the surface produced by surface tension, because surface tension is important only if the curvature radius of the surface is, say, <1 cm. Thus, the amplitude of the combined last two terms in (4.17) is zero. Using (4.16) for z_s and (4.15) for ϕ, this results in the condition

$$\rho g A - \rho c \omega A = 0. \tag{4.18}$$

With $c = \omega/k$ (corresponding to Eq. (4.12)) follows

$$k = \omega^2/g. \tag{4.19}$$

This so-called dispersion relation defines the relation between, on the one hand, wave number and wave length, and, on the other hand, wave frequency (at a fixed point) and phase velocity:

$$c = \frac{\omega}{k} = \frac{g}{\omega} = \sqrt{g/k} = \sqrt{g\lambda/(2\pi)}. \tag{4.20}$$

4.1.3 Corollaries

Equation (4.16) shows. If A is real, then at time $t = 0$ there is a wave trough (maximum of z_s) at $(x, y) = (0, 0)$.

The fluid velocity follows from (4.2):

$$\mathbf{u} = \nabla\phi = \mathrm{Re}\left((-\cos\mu, \sin\mu, i)^T \omega A e^{-kz+i\chi}\right). \tag{4.21}$$

It is called 'orbital velocity' because the fluid particles move on circular paths (Latin orbis means circle).

For a fixed point at position $\mathbf{x} = (x, y, z)^T$ (T indicates transposition) below the water surface, Bernoulli's equation together with the above equations give the pressure as

$$p = p_0 + \rho g z - \mathrm{Re}(\rho g A e^{-kz+i\chi}). \tag{4.22}$$

For a point moving with velocity u in positive x-direction (for instance, a point fixed to a ship with speed u ahead), the surface depth is

$$z_s(x, y, t) = \mathrm{Re}\left(A e^{i(\omega t - k[x+ut]\cos\mu + ky\sin\mu)}\right) = \mathrm{Re}\left(A e^{i([\omega - ku\cos\mu]t - kx\cos\mu + ky\sin\mu)}\right). \tag{4.23}$$

Thus, at the moving point the oscillation of z_s, as well as that of any other linear response, varies with the 'encounter angular frequency'

$$\omega_e = \omega - ku\cos\mu. \tag{4.24}$$

A superposition of two waves of equal frequency and amplitude, but of opposite direction, results in 'standing waves':

$$z_s = \mathrm{Re}\left(A[e^{i(\omega t - k\xi)} + e^{i(\omega t + k\xi)}]\right) = \mathrm{Re}\left(A e^{i\omega t}\right) 2\cos(k\xi). \tag{4.25}$$

Here, wave troughs and crests remain fixed in space; their depth coordinate z_s oscillates over time.

4.1.4 Stokes Waves

The above formulae are (almost) exact in the limit of vanishing wave steepness $2A/\lambda$. For moderate steepness, approximation errors can be decreased by relatively simple modifications. Stokes [2] developed such modifications using perturbation theory. His third-order formulae for a regular progressive wave are derived here; that may serve as an introduction to methods applied in potential flow theory.

As for linear waves, Laplace's equation

$$\Delta\phi(\xi, z, t) = \phi_{\xi\xi} + \phi_{zz} = 0 \tag{4.26}$$

must be satisfied for all time t and everywhere in the fluid domain. As above, ξ is the coordinate in the direction of wave propagation. To determine ϕ and the initially unknown surface depression $z_s(\xi, t)$, we need two boundary conditions at the water surface. There may be no flow through the surface (kinematic condition), and the pressure must be constant (dynamic condition).

Compared to (4.17), the Bernoulli equation must be complemented here by the term involving the squared fluid velocity:

$$p - p_0 = \rho\left(gz - \phi_t - \frac{1}{2}\nabla\phi \cdot \nabla\phi\right) = 0. \tag{4.27}$$

The equation must be satisfied on the fluid surface $z = z_s(\xi, t)$. Thus, it gives an expression for z_s:

$$z_s = -\frac{1}{g}\left(\phi_t + \frac{1}{2}\nabla\phi \cdot \nabla\phi\right). \tag{4.28}$$

A 'substantial time derivative' is a time derivative at a point moving with the fluid:

$$\frac{D}{Dt} = \frac{\partial}{\partial t} + \nabla\phi \cdot \nabla. \tag{4.29}$$

From the kinematic condition follows that a fluid particle on the fluid surface will remain on the surface. Thus, also D/Dt of Eq. (4.27) is zero:

$$\left(\frac{\partial}{\partial t} + \nabla\phi \cdot \nabla\right)\left(gz - \phi_t - \frac{1}{2}\nabla\phi \cdot \nabla\phi\right) = g\phi_z - \left(\frac{\partial}{\partial t} + \nabla\phi \cdot \nabla\right)\left(\phi_t + \frac{1}{2}\nabla\phi \cdot \nabla\phi\right) = 0. \tag{4.30}$$

Evaluating the derivatives gives

$$\phi_{tt} + 2\nabla\phi \cdot \nabla\phi_t - g\phi_z + \frac{1}{2}(\nabla\phi \cdot \nabla)(\nabla\phi \cdot \nabla\phi) = 0. \tag{4.31}$$

The purpose of the above re-formulation of the free-surface boundary condition is to eliminate the unknown function $z_s(\xi, t)$. However, z_s is implicitly contained in (4.31) because ϕ is short for $\phi(\xi, z_s, t)$. Because we search for an approximation for moderate wave amplitude A, also $|z_s|$ is moderate. A Taylor expansion for ϕ around $z = 0$ will converge for sufficiently small A:

$$\phi(x, z_s, t) = \phi(x, 0, t) + z_s\phi_z(x, 0, t) + \frac{1}{2}z_s^2\phi_{zz}(x, 0, t) + \cdots \tag{4.32}$$

Higher order terms of this expansion are not necessary for the third-order approximation in A. Inserting here for z_s the expression (4.28), and further inserting the result into the boundary condition (4.31) gives an approximate boundary condition, where terms proportional to A^4 and higher powers of A have been neglected. This condition contains z_s neither explicitly nor implicitly:

$$\underbrace{\phi_{tt} - g\phi_z}_{\text{first order in } A} + \underbrace{2\nabla\phi \cdot \nabla\phi_t + \frac{1}{g}\phi_t(\phi_{tt} - g\phi_z)_z}_{\text{2nd order in } A}$$

$$+ \frac{1}{2}(\nabla\phi \cdot \nabla)(\nabla\phi \cdot \nabla\phi) + \frac{1}{2g}(\nabla\phi \cdot \nabla\phi)(\phi_{tt} - g\phi_z)_z + \frac{1}{2g^2}(\phi_t)^2(\phi_{tt} - g\phi_z)_{zz}$$

$$+ \frac{2}{g}\phi_t(\nabla\phi_z \cdot \nabla\phi_t + \nabla\phi \cdot \nabla\phi_{zt}) + \frac{1}{g^2}\phi_t\phi_{zt}(\phi_{tt} - g\phi_z)_z = 0. \tag{4.33}$$

Here, ϕ is short for $\phi(\xi, 0, t)$. Such complicated formulas are typical for perturbation methods of higher than first- or second-order.

The following is simplified by restricting the derivation to real wave amplitude A, i.e., to waves having their trough at time $t = 0$ at the coordinate origin. Equation (4.15) can then be expressed in the form

$$\phi = cAe^{-kz}\sin(\omega t - k\xi). \tag{4.34}$$

We test this first-order approximation by inserting it into (4.33); however, with a modified dispersion relation

$$\omega^2 = gk + fA^2, \tag{4.35}$$

where f is a real factor. This modification has no effect on second- and third-order terms because it generates $\geq$ fourth-order terms which are omitted. But in connection with the linear term in the boundary condition (4.33), it gives rise to a third-order term which cancels a third-order term in (4.33) for a suitable selection of factor f. The remaining second- and third-order terms cancel automatically for $\omega^2 = gk$. Thus, the third-order boundary condition is satisfied by the first-order wave potential (4.15) or (4.34) if the following modified dispersion relation is applied:

$$\omega^2 = gk(1 + k^2A^2). \tag{4.36}$$

It gives rise to a higher phase speed of the third-order Stokes wave compared to the linear Airy wave of same length:

$$c = \frac{\lambda}{T} = \frac{\omega}{k} = \sqrt{\frac{g}{k}}\sqrt{1 + k^2A^2} \approx \sqrt{\frac{g}{k}}\left(1 + \frac{1}{2}k^2A^2\right), \tag{4.37}$$

where $\approx$ means equal up to third order in A.

The surface depression z_s is derived from (16.1). On the right-hand side, ϕ_t is short for $\phi_t(\xi, z_s, t)$; correspondingly, for $\nabla\phi$. Again a Taylor expansion around $z = 0$ is made, and (4.28) is once more applied to remove z_s from the Taylor expansion. The resulting third-order approximation for z_s is

$$z_s = \frac{1}{g}\left[\phi_t + \frac{1}{g}\phi_t\phi_{tz} + \frac{1}{g^2}(\phi_{tz})^2\phi_t + \frac{1}{2g}\phi_{tz}(\nabla\phi)^2 + \frac{1}{2g^2}(\phi_t)^2\phi_{tzz} + \frac{1}{2}(\nabla\phi)^2 + \frac{1}{g}\phi_t\nabla\phi\cdot\nabla\phi_z\right], \tag{4.38}$$

where ϕ is now short for $\phi(\xi, 0, t)$. Inserting (4.34) and using the relations

$$\cos^2\alpha = \frac{1}{2}(1 + \cos 2\alpha) \tag{4.39}$$

and

$$\cos^3\alpha = \frac{1}{4}(\cos 3\alpha + 3\cos\alpha) \tag{4.40}$$

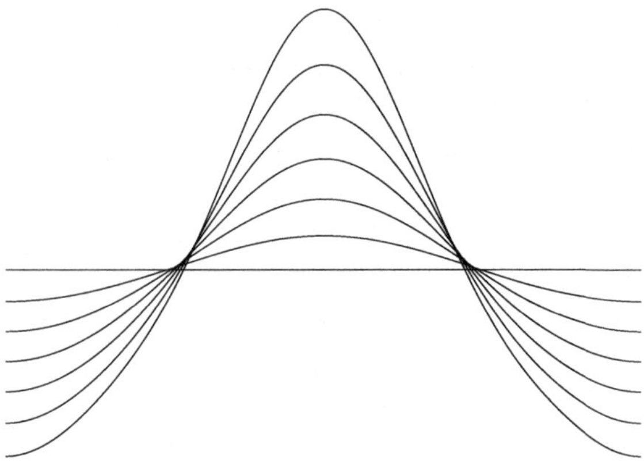

Fig. 4.1 Wave contours of third-order Stokes waves of nominal steepness $2A/\lambda = 0.02$ to 0.12 in steps of 0.02. Vertical coordinates magnified by factor 5

gives the following 3rd-order approximation for the surface depression:

$$z_s = A\left[\left(1+\frac{5}{8}k^2A^2\right)\cos(\omega t - k\xi) - \frac{1}{2}kA\cos 2(\omega t - k\xi) + \frac{3}{8}k^2A^2\cos 3(\omega t - k\xi)\right]. \tag{4.41}$$

This results in a wave height (measured from crest at $\omega t - k\xi = \pi$ to trough at $\omega t - k\xi = 0$) of

$$h = 2A(1 + k^2A^2). \tag{4.42}$$

The wave steepness is

$$s = \frac{h}{\lambda} = \frac{1}{\pi}kA(1 + k^2A^2) = s_n(1 + k^2A^2), \tag{4.43}$$

where s_n is the *nominal* steepness kA/π.

Figure 4.1 shows z_s for 6 different values of nominal wave steepness up to 0.12. The actual steepness of the steepest wave of Fig. 4.1 is about 0.137, which is near to the maximum steepness of progressive waves before wave breaking occurs. The figure shows that third-order Stokes waves approximate (but under-predict) the higher curvature of actual waves at the wave crest compared to that at the wave trough. Another often encountered property of actual wind waves is not shown by the Stokes approximation. The higher slope ahead of the wave crest compared to the slope behind the crest.

In the relevant literature, many different formulae are given for z_s of the third-order Stokes wave. They are caused not only by different definitions of coordinate direction, phase angle, and nominal amplitude, but some published formulas are even incorrect. Thus, it is recommended to test such formulas by adapting the short program of Appendix A to the formula to be tested, to determine whether third-order accuracy is really attained.

At fixed z coordinates below the wave trough, the wave potential generates a fluid speed varying harmonically about the average value 0. In the depth region between crest and trough height, however, the periodic emergence causes a mean water flow in the direction of wave propagation. The average volume flow generated by this effect is $gkA^2/(2\omega)$ per breadth in crest direction. In a closed basin (e.g., a wave tank), this average flow due to the wave is canceled by a steady back flow. For moderate fluid depth, the back flow will be rather constant over depth. It adds on to the flow speed described by the wave potential.

4.1.5 Linear Regular Waves in Deep and Shallow Water

Contrary to the above cases, here a horizontal impermeable water bottom is assumed at a depth d below the average water level. The case of deep water is handled by the procedure derived in the following either by using a large (compared to the wavelength) depth or by taking the limit for $d \to \infty$.

The general dependence on z of the linear wave potential, which is necessary to satisfy the Laplace equation, was given already in (4.8). In deep water, it simplifies to $f(z) = e^{-kz}$ as shown in (4.9). In shallow water, the z dependence must be changed to

$$f(z) = \frac{\cosh[k(z-d)]}{\cosh(kd)}. \tag{4.44}$$

This is a special case of (4.8); it results in $f(z) = 1$ at the mean water level $z = 0$, corresponding there to $f(z)$ for deep water; and at the water bottom $z = d$, it results in $f'(z) = 0$; thus, there is no vertical velocity at the bottom.

For large water depth, $\cosh(kd)$ may be out of the range of admissible floating-point numbers. Therefore, in computer programs the factor $\cosh[k(z-d)]/\cosh(kd)$ should be evaluated as

$$\frac{\cosh[k(z-d)]}{\cosh(kd)} = \cosh(kz) - \sinh(kz)\tanh(kd). \tag{4.45}$$

The equation follows from the formula for a sum of arguments of a hyperbolic cosine function. A similar expression occurs in partial derivatives of the wave potential and should also be evaluated as

$$\frac{\sinh[k(z-d)]}{\cosh(kd)} = \sinh(kz) - \cosh(kz)\tanh(kd). \tag{4.46}$$

The consideration leading to the dispersion relation for low-amplitude waves in deep water given in (4.19) can be repeated for shallow water using (4.44). This results in the dispersion relation for low-amplitude waves in shallow water

$$k\tanh(kd) = \omega^2/g. \tag{4.47}$$

From this follows the phase velocity of a low-amplitude regular wave in shallow water:

$$c = \frac{\omega}{k} = \frac{g \tanh(kd)}{\omega} = \sqrt{g \tanh(kd)/k} = \sqrt{g\lambda \tanh(kd)/(2\pi)}. \tag{4.48}$$

Thus, in shallow water the phase velocity is smaller than in deep water for the same wave length, and for the same wave frequency as well.

Correspondingly, the linear flow potential of a wave in shallow water is, analogous to (4.15):

$$\phi(x, y, z, t) = \mathrm{Re}\left(-icA\frac{\cosh[k(z-d)]}{\cosh(kd)}e^{i\chi}\right). \tag{4.49}$$

The wave contour is given by (4.16) as for the case of deep water.

4.1.6 Nonlinear Regular Waves in Deep and Shallow Water

In Sect. 4.1.4, the case of a steep wave in deep water was handled using a perturbation method. It generates a relatively simple formula by complicated formula manipulations. Here, for the case of shallow water (including arbitrarily large water depth), we will derive a system of nonlinear equations to compute the wave potential and the surface contour up to an arbitrary order of approximation. Instead of complicated formula manipulations, a general procedure for solving nonlinear equation systems is used to do the necessary work numerically for the actual case. The chapter is based on and corresponds largely to [3, 4].

At first, formula (4.49) for the first-order wave potential is confined to the case that a wave trough is at the coordinate origin at t = 0, i.e., that the wave amplitude A is real:

$$\phi(x, y, z, t) = cA\frac{\cosh[k(z-d)]}{\cosh(kd)}\sin\chi. \tag{4.50}$$

To take account of higher harmonics, the formula is extended up to order N:

$$\phi(x, y, z, t) = \sum_{n=1}^{N} B_n \frac{\cosh[nk(z-d)]}{\cosh(nkd)}\sin n\chi. \tag{4.51}$$

Every term in the sum satisfies the Laplace equation (to model an incompressible fluid) and the boundary condition at the water bottom. The constants B_n will be used to satisfy the other conditions.

To further simplify the following, a new coordinate system moving with the phase speed c in wave propagation direction will be used. In this system, the wave is stationary. Its horizontal coordinate is

$$X = \xi - ct. \tag{4.52}$$

The potential expressed in this system becomes

$$\phi(X, z) = -cX + \sum_{n=1}^{N} B_n \frac{\cosh[nk(z-d)]}{\cosh(nkd)} \sin(nkX). \tag{4.53}$$

The first term on the right-hand side gives the back flow of water relative to the moving coordinate system.

To derive conditions for the B_n, we introduce the stream function $\psi(X, z)$. It is defined as the flux (volume per time and per breadth in wave crest direction) between the point (X, z) and the bottom. (For any curve between the point and the bottom, the flux of an incompressible fluid crossing the curve has the same value $\psi(X, z)$.) The partial derivatives of the stream function define the fluid speed components U (in X-direction) and w (in z direction), however, different from the corresponding definitions for the flow potential ϕ:

$$U = \partial\psi/dz = \partial\phi/dX; \quad w = -\partial\psi/dX = \partial\phi/dz. \tag{4.54}$$

These relations between ψ and ϕ are satisfied everywhere if the following modification of (4.53) is chosen for ψ:

$$\psi(X, z) = -c(z-d) + \sum_{n=1}^{N} B_n \frac{\sinh[nk(z-d)]}{\cosh(nkd)} \cos(nkX). \tag{4.55}$$

We choose $N + 1$ points per half wavelength on the water surface to define the wave contour. Their equidistant horizontal coordinates are

$$X_j = j\pi/(Nk) \tag{4.56}$$

for $j = 0$ to N. Their vertical coordinates z_j satisfy the implicit equations following from (4.55):

$$-c(z_j - d) + \sum_{n=1}^{N} B_n \frac{\sinh[nk(z_j - d)]}{\cosh(nkd)} \cos(nj\pi/N) = \psi_s \tag{4.57}$$

for $j = 0$ to N. ψ_s is the total flux between bottom and water surface; it is the same for all j.

To obtain constant pressure at the chosen points on the water surface, Bernoulli's equation is written at these points:

$$-\frac{1}{2}[\partial\psi(X_j, z_j)/dz]^2 - \frac{1}{2}[\partial\psi(X_j, z_j)/dX]^2 + gz_j = b \tag{4.58}$$

for $j = 0 \ldots N$. The 'Bernoulli constant' b is also independent of j. Inserting here ψ from (4.55) results in the equations:

$$-\frac{1}{2}\left(-c + \sum_{n=1}^{N} nkB_n \frac{\cosh[nk(z_j - d)]}{\cosh(nkd)} \cos(nj\pi/N)\right)^2$$

$$-\frac{1}{2}\left(\sum_{n=1}^{N} nkB_n \frac{\sinh[nk(z_j - d)]}{\cosh(nkd)} \sin(nj\pi/N)\right)^2 + gz_j = b \tag{4.59}$$

for $j = 0 \ldots N$.

Equations (4.57) and (4.59) are together $2N + 2$ equations to determine the $N + 1$ unknowns z_j and the N unknowns B_n. Further unknowns are the phase speed c, the flux ψ_s, and the Bernoulli constant b. Thus, we need two additional equations. These follow from definitions introduced above: For the wave height

$$h = z_0 - z_N, \tag{4.60}$$

and for the average surface coordinate

$$\frac{1}{2}(z_0 + z_N) + \sum_{j=1}^{N-1} z_j = 0. \tag{4.61}$$

(The average within half a wavelength is determined using the trapezoidal rule, which is the optimum approximation here.)

The nonlinear equation system can be solved using a standard iterative procedure based on Newton's method. For the $2N + 4$ unknowns, initial values are determined from linear theory ($N = 1$), except for the unknowns B_2 to B_N, for which zero is assumed as initial value. Convergence proved to be no problem at least up to a wave height of 92% of the maximum physically possible value. For very shallow water ($d/\lambda < 3\%$), the iteration may converge to a solution having more than one wave crest per given wavelength; but for seakeeping of ships this case is not relevant.

For the approximation order N, a value of 20 can be recommended. The reduced computer time for smaller N is minimal, and higher N values hardly change results.

The X derivative of the potential (4.53) is the horizontal water velocity in the coordinate system moving with the wave. To obtain the wave potential in the earth-fixed reference system, the first term $-cX$ must be deleted. Thus, the velocities are defined alone as partial derivatives of the second term. It gives a time-averaged horizontal velocity of zero at any fixed point below the wave trough level. At points between trough and crest level, however, there is no water flow when the point is emerged (e.g., above a wave trough), while the flow is mostly in wave propagation direction when the point is immersed (below a wave crest). Thus, the wave generates an average flux of water in wave propagation direction given here by ψ_s. In a closed basin, or when the waves are running perpendicularly to a shoreline, the flux is canceled by a superimposed steady current opposite to the wave propagation direction. The depth-averaged velocity of this back flow is ϕ_s/d. It reduces the propagation of waves relative to the ground. If we assume the back flow to be constant over z (otherwise, it would be a shear flow having no potential), the wave flow in the earth-fixed system has the potential

$$\phi(X, z) = (-c + \psi_s/d)X + \sum_{n=1}^{N} B_n \frac{\cosh[nk(z-d)]}{\cosh(nkd)} \sin(nkX). \tag{4.62}$$

Where c is then the wave phase speed not relative to the earth, but relative to the average position of water particles below the wave trough level. The possible presence of the back flow in shallow water and its effect on the phase speed was already discussed by Stokes [2].

4.2 Natural (Irregular) Waves

4.2.1 Linear Superposition

To simplify the following, a 'complex vector wave number' $\boldsymbol{\nu}$ is introduced:

$$\boldsymbol{\nu} = ik(-\cos\mu, \sin\mu, i). \tag{4.63}$$

Using this abbreviation, formulas (4.15) and (4.16) become

$$\phi(\mathbf{x}, t) = \mathrm{Re}(-icAe^{i\omega t + \boldsymbol{\nu}\cdot\mathbf{x}}) \tag{4.64}$$

and

$$z_s(x, y, t) = \mathrm{Re}(Ae^{i\omega t + \boldsymbol{\nu}\cdot(x,y,0)^T}). \tag{4.65}$$

To simulate floating bodies in a natural seaway, often a linear superposition of Airy (linear) waves is applied:

$$\phi(\mathbf{x}, t) = \sum_{j=1}^{J} \mathrm{Re}(-ic_j A_j e^{i\omega t + \boldsymbol{\nu}\cdot_j \mathbf{x}}), \tag{4.66}$$

$$z_s(x, y, t) = \sum_{j=1}^{J} \mathrm{Re}(A_j e^{i\omega_j t + \boldsymbol{\nu}_j\cdot(x,y,0)^T}). \tag{4.67}$$

If the wave angles μ_j, which are contained in $\boldsymbol{\nu}_j$, are the same for all j, the resulting sum describes a long-crested seaway, in which wave crests (as well as troughs) are of infinite length perpendicular to the wave propagation direction μ. A better model of a natural wind-generated seaway is obtained when waves of different direction are superimposed.

To model natural wind-generated seaways, (4.66) and (4.67) cannot always be recommended. The reason is that wind seas contain waves spanning a broad range of wave numbers. Each term in the sum for ϕ varies with vertical coordinate by a factor e^{-kz}, which is contained in the scalar product $\boldsymbol{\nu}_j \cdot \mathbf{x}$. This is correct within the range of typical wave amplitudes *belonging to the wave number* k, i.e., up to about 1/15 wavelength. But it is grossly in error if the factor is used in short waves, while z corresponds to the maximum height of superimposed longer and higher waves. The factor e^{-kz} has the effect that short waves have higher amplitudes on wave crests of the total seaway than in wave troughs. If the steepness of the short wave exceeds about 1/7, the wave will break, dissipating much of its energy and transferring the rest to longer waves, increasing their amplitude.

The fact that the factor e^{-kz} may cause unreasonably high orbital velocities at the wave crest has been observed by several authors. Typical corrections proposed are

- Amplitude stretching, in which ϕ and the orbital velocity are determined from the linear superposition formulae at $z = 0$ instead of at the actual z coordinate above (and in some proposals also below) the average height $z = 0$.
- Limiting k_j by a 'cut-off wave number', omitting the contribution of shorter waves.

Both remedies are quite rough and may give substantial deviations from reality for the wave properties, especially for the fluid velocity at the wave crest (but not necessarily for responses of a floating structure). Results of the 'cut-off wave number method' are also sensitive to the cut-off limit.

This motivates another remedy to the problem of superimposing regular waves. We introduce another position vector $\mathbf{x}_0$. It gives the position where a fluid particle, which is actually at position $\mathbf{x}$, would come to rest if the wave amplitudes were decreased to zero. In regular waves, it would correspond to the center of the orbital motion of a fluid particle. For a given position $\mathbf{x}_0$, the properties of the wave superposition are determined at another position $\mathbf{x}$. The regular elementary waves are evaluated at a location that differs from $\mathbf{x}_0$ by the influence of all shorter elementary waves, but not by that of the longer waves. This is expressed by the formulas

$$\phi(\mathbf{x}, t) = \sum_{j=1}^{J} \mathrm{Re}(-ic_j A_j e^{i\omega_j t + \boldsymbol{\nu}_j \cdot \mathbf{x}_j}), \tag{4.68}$$

where

$$\mathbf{x}_j = \mathbf{x}_0 + \sum_{l=j+1}^{J} \mathrm{Re}(A_l e^{i\omega_l t + \boldsymbol{\nu}_l \cdot \mathbf{x}_l}) \tag{4.69}$$

and

$$\mathbf{x}_J = \mathbf{x}_0, \quad \mathbf{x} = \mathbf{x}_1. \tag{4.70}$$

Here, it is supposed that the wave numbers k_j monotonically increase with j. The surface depression is obtained from (4.69) for $\mathbf{x}_0$ vectors having z-coordinate 0. The equations involve an error proportional to A^2 of the Laplace equation, additionally to the error of the free-surface condition, which remains of second order.

The evaluation of formulas (4.68) and (4.69) is easily done in a loop running backward from J (shortest wave) to 1 (longest wave). However, evaluations for a given location $\mathbf{x}$ require an iteration to determine $\mathbf{x}_0$. Compared to (4.66) and (4.67), (4.68) and (4.69) modify (often quite substantially) the short waves contained in the seaway. However, if short waves (relative to the most significant waves within the seaway) are irrelevant for the structure of interest, the application of (4.66) and (4.67) combined with the cut-off wave number method is recommended to avoid the iteration.

4.2.2 Nonlinear Natural Seaway

If linear wave superposition to model natural seaways appears too inaccurate, higher order approximations may be applied. At first, the second-order approximation is given. Whereas in linear waves the complex form of formulas appears simpler both for theory and programming, in nonlinear waves the real form (which was used already for Stokes third-order waves) appears often more suitable.

In a similar way as Stokes third-order regular waves were derived, one can derive formulas equivalent to (4.66) and (4.67) that are accurate up to second order:

$$z_s(x, y, t) = \sum_{j=1}^{J} A_j \cos\chi_j - \frac{1}{2}\sum_{j=1}^{J} k_j A_j^2 \cos(2\chi_j)$$

$$-\frac{1}{2}\sum_{j=1}^{J-1}\sum_{l=j+1}^{J} A_j A_l[(k_j + k_l)\cos(\chi_j + \chi_l) + (k_j - k_l)\cos(\chi_j - \chi_l)]; \tag{4.71}$$

$$\phi(x, y, z, t) = \sum_{j=1}^{J} c_j A_j e^{-k_j z} \sin\chi_j - \sum_{j=1}^{J-1}\sum_{l=j+1}^{J} \omega_l A_j A_l e^{(k_j - k_l)z} \sin(\chi_j - \chi_k). \tag{4.72}$$

The formulas require that k_j are ordered: Greater j implies greater k_j and, thus, greater $\omega_j = \sqrt{gk_j}$. Furthermore the following abbreviation was used:

$$\chi_j = \omega_j t - k_j x \cos\mu_j + k_j y \sin\mu_j + \epsilon_j, \tag{4.73}$$

where ϵ_j is an arbitrary phase angle. Also here, before using published formulae it is recommended to test them numerically for satisfying the dynamic and kinematic free-surface condition up to second order (i.e. errors proportional to A^3), because sometimes wrong formulae have been published. As discussed in the linear case, also for second-order wave superposition, contributions of short waves, when evaluated high above the still water level, may produce large errors if no appropriate precautions are made.

To simulate natural seaways more accurately, including exceptionally high (freak or rogue) waves, various numerical methods are applicable. Especially suitable due to its versatility, accuracy, and short computing time appears the 'higher-order spectral' (HOS) method [5] realized in a public-domain program. Starting from given initial fields of z_s and ϕ at the water surface (computed, e.g. using the linear approximation (4.66) and (4.67)), the further development of the wavefield is computed in a rectangular domain, assuming that waves leaving one boundary enter at the opposite boundary. Arbitrary orders of approximation may be selected. The method uses a Taylor expansion of ϕ around $z = 0$, the Fast Fourier transformation over the horizontal plane, and Runge-Kutta time integration.

4.2.3 Wave Spectrum

A natural seaway is characterized by the distribution of its energy per unit area, $\rho g S(\omega, \mu)$, over angular frequency ω and propagation direction μ. S is called the wave spectrum. The energy of a regular wave per unit area is $\frac{1}{2}\rho g A^2$. Correspondingly, if a natural seaway is approximated by a finite number of regular waves (as described in Sect. 4.2.1), the amplitudes of the latter are derived from the wave spectrum as

$$A = \sqrt{\int_{\omega_1}^{\omega_2} \int_{\mu_1}^{\mu_2} 2S(\omega, \mu) d\omega d\mu}, \tag{4.74}$$

where ω_1, ω_2, μ_1 and μ_2 are the limits of the (ω, μ)-Region which is modeled by one of those regular waves.

Often a seaway contains contributions from the actual wind (the wind sea) and of previous wind fields of higher velocity (called swell). Swell generated by lower than the actual wind speed combines with the actual wind sea after a short while. Thus, in extreme seas swell is rather unimportant.

Whereas swells may have various forms of energy distribution, the spectra of wind seas closely approximate a common shape family, the JONSWAP (for Joint North Sea Wave Project; [6]) spectrum (Fig. 4.2):

$$S(\omega, \mu) = \frac{\alpha g^2}{\omega^5} \exp\left(-\frac{5}{4}(\omega_p/\omega)^4\right) \gamma^\Gamma D(\omega, \mu) \tag{4.75}$$

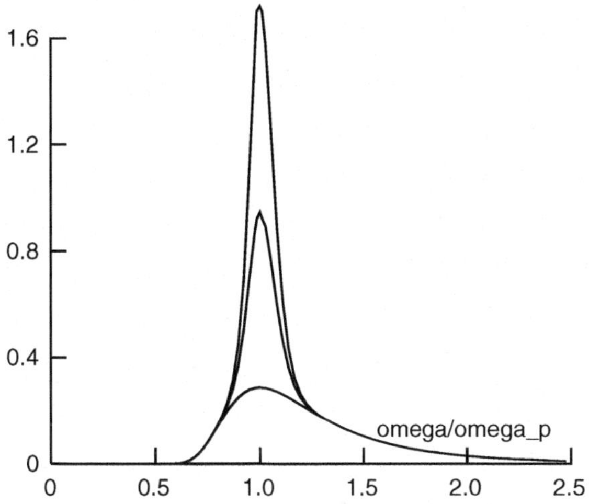

Fig. 4.2 Non-dimensional Jonswap spectrum $S\omega_p^5/(\alpha g^2)$ for peak enhancement factors 1, 3.3 and 6

Here, ω_p is the peak frequency (maximum of S over ω; independent of μ). $\gamma \geq 1$ is called 'peak enhancement factor' because γ^Γ enhances the peak of the spectrum by the factor γ, but lets S nearly unchanged at substantially lower and higher frequencies. This is due to the exponent

$$\Gamma = \exp\left(-\frac{(\omega - \omega_p)^2}{2\sigma^2\omega_p^2}\right) \tag{4.76}$$

with

$$\sigma = \big(0.07 \text{ if } \omega < \omega_p \text{ else } 0.09\big)\,. \tag{4.77}$$

For $\gamma = 1$, the spectrum is called Pierson-Moskowitz spectrum. It approximates the wind sea spectrum under long stationary wind conditions. In other cases, a reasonable approximation is

$$\gamma = 1.7 + 6\log_{10}(U/c_p), \tag{4.78}$$

where U is the wind speed 10 m above the mean sea level, and c_p is the phase speed for wave frequency ω_p. It is related to the phase speed c_p and the wave number k_p of those waves by the approximate relation

$$c_p = \frac{\omega_p}{k_p} \approx 1.03\sqrt{\frac{g}{k_p}} = 1.06\frac{g}{\omega_p}, \tag{4.79}$$

where the factors 1.03 and 1.06 take account of nonlinear effects. In many papers the peak enhancement factor $\gamma = 3.3$ is used because this was the average value measured in the wave measuring project [6]. $U/c_p > 1$ and thus $\gamma > 1$ occurs when the seaway did not have enough space or time to fully adapt to the actual wind.

The factor α in (4.75) depends on the 'significant wave height' H_s, which is, loosely defined, the mean of the 1/3 largest wave heights (from crest to trough). From the exact definition

$$H_s = 4\sqrt{\int_0^\infty \int_0^{2\pi} S(\omega, \mu) d\mu d\omega} \tag{4.80}$$

and the given empirical formulae follows the approximation

$$\alpha = \left(\frac{\omega_p^2 H_s}{g}\right)^2 / (2.64 + 0.645\gamma). \tag{4.81}$$

The above empirical relations were derived from various sources, most notably from [6–8]. This holds also for the factor $D(\omega, \mu)$ describing the energy distribution over wave direction μ. Of the various, largely equivalent formulae given for D we present here that given in [8] (Fig. 4.3):

$$D(\omega, \mu) = \frac{0.5\beta}{\cosh^2[\beta(\mu - \mu_0)]} \tag{4.82}$$

where μ_0 is the direction of maximum wave energy, corresponding to the mean wind direction. β determines the 'breadth' of the distribution. An empirical approximation is (Fig. 4.4):

$$\beta = \begin{pmatrix} 2.61(\omega/\omega_p)^{1.3} & \text{if} \quad \omega/\omega_p < 0.95 \\ 2.28(\omega/\omega_p)^{-1.3} & \text{if} \quad \omega/\omega_p \geq 0.95 \\ \text{at least} & 1.24 \end{pmatrix}. \tag{4.83}$$

4.2.4 Relations Between Wind and Seaway

Wave spectra can be predicted from the present and past wind field by numerical methods. Here, only approximations for idealized conditions are given. It is assumed that starting from a calm sea, a seaway is produced by a wind speed U (measured 10 m above the mean sea level), where U is constant over a time T and a fetch F (length of wind field upwind from the actual location). Then the spectrum parameters may be approximated by

$$U/c_p = \text{Max}\left(0.95,\ 18(gF/U^2)^{-0.3},\ 110(gT/U)^{-3/7}\right) \tag{4.84}$$

and

$$\alpha = 0.009(U/c_p)^{0.55}. \tag{4.85}$$

U is related to the Beaufort wind scale B_F:

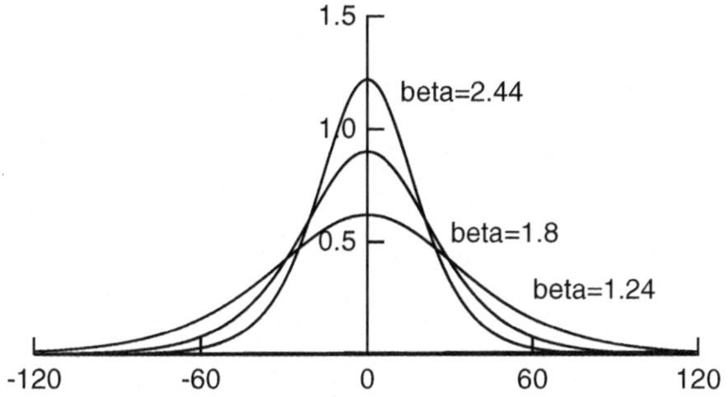

Fig. 4.3 Angular distribution D of wave energy over $\mu - \mu_0$

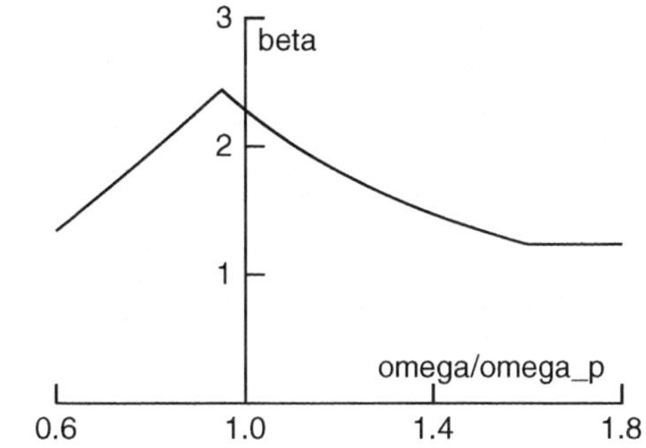

Fig. 4.4 Dependence of β on ω/ω_p

$$U = 0.836 B_F^{1.5}. \tag{4.86}$$

These approximations result in a significant wave height according to Fig. 4.5 and a peak period $2\pi/\omega_p$ according to Fig. 4.6. The relation between ω_p and the mean angular frequency ω_0 of up-crossing zeros is

$$\omega_0 \approx \omega_p \sqrt{1 + \gamma^{-0.4}}. \tag{4.87}$$

If wind speed U, fetch F and duration T are given, approximate seaway parameters and the corresponding wave spectrum can be determined. Equation (4.84) gives the 'maturation parameter' U/c_p; from this follow the peak phase speed c_p, the 'Phillips constant' α using (4.85), and the peak enhancement factor γ using (4.78). The significant wave height H_s follows from α using (4.81). The peak angular frequency ω_p follows from c_p using (4.79), giving also the peak period $T_p = 2\pi/\omega_p$. Equation (4.87) gives the zero-upcrossing angular frequency ω_0, from which the zero-upcrossing period T_0 follows correspondingly. Finally, the angular spreading of wave energy follows from (4.82) together with (4.83). Thus all quantities required to evaluate the seaway spectrum given by Eqs. (4.75)–(4.77) are known.

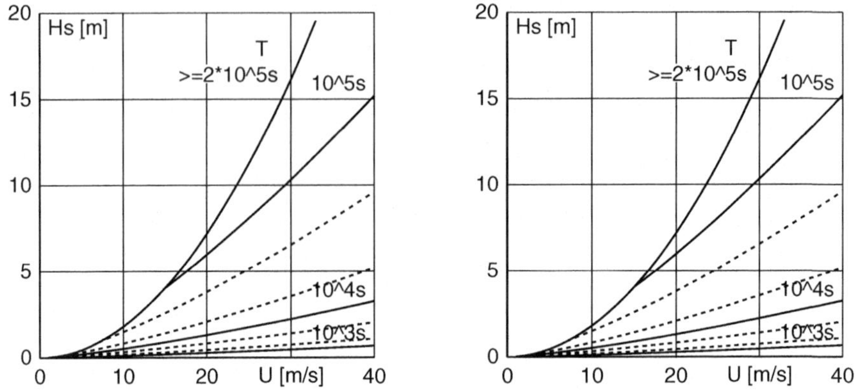

Fig. 4.5 Significant wave height H_s depending on wind speed U and either wind duration T (left) or fetch F (right). Broken lines are for parameter values 2 times and 0.5 times of those given for the next continuous curve. The lower value of both diagrams should be taken

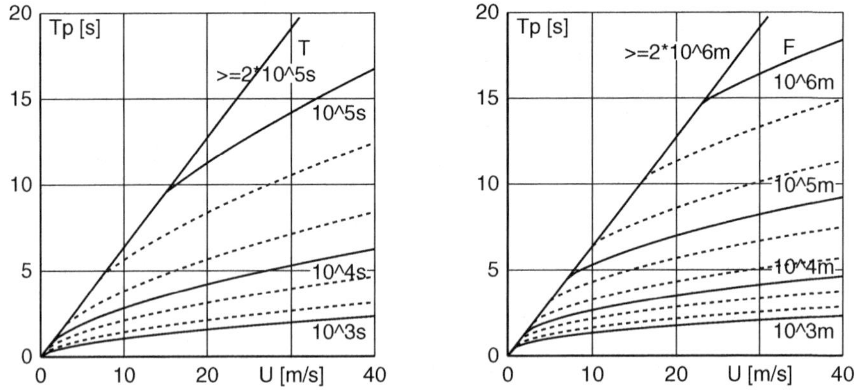

Fig. 4.6 Like Fig. 4.5, however, for peak period $T_p = 2\pi/\omega_p$

4.2.5 Simulation of Natural Seaways

To model nonlinear responses of fixed or floating bodies in a natural seaway, numerical simulation is the most appropriate tool. To simulate a seaway with a given spectrum $S(\omega, \mu)$ using linear superposition, for each elementary wave, we have to determine its amplitude A, angular frequency ω, direction μ, and phase angle ϵ.

The phase angle ϵ should be selected as a random number equally distributed between 0 and 2π and uncorrelated to the phase angles of the other elementary waves. To determine A, the range of ω and μ values is subdivided into a reasonable number of intervals, using smaller intervals where S is larger, and determining A from (4.74). (Another possibility is to select the uncorrelated quantities $A \cos \epsilon$ and $A \sin \epsilon$ from their probability distributions.) For each elementary wave, it is recommended to determine ω and μ also as random numbers equally distributed over the representative (ω, μ) interval. This avoids incorrect statistical properties of the seaway which can occur for a more deterministic selection of ω and μ. For instance, if the ω of the different elementary waves are integer multiples of a common interval $\Delta\omega$, then the total wavefield repeats after a time $2\pi/\Delta\omega$. Thus, a simulation extended over more than half of the repetition time would involve erroneous correlations. (The selection of equidistant frequencies is advantageous because it allows to reduce the computing time by applying the Fast Fourier Transform algorithm; however, the wrong statistics exclude this possibility.)

Linear superposition of elementary Airy waves has the same deficiencies as Airy waves themselves; especially the wider wave troughs and sharper wave crests are not shown in the linear superposition. To use nonlinear Stokes waves as 'elementary waves' would not be correct, because the nonlinearity would disappear in the limit of many elementary waves, each of which approach zero steepness. A second-order superposition according to Sect. 3.2 may be recommended, at least for testing whether it gives substantially different results for the response of interest. If not, linear superposition would be preferable as it is computationally faster because it lacks the double sums over elementary waves.

4.2.6 Statistics of Seaway Parameters

The main parameters characterizing a seaway are significant height H_s, mean period T_m, and direction μ_0. If we designate as the ith spectrum moment the quantity

$$m_i = \int\int \omega^i S(\omega, \mu) d\omega d\mu, \tag{4.88}$$

the quantity T_m may be defined as either $2\pi m_0/m_1$ or $2\pi\sqrt{m_0/m_2}$. In the following, we use the latter definition. It corresponds to the previously defined mean zero-up-crossing frequency ω_0 of (4.87), i.e.

$$T_m = 2\pi/\omega_0. \tag{4.89}$$

The same definition is used also by the 'KNMI/ERA-40 Wave Atlas' [9, 10]. It is an extremely comprehensive, freely accessible source of global wave data based on, besides others, hind-cast seaway spectra computed from the global wind field between the years 1957 and 2002. Table 4.1 shows an example taken from the atlas. A corresponding table of IACS (International Association of Classification Societies) contained in Recommendation 34 Revision 1 of 2001, which is still often used and recommended by ISSC (International Ship and Offshore Structures Congress), is definitely erroneous by giving up to 5 m too high H_s for a given T_m.

Table 4.1 10^4 times probability of combinations of H_s and T_m for all wave directions and months at a position in the North Atlantic (longitude 324.33°, latitude 54.63°)

$H_s[m]$ $T_m[s]$	0-1	1-2	2-3	3-4	4-5	5-6	6-7	7-8	8-9	9-10	10-11	11-12	12-13	13-14	14-33	Sum
0–3	0	0	0	0	0	0	0	0	0	0	0	0	0	0	0	0
3–4	4	22	0	0	0	0	0	0	0	0	0	0	0	0	0	25
4–5	24	501	123	1	0	0	0	0	0	0	0	0	0	0	0	648
5–6	65	725	902	224	7	0	0	0	0	0	0	0	0	0	0	1921
6–7	24	702	697	815	384	31	1	0	0	0	0	0	0	0	0	2655
7–8	1	192	579	484	513	428	113	8	0	0	0	0	0	0	0	2318
8–9	0	14	195	365	265	263	266	145	28	2	0	0	0	0	0	1543
9–10	0	0	11	90	151	115	76	95	90	35	5	0	0	0	0	669
10–11	0	0	0	3	22	42	30	18	18	25	19	7	1	0	0	186
11–12	0	0	0	0	0	3	7	5	2	2	3	5	4	1	0	31
12–13	0	0	0	0	0	0	0	1	1	0	0	0	0	0	1	3
Sum	117	2156	2507	1982	1342	881	493	273	138	63	27	12	5	2	1	10000

4.3 Appendix A: Fortan90 Program for Testing Formula (4.41)

```
program wave !Stokes third-order
implicit none
real(kind=8),parameter:: pi=4*atan2(1.,1.)
integer:: i
real(kind=8)::g=10.,a(2)=(/0.1,0.01/),k=0.1,psi,ka(2),omega(2),z(2),phit(2),pdivrho(2)
ka=k*a
omega=sqrt(g*k*(1+ka**2))
For9Phases: do i=0,8
 psi=pi*i/8.0      !stands for: omega t - k xi
 z=a*( (1+5./8.*ka**2)*cos(psi) - 0.5*ka*cos(2*psi) + 3./8.*ka**2*cos(3*psi))
 phit=omega**2/k*a*cos(psi)*exp(-k*z)
 pdivrho=g*z-phit-0.5*(omega*a*exp(-k*z))**2
 print *,psi*180/pi,pdivrho  !The ratio of two printed p/rho
  values should be 10**4!
 !One should test correspondingly also the kinematic condition
  z_t + phi_x z_x -phi_z = 0!
enddo For9Phases
end program wave
```

References

1. G.B. Airy, Tides and waves, *Encyclopaedia Metropolitana. B. Fellows* (1841), pp. 241–396
2. G.G. Stokes, On the theory of oscillatory waves. Trans. Camb. Philos. Soc. **8**, 441–455 (1847)
3. J.D. Fenton, Use of the programs FOURIER, CNOIDAL and STOKES for steady waves, *The Sea*, vol. 9, ed. by B. Le Mehaute, D.M. Hanes (1990)
4. J.D. Fenton, Nonlinear wave theories (2018), http://johndfenton.com/Steady-waves/Instructions.pdf
5. G. Ducrozet, F. Bonnefoy, P. Ferrant, *HOS-Ocean: Open-Source Solver for Nonlinear Waves in Open Ocean Based on High-Order Spectral Method* (Elsevier, Amsterdam, 2016)
6. K. Hasselmann, Measurements of wind-wave growth and swell decay during the Joint North Sea Wave 543 project (JONSWAP). Ergänzungsheft zur Deutschen Hydrographischen Zeitschrift Reihe **12**, 95 (1973)
7. K. Richter, Seaway as a basis for predicting hydrodynamic loads (in German). Jahrbuch der Schiffbautechnischen Gesellschaft **75**, 247–271 (1981)
8. M.A. Donelan, J. Hamilton, W. Hui, Directional spectra of wind-generated waves. Proc. Philos. Trans. R. Soc. A315 (1985)
9. S. Caires, A. Sterl, G. Komen, V. Swail, The KNMI/ERA-40 wave atlas derived from 45 years of ECMWF reanalysis data (2013), http://www.knmi.nl/waveatlas
10. S.M. Uppala, The ERA-40 re-analysis. Meteorol. Soc. **131**, 2961–3012 (2005)

Chapter 5
Strip Methods

Abstract This chapter describes the traditional method for seakeeping analyses. Seventy years after the pioneering work of Ursell to determine added mass and damping of ship sections oscillating at the water surface, this book describes a highly accurate, mathematically simple, and computationally fast method for this. For the sake of brevity, the step from ship sections in forced motion to free motions of a total ship with speed ahead is made by heuristic considerations. A special feature of this chapter is the speed-dependent hydrodynamic interaction between hulls, which is important for catamarans and trimarans.

5.1 History

In early attempts to compute ship motions in waves, the hydrodynamic force and moment on the hull were determined from the pressure distribution in a regular wave, undisturbed by the ship. This resulted in the so-called Froude-Krylov force and moment, which neglect the change of the pressure distribution by the presence of the hull. It turned out that this gave unreasonable results because the effects of the added mass of the water on ship motions, the damping by waves generated by the oscillating body, and the diffraction of incoming waves by the body are essential to obtain correct results.

Korvin-Kroukovski [1] proposed a better, simple method to compute the vertical (heave) force and the pitch moment (about the transverse axis) acting on a ship sailing in a regular wave. From the force and moment follows the combined heave and pitch motion of the ship. Force and moment were derived by integrating over ship length the force per length on ship sections performing heave motions relative to the water surface. For semicircular ship sections, Ursell [2] published a potential method to compute these forces, including added mass and damping effects. Wave effects were dealt with by assuming that the relative motion between water surface and body section determined the hydrodynamic force on the section. Grim [3, 4] extended Ursell's method to more general section shapes, and he considered body motion and wave motion separately.

B. O. el Moctar et al., *Numerical Methods for Seakeeping Problems*,
https://doi.org/10.1007/978-3-030-62561-0_5

The original publication of Korvin-Kroukovski was improved substantially by Korvin-Kroukovski and Jacobs [5] with respect to forward speed effects. Their method, which was widely applied later on, is often called OSM (ordinary strip method). Further improvements followed; here only two of them, which were and are widely applied even today, are the methods of Söding [6] and Salvesen, Tuck and Faltinsen [7].

Later extensions dealt with motions in all six degrees-of-freedom of a rigid body, including shear forces and bending moments, second-order drift forces, multi-hulls, asymmetric bodies (asymmetrical catamarans, heeled ships), surface-effect ships and their air cushion, with interaction effects between several bodies near each other, nonlinear effects, etc.

The accuracy of the results of strip methods is less than the accuracy obtainable with three-dimensional potential flow methods and finite-volume methods (Euler or RANSE solvers). Nevertheless, strip methods are used even today because their accuracy appears sufficient for many purposes and because they require less computing time than other methods. This is important, especially for optimizations.

The following elaboration is based mainly on the 'public-domain strip method' PDSTRIP (Bertram et al. [8–10]).

5.2 Computing Added Mass, Damping, and Excitation in Two-Dimensional Flow

The strip method requires means of computing force, moment, and pressures on transverse ship sections oscillating at the water surface disturbed by regular waves. The flow is assumed to be confined to the y, z plane, where y points to starboard, z downward. Except when propagating in $\pm y$ direction, the incoming wave is not confined to the y, z plane. Therefore, one combines the flow from that of the wave (described by the Airy potential) and a disturbance. Only the latter is assumed to be confined to the y, z plane. In reality, the disturbance has also longitudinal components. Because of this, strip methods are not applicable in short waves, say for wave lengths $\leq$0.4L, except when using certain tricks not described here, or for waves running in the transverse direction ($\mu = \pm 90°$). Also, in long waves the neglect of longitudinal components of the disturbance flow leads to inaccuracies, but these are less relevant because the disturbance flow is less important than hydrostatic pressure and wave pressure in long waves.

Typically, strip methods use, potential methods to determine added mass, damping and excitation force, and moment on ship sections or, equivalently, on cylindrical three-dimensional bodies. There are various methods to compute these responses. In the following, a Rankine source method is described based on Söding [11, 12], here, however, restricted to responses depending linearly on motion or wave amplitude. The linearization allows separating added mass and damping, from wave excitation.

The method is described below for the case of deep water. However, it may be extended easily to finite water depth by
(a) using the finite-depth potential of the incoming wave,
(b) applying the relation between wave frequency and wave number for finite water depth, and
(c) using, instead of a point source, a pair of an original source and its mirror image below the water bottom. If both sources have the same strength, the condition on a flat bottom (no flow through it) is satisfied.

5.2.1 Fundamental Equations

The water is approximated as an ideal (i.e., incompressible, non-viscous) fluid. The flow is assumed irrotational, thus, it can be described by a potential $\phi(\mathbf{x}, t) = \phi(y, z, t)$ depending on location $\mathbf{x}$ and time t. Coordinate axes point normal to the flow plane (x), to starboard (y), and downward (z). The coordinate origin is assumed within the cylindrical body (axis in x-direction) in height of the undisturbed water surface. $\zeta(y, t)$ designates the z coordinate of the free surface.

Because of the incompressibility, the Laplace equation

$$\Delta\phi = \phi_{yy} + \phi_{zz} = 0 \tag{5.1}$$

must be satisfied everywhere within the fluid. Indices y and z, and later t, indicate partial derivatives.

There is no flux through the cross section. This is described by the boundary condition

$$\nabla\phi \cdot \mathbf{n} = (\boldsymbol{\tau}_t + \boldsymbol{\varphi}_t \times \mathbf{x}) \cdot \mathbf{n}, \tag{5.2}$$

where $\mathbf{n}$ is the unit normal vector (pointing into the body) on the section contour. $\boldsymbol{\tau}(t)$ is the time-dependent translation vector (within the yz plane), and $\boldsymbol{\varphi}(t)$ is the rotation vector of the section (in x direction) about the coordinate origin. The condition must be satisfied along with the instantaneous position of the moving section contour. By means of a Taylor expansion of all terms around the average contour, it can be converted to a condition to be satisfied along the average contour. Both conditions differ only with respect to terms of higher than first order in motion amplitude. To determine first-order responses, nonlinear terms can be omitted; thus, the condition needs to be satisfied only on the non-moving average body contour.

Constant pressure on the water surface gives the dynamic free-surface boundary condition

$$g\zeta - \phi_t - \frac{1}{2}(\nabla\phi)^2 = 0, \tag{5.3}$$

which is to be satisfied at $z = \zeta(y)$ outside of the body. Here g is the gravity acceleration. As above for the body boundary condition, to determine only first-order responses the nonlinear term $\frac{1}{2}(\nabla\phi)^2$ can be omitted, and the resulting condition

$$g\zeta - \phi_t = 0 \tag{5.4}$$

can be satisfied using $\phi_t(y, 0, t)$ instead of $\phi(y, \zeta, t)$.

Water particles on the free surface remain on the free surface. This kinematic condition is expressed by

$$\frac{D}{Dt}(\zeta - z) = \left(\frac{\partial}{\partial t} + \nabla\phi \cdot \nabla\right)(\zeta - z) = 0, \tag{5.5}$$

where D/Dt indicates the substantial time derivative (i.e. at a point moving with the fluid). Evaluating the partial derivatives of z and ζ gives

$$\zeta_t + \phi_y\zeta_y - \phi_z = 0. \tag{5.6}$$

Again, for responses up to first order, the nonlinear term $\phi_y\zeta_y$ can be omitted, and the resulting equation

$$\zeta_t - \phi_z = 0 \tag{5.7}$$

can be satisfied at the average free surface $z = 0$.

For $z \to \infty$, the fluid velocity should tend to zero in deep water (bottom condition). Furthermore, we want to compute the flow generated alone by the incoming wave and the (moving or stationary) body, not by other bodies, by wave reflections at walls, etc. This is usually expressed by considering, besides the incident wave, far from the body only out-going waves, no waves running to the body. This is the so-called radiation condition.

5.2.2 *Preparation of the Flow Potential*

As the flow is assumed to depend linearly on the sinusoidal wave flow, it oscillates harmonically over time:

$$\phi(y, z, t) = \mathrm{Re}[\hat{\phi}(y, z)e^{i\omega_e t}]. \tag{5.8}$$

The hat symbol is used to designate complex amplitudes, here the amplitude of the potential ϕ. ω_e is the encounter angular frequency given in (4.24). Corresponding equations hold for the potential of the incident wave, for the body motions and all other linear responses: oscillate harmonically with frequency ω_e. Thus, we will deal in the following not directly with time functions, but with their complex amplitudes because the time functions follow from them simply by multiplication by $e^{i\omega_e t}$ and taking the real part of the product. From (5.8) follows that derivatives $\partial/\partial t$ for time functions correspond to a multiplication by $i\omega_e$ for complex amplitudes.

ϕ is the sum of contributions from the wave (Airy) potential and the disturbance potential ϕ_D. Thus, its complex amplitude is, according to (4.15),

$$\hat{\phi}(y, z) = -icAe^{-kz-ikx\cos\mu+iky\sin\mu} + \hat{\phi}_D. \tag{5.9}$$

Here, $c = \omega/k$ is the phase velocity, ω the angular frequency, and k the wave number of the incoming wave. A is the amplitude of the incoming wave. Here, we assume that

$$Ae^{-ikx\cos\mu} = 1, \tag{5.10}$$

where x is the longitudinal coordinate of the section. Later, results will be scaled to account for the actual amplitude. Combining (5.9) and (5.10) gives

$$\hat{\phi}(y, z) = -ice^{-kz+iky\sin\mu} + \hat{\phi}_D. \tag{5.11}$$

The disturbance potential ϕ_D is further subdivided into contributions ϕ_2 caused by the sway motion of the section, ϕ_3 by heave, ϕ_4 by roll, and ϕ_7 caused by the interaction between the incoming wave and the non-moving section. Thus, the complex amplitude of the total potential ϕ is composed of 5 contributions:

$$\hat{\phi} = -ice^{-kz+iky\sin\mu} + \hat{\phi}_2 + \hat{\phi}_3 + \hat{\phi}_4 + \hat{\phi}_7. \tag{5.12}$$

Here ϕ_2 to ϕ_4 are called radiation potentials because they contain the waves radiated by the moving body, and ϕ_7 is called diffraction potential because it models the diffraction of the incoming wave by the body at rest. The three radiation potentials are determined here for unit motion amplitudes; the actual amplitudes will be taken into account later by scaling the results.

The body boundary condition for the complex amplitude of the sway motion follows from (5.2):

$$\nabla\hat{\phi}_2 \cdot \mathbf{n} = (0, 1, 0) \cdot \mathbf{n} = n_2, \tag{5.13}$$

where n_2 is the y-component of the unit normal vector $\mathbf{n}$. Here and in the following, three terms separated by a comma and enclosed between parentheses designate a three-dimensional vector defined by its three components.

Correspondingly, the body boundary conditions for ϕ_3 and ϕ_4 are

$$\nabla\phi_3 \cdot \mathbf{n} = n_3 \tag{5.14}$$

and

$$\nabla\hat{\phi}_4 \cdot \mathbf{n} = [(1, 0, 0) \times \mathbf{x}] \cdot \mathbf{n} = -zn_2 + yn_3. \tag{5.15}$$

The body boundary condition for the complex potential of the diffraction potential ϕ_7 follows from (5.2) and (5.11):

$$\nabla\hat{\phi}_7 \cdot \mathbf{n} = \nabla(ice^{-kz+iky\sin\mu}) \cdot \mathbf{n}. \tag{5.16}$$

The free surface conditions follow from (5.4) and (5.7):

$$g\hat{\zeta} = i\omega_e\hat{\phi}; \quad i\omega_e\hat{\zeta} = \hat{\phi}_z. \tag{5.17}$$

Eliminating $\hat{\zeta}$ and using the dispersion relation (4.35) gives

$$k\hat{\phi} + \hat{\phi}_z = 0. \tag{5.18}$$

This homogeneous condition holds for each of the potentials ϕ_2, ϕ_3, ϕ_4 and ϕ_7 on $z = 0$ outside of the body.

5.2.3 Numerical Method

$\hat{\phi}_2$ to $\hat{\phi}_7$ are approximated as a superposition of point sources at locations $\mathbf{x}_j$:

$$\hat{\phi}_D(\mathbf{x}) = \sum_j G(\mathbf{x}, \mathbf{x}_j)\hat{q}_j \tag{5.19}$$

with Green functions

$$G(\mathbf{x}, \mathbf{x}_j) = \log|\mathbf{x} - \mathbf{x}_j|. \tag{5.20}$$

Each Green function is the potential of a two-dimensional point source of strength 2π. The superposition factors $\hat{q}_j$ determine the source strengths. They are the complex amplitudes of the ejected fluid area per time divided by 2π. They will be determined so that the body and free-surface boundary conditions are approximately satisfied.

The Green functions satisfy the two-dimensional Laplace equation (5.1). As the Laplace equation is linear, it is satisfied also by the superposition (5.19) of Green functions. The Airy potential, i.e., the first term on the right-hand side of (5.11), satisfies the three-dimensional Laplace equation

$$\phi_{xx} + \phi_{yy} + \phi_{zz} = 0. \tag{5.21}$$

Thus, the two-dimensional Laplace equation is violated by the term ϕ_{xx}. This is, again, an error accepted in strip methods for simplification.

Contrary to the more usual panel methods, which use distributed sources, here point sources will be used. That necessitates to satisfy the boundary conditions not at collocation points, but on the average over each panel (contour segment). This 'patch method' is described in depth in Sect. 3.8. The flux $\int \nabla G\,\mathbf{n}\,ds$ through a panel generated by a single source j of strength 2π is the angle under which the panel is seen from the source location $\mathbf{x}_j$. If the end points of the panel are $\mathbf{x}_a$ and $\mathbf{x}_b$, this angle is

$$\beta_j = \arctan\frac{[(\mathbf{x}_a - \mathbf{x}_j) \times (\mathbf{x}_b - \mathbf{x}_j)]_1}{(\mathbf{x}_a - \mathbf{x}_j) \cdot (\mathbf{x}_b - \mathbf{x}_j)}. \tag{5.22}$$

The index 1 in the numerator designates the x-component of the vector product. β_j has the correct sign (i.e., β_j is positive if the flux has the direction of $\mathbf{n}$) if, seen in $+x$ direction, the vector $\mathbf{x}_b - \mathbf{x}_a$ points anti-clockwise about $\mathbf{x}_j$.

The submerged part of the cross section is described by offset points starting at the port and ending at the starboard intersection of the cylinder with the rest waterline $z = 0$. Between offset points, the contour is approximated as a straight line called panel. Fully submerged sections, and sections consisting of more than one body, should also be accounted for in programs to model ship sections of a bulbous bow located in front of the forward perpendicular.

On the mean free surface $z = 0$, n_f panels are generated on either side of the body. For each panel the condition (5.18), integrated over the panel length, is satisfied. $n_f = 40$ is, generally, enough to compute accurate results for first-order force and moment. The length of free-surface panels adjacent to the body is chosen equal to the length of the uppermost body panels. The next outer panels increase in length by a constant factor up to a maximum length of 1/15 of the wave length. Farther outside, the panel length is held constant.

The discretized boundary conditions are as follows:

- For body panels:

$$\sum_j \int \nabla G \cdot \mathbf{n}\, ds\, \hat{q}_j = \sum_j \beta_j \hat{q}_j = \int I\, ds, \tag{5.23}$$

  where the integrals over arc length s (also in the following) cover just one panel. I is the in-homogeneous part of the boundary conditions (5.13)–(5.16).

- For free surface panels:

$$\sum_j \int (kG + G_z)\, ds\, \hat{q}_j = 0. \tag{5.24}$$

For the diffraction potential, the right-hand side of (5.23) is evaluated analytically:

$$\int \nabla(ice^{-kz+iky\sin\mu})\mathbf{n}\, ds = ce^{-kz_a+iky_a\sin\mu}(e^{ik\Delta y\sin\mu - k\Delta z)} - 1), \tag{5.25}$$

where $\Delta y = y_b - y_a$ and $\Delta z = z_b - z_a$ depend on the panel end points a and b.

Equation (5.24) contains the integral over a source potential and its normal derivative. The first of these integrals is evaluated analytically between $(y_a, 0)$ and $(y_b, 0)$:

$$\int_{y_a}^{y_b} G\, dy = \int_{y_a}^{y_b} \frac{1}{2}\log(y^2 + z_j^2)\, dy - \left[\frac{1}{2} y \log(y^2 + z_j^2) + z_j \arctan\left(\frac{y}{z_j}\right) - y\right]_{y_a - y_j}^{y_b - y_j}. \tag{5.26}$$

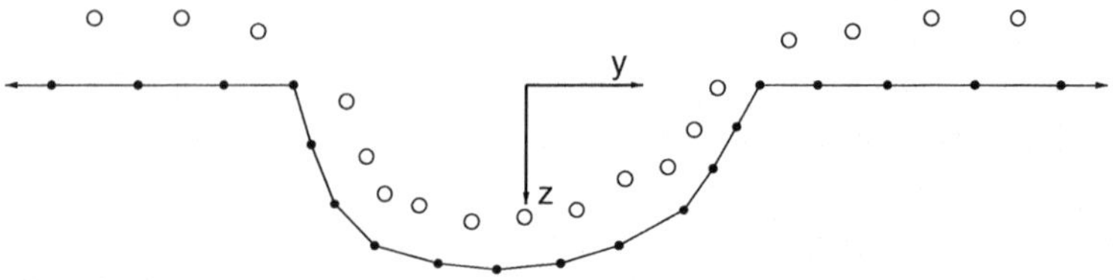

Fig. 5.1 Panel and source arrangement along water surface and body contour

The integral over the normal derivative of the source potential in (5.24) is evaluated using (5.22). The panel average of the derivative of G in the tangential direction is simply the difference between the G values at the ends of the panel divided by panel length.

The bottom condition $\lim_{z\to\infty} \nabla\phi = 0$ may be satisfied for the disturbance potential ϕ_D by requiring that the sum of source strengths is 0. (An additional source must then be arranged to obtain an equal number of conditions and unknown source strengths.) However, it was found that this is not necessary here because that condition is satisfied automatically with a high degree of accuracy, and that it did not improve the consistency of numerical results.

The accuracy of results is quite sensitive, however, to the details of satisfying the radiation condition. Here, a method is used that is simple and accurate, but perhaps not optimal with respect to computing time. In the free-surface condition, the wave number $k = \omega^2/g$ is computed not with the real gravity acceleration g, but with its complex analog

$$\tilde{g} = g[1 + i\alpha(y)]. \tag{5.27}$$

A positive imaginary part of $\tilde{g}$ dampens the waves. It proved advantageous to increase α proportional to the square of the distance to the near side of the body up to 0.5 at the outer end of the discretized free surface.

Sources are arranged within the body, typically 75% of the panel length shifted inside in normal direction from each panel midpoint (Fig. 5.1). However, if a source is outside of the body (it may occur in narrow parts of the section, e.g., at a skeg), or near any other panel, it is shifted nearer to the associated panel midpoint. The same holds if the distance between a source and the associated panel midpoint is larger than half the local curvature radius of the contour. Along the free surface, sources are arranged, correspondingly, at 0.75 panel length above the panel midpoints.

Equations (5.23) and (5.24) form a linear equation system to determine the unknowns $\hat{q}_j$. It has four right-hand sides and thus four sets of $\hat{q}_j$ to generate the potentials ϕ_2, ϕ_3, ϕ_4, and ϕ_7. In our programs, it is solved using the subroutine `cgesv` (LU decomposition for a fully occupied complex matrix) of the LAPACK system. The complex amplitudes of the four potentials are then computed at all body offset points using (5.11), (5.19), and (5.20).

In case of symmetrical section shapes, it would suffice to discretize only one symmetry half of the body and the free surface and to use, instead of each source, a pair of sources arranged at corresponding locations on both sides of the symmetry plane. The source strength of original and mirror source should be the same for heave motion, and of opposite sign for sway and roll motion. To determine the diffraction potential ϕ_7. The body boundary condition has to be split into two conditions which can be satisfied by

a) equal source strengths, and
b) source strengths of opposite sign, on port and starboard side.

If computer time is not relevant, it seems better to avoid this complexity.

5.2.4 Determination of Pressure, Force, and Moment

The pressure p (difference to air pressure) results from Bernoulli's equation:

$$p = \rho\left(gz - \dot{\phi} - \frac{1}{2}(\nabla\phi)^2\right). \tag{5.28}$$

As before, the last term is omitted because it is of second order. The complex amplitude of the first-order pressure oscillation at a body-fixed point is, thus,

$$\hat{p} = \rho(g\hat{z} - i\omega_e\hat{\phi}). \tag{5.29}$$

To determine the complex amplitude of the force $\hat{F}$ on the section, and of the moment $\hat{M}$ about the coordinate origin, we need the average of $\hat{p}_m$ over each body panel m. It is approximated as the mean of the values at both panel ends; more involved approximations do not improve the accuracy. From $\hat{p}_m$ follows

$$\hat{\mathbf{F}} = \sum_m \hat{p}_m \mathbf{f}_m \tag{5.30}$$

and

$$\hat{\mathbf{M}} = \sum_m \hat{p}_m \mathbf{x}_m \times \mathbf{f}_m. \tag{5.31}$$

Here, $\mathbf{f}_m$ is the panel length vector: normal on the panel, pointing into the body, vector length (or magnitude) = panel length. $\mathbf{x}_m$ is the position vector of the panel midpoint. Only the x component of $\hat{\mathbf{M}}$ is non-zero.

5.2.5 Verification

Figure 5.2 shows the amplitude of the fluid force on a heaving semicircle depending on oscillation frequency. It demonstrates (a) that 20 body panels are fully sufficient in this case and (b) that the method gives practically the same results as the quite

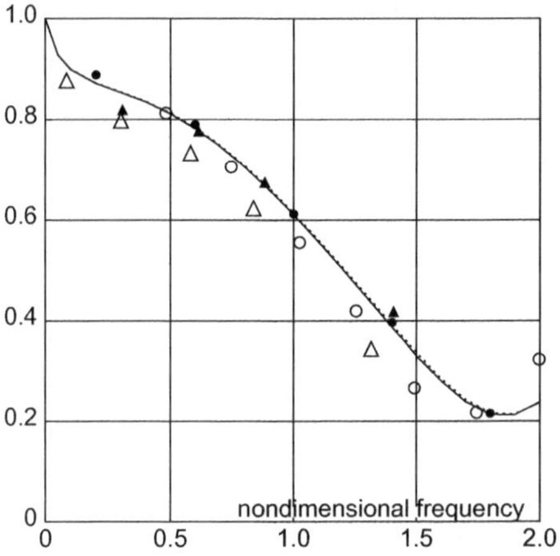

Fig. 5.2 Amplitude of vertical first-order force divided by $2\rho gAR$ on a semicircle of radius R heaving with amplitude A. Abscissa: $FP = \omega^2 R/g$. Curves: present method for 99 (continuous) and 20 (dotted) body panels. Markers computed by Papanikolaou ([13]; dots, potential method) and Gentaz et al. ([14], circles, Navier-Stokes) and measured by Tasai and Koterayama ([15], amplitude 0.2R, black triangles; and 0.4R, open triangles)

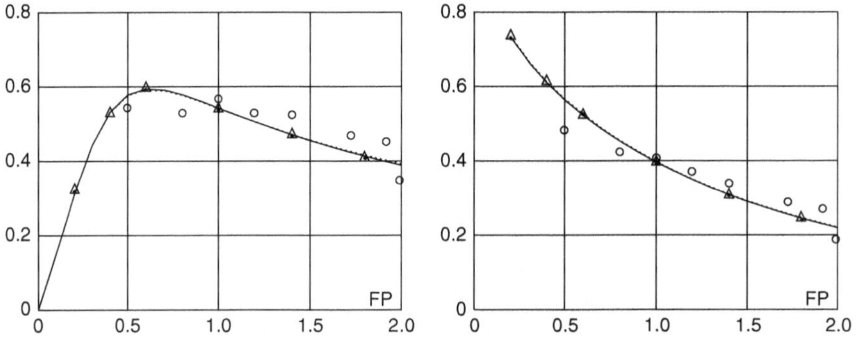

Fig. 5.3 Amplitude of horizontal (left) and vertical (right) first-order force on a non-moving semicircle of radius R in a regular transverse wave of amplitude A. Abscissa: $FP = \omega^2 R/g$; ordinate: force divided by $2\rho gAR$. Curves: present method for 99 (continuous) and 20 (dotted) body panels. Markers computed by Papanikolaou ([13]; dots, potential method) and Cheung ([16]; triangles; potential method) and measured by Kyozuka ([17]; circles)

different potential method of Papanikolaou [13], and also as experiments by Tasay and Koterayama for small heave amplitudes. The latter fact seems to indicate that the larger deviations of the viscous calculations of Gentaz et al. [14] are not caused by viscous effects.

Figure 5.3 shows amplitudes of the horizontal and vertical forces in a regular transverse wave. Also, the difference between results using 20 and 99 body panels

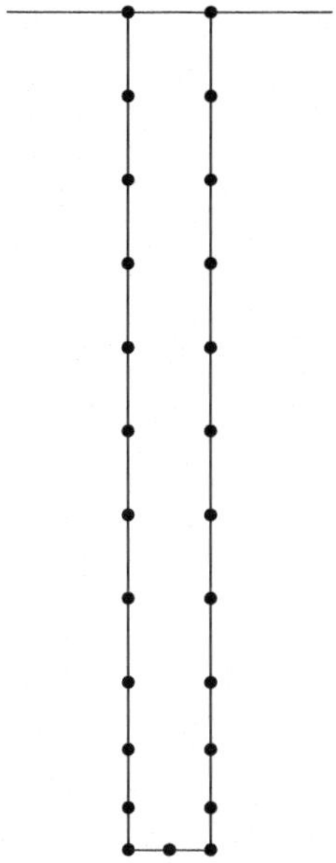

Fig. 5.4 Test section modeled by 24 panels

is hardly visible. The differences between the present results and those of different potential calculations of Papanikolaou [13] and of Cheung [16] are negligible.

A semi-circle is, probably, the 'easiest' shape. A thin rectangle (Fig. 5.4) used by Sutulo and Guedes Soares [18] for comparisons causes larger numerical errors because (a) the sources must be arranged nearer to the section contour and (b) the pressure varies extremely at the corners of the rectangle. Furthermore, potential methods do not model the flow separation at the corners occurring in real flow. In Fig. 5.5, however, only results of potential methods are compared.

Whereas in Figs. 5.2 and 5.3 the phase of the force was ignored, Fig. 5.5 gives the non-dimensional added mass m and damping d for a sway motion with complex translation amplitude $\hat{\tau}$. Here, m and d are defined by the equation

$$\hat{F} = -(-\omega_e^2 \hat{\tau} M + i\omega_e \hat{\tau} D) = -\left(-\omega_e^2 \underbrace{m \cdot \rho\pi T^2}_{M} + i\omega_e \underbrace{d \cdot \rho\pi T^2 \omega_e}_{D}\right)\hat{\tau}, \quad (5.32)$$

where M and D are the added mass and damping, respectively, per cylinder length, and m and d are their non-dimensional equivalents. T is the section draft. m and d follow, thus, from the real and imaginary, respectively, parts of $\hat{F}$. (The corresponding equation for heave contains an additional hydrostatic term $\rho g B\hat{\tau}$.)

Figure 5.5 shows that, for this 'difficult' section, 24 body panels give inaccurate results: However, for 48 panels, results of the present method coincide with those of the quite different potential method of Sutulo and Guedes-Soares.

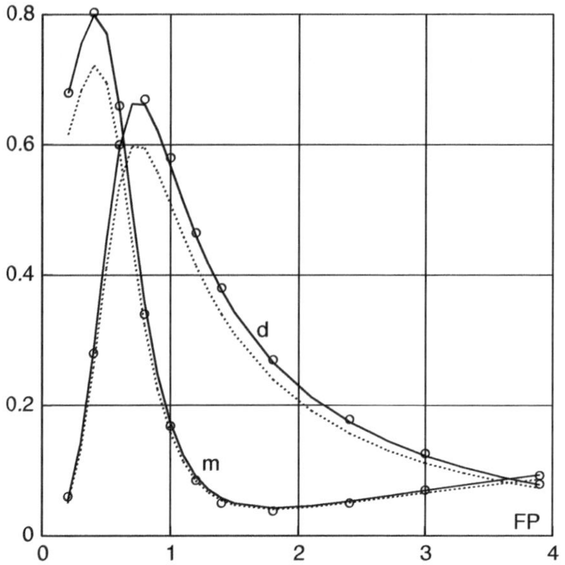

Fig. 5.5 Non-dimensional added mass (m) and damping (d) of a thin rectangular section (Fig. 5.4) in horizontal motion. Abscissa: $FP = \omega^2/g \cdot$ section draft. Curves: present method for 48 (continuous) and 24 (dotted) body panels. Markers: results of Sutulo and Soares [18]

5.3 Determination of Ship Motions in a Regular Wave

Like for the two-dimensional case, we deal here only with responses depending linearly on wave amplitude. In the following it is assumed to be unity. The immersed ship hull is described by a sufficient number of transverse sections. In case of a steady trim angle, the ship sections are shifted vertically, depending on their longitudinal position. Also the mass matrix of the body must be adapted to the trim; that means: it refers to horizontal and vertical axes while the ship is in its time-averaged position. A steady wave field generated by forward speed may be taken into account by arranging the section waterline above or below the undisturbed water surface.

In the following, the hull is assumed rigid.

5.3.1 Coordinate Systems

We need two-coordinate systems:

- An inertial system x, y, z, the origin of which follows the average forward motion of the ship. Its axes point forward (x), horizontally to starboard (y), and downward (z).
- A body-fixed system $\underline{x}$, $\underline{y}$, $\underline{z}$. Its origin and its axes coincide, in the time average, with the inertial system.

Ship motions relative to the inertial system are designated by $\mathbf{u}$ (translational motion of the coordinate origin) and $\boldsymbol{\alpha} = (\varphi, \theta, \psi)$ (rotation). Components 1–3 of $\mathbf{u}$ are the translations surge, sway, and heave (in x, y, z direction, respectively), and components 1–3 of $\boldsymbol{\alpha}$ are the rotations roll, pitch, and yaw (right-hand around the x, y, and z axes, respectively).

Omitting nonlinear terms, the following matrix equation relates inertial and body-fixed coordinates of the same point:

$$\begin{pmatrix} x \\ y \\ z \end{pmatrix} = \begin{pmatrix} 1 & -\alpha_3 & \alpha_2 \\ \alpha_3 & 1 & -\alpha_1 \\ -\alpha_2 & \alpha_1 & 1 \end{pmatrix} \begin{pmatrix} \underline{x} \\ \underline{y} \\ \underline{z} \end{pmatrix} + \begin{pmatrix} u_1 \\ u_2 \\ u_3 \end{pmatrix}. \tag{5.33}$$

The equation follows from (2.30) and (2.47).

As in the two-dimensional case, we use complex amplitudes (designated by the hat symbol) instead of time functions. The relation between both is, e.g., for the motion $\mathbf{u}$:

$$\mathbf{u}(t) = \mathrm{Re}(\hat{\mathbf{u}}e^{i\omega_e t}). \tag{5.34}$$

The absolute value of the complex amplitude is the (real) amplitude, also called response amplitude operator RAO (as the wave amplitude is assumed as 1). The ratio between imaginary and real part of the complex amplitude describes the phase shift between the response and the exciting wave. We will assume the exciting wave having a wave trough at the coordinate origin at time $t = 0$. The real part of a complex amplitude is the response at time $t = 0$, and the imaginary part is the response a quarter encounter period earlier or later if the encounter frequency ω_e is positive (normal case) or negative, respectively.

5.3.2 *Equation of Motion*

The following equations do not suppose that weight and buoyancy are in equilibrium, and they are suitable not only for the total ship but also for parts of a ship (or a single hull) in front of a certain x-coordinate if the integrations start at this x. This is required for computing section forces and moments in transverse sections of the body.

We want to compute the complex amplitude of the generalized motion vector $(\hat{\mathbf{u}}, \hat{\boldsymbol{\alpha}})^T$. The complex amplitude of the motion acceleration is $-\omega_e^2(\hat{\mathbf{u}}, \hat{\boldsymbol{\alpha}})^T$, where ω_e is the encounter angular frequency. The motion equation is written in matrix form:

$$-\omega_e^2 \mathbf{M} \begin{pmatrix} \hat{\mathbf{u}} \\ \hat{\boldsymbol{\alpha}} \end{pmatrix} = \begin{pmatrix} \hat{\mathbf{F}} \\ \hat{\mathbf{M}} \end{pmatrix}. \tag{5.35}$$

Here, $\hat{\mathbf{F}}$ and $\hat{\mathbf{M}}$ are the complex amplitudes of the periodical force and moment acting on the ship. $\mathbf{M}$ is the mass matrix:

$$\mathbf{M} = \begin{pmatrix} m & 0 & 0 & 0 & mz_G & -my_G \\ 0 & m & 0 & -mz_G & 0 & mx_G \\ 0 & 0 & m & my_G & -mx_G & 0 \\ 0 & -mz_G & my_G & \theta_{xx} & -\theta_{xy} & -\theta_{xz} \\ mz_G & 0 & -mx_G & -\theta_{xy} & \theta_{yy} & -\theta_{yz} \\ -my_G & mx_G & 0 & -\theta_{xz} & -\theta_{yz} & \theta_{zz} \end{pmatrix} \tag{5.36}$$

with:
m ship's mass,
x_G, y_G, z_G mass center of gravity,
$\theta_{xx}, \ldots$ mass moments of inertia.
The mass moments of inertia are defined with respect to the coordinate origin, e.g.:

$$\theta_{xx} = \int (y^2 + z^2)\, dm, \tag{5.37}$$

$$\theta_{xy} = \int xy\, dm, \tag{5.38}$$

where $\theta_{xy} = \theta_{yz} = 0$ for mass distributions which are symmetrical to $y = 0$. Here, we need not distinguish between ship-fixed and inertial coordinates, because errors of second order in motion amplitudes are tolerated in (5.35). Compared to (2.40) and (2.41), there are also differences of first order; these occur because here $\hat{\mathbf{u}}$ and $\hat{\mathbf{M}}$ do not refer to the center of gravity.

The most important contribution to the force and moment vector $(\hat{\mathbf{F}}, \hat{\mathbf{M}})^T$ in (5.35) is due to the water pressure on the hull. It consists of a hydrostatic part ($-\mathbf{S}$ times motion amplitude), a hydrodynamic part ($\mathbf{B}$ times motion amplitude) depending on motion velocity and acceleration, and a contribution due to the incident wave and its disturbance (diffraction) by the ship. The latter contribution is designated by index e for excitation. This yields the fundamental equation of motion amplitude in a regular wave:

$$(-\omega_e^2 \mathbf{M} - \mathbf{B} + \mathbf{S}) \begin{pmatrix} \hat{\mathbf{u}} \\ \hat{\boldsymbol{\alpha}} \end{pmatrix} = \begin{pmatrix} \hat{\mathbf{F}}_e \\ \hat{\mathbf{M}}_e \end{pmatrix}, \tag{5.39}$$

where $\mathbf{M}$ and $\mathbf{S}$ are real 6 by 6 matrices, $\mathbf{B}$ is a complex 6 by 6 matrix, and $(\hat{\mathbf{F}}, \hat{\mathbf{M}})_e^T$ is a six-component complex column vector.

5.3.3 Restoring Forces

Formulas for restoring forces will not be derived here, because the derivation is straightforward; only the result will be given. However, one should be aware that the restoring force and moment are expressed here in ship-fixed coordinates. For the total ship the result is, up to first order, the same as that expressed in inertial coordinates;

thus, both coordinate systems are appropriate. However, the same expressions—and the same numbers during the computation—will be used also to determine sectional forces and moments at transverse sections of the body. For instance, the static vertical shear force in a transverse section at x gives a first-order contribution to the 'horizontal' shear force in ship-fixed coordinates if the ship is rolling. Practically, more important is the corresponding effect of the static vertical bending moment on the first-order horizontal bending moment for a rolling ship. Sectional force and moments make much more sense in ship-fixed than in inertial coordinates because the structural data are time-independent only in a ship-fixed system. That is the reason for computing **S** in body-fixed coordinates.

For the general case where the ship is not symmetric to $\underline{y} = 0$, the restoring matrix **S** is

$$\mathbf{S} = \begin{pmatrix} 0\;0 & \rho g A_{tr} & \rho g \underline{y}_{tr} A_{tr} & \begin{array}{c} -\rho g \underline{x}_{tr} A_{tr} + gm \\ -\rho g \int A \, d\underline{x} \end{array} & 0 \\ 0\;0 & 0 & \rho g \int A \, d\underline{x} - gm & 0 & 0 \\ 0\;0 & \rho g \int B_w \, d\underline{x} & \rho g \int \underline{y}_w B_w \, d\underline{x} & -\rho g \int \underline{x} B_w \, d\underline{x} & 0 \\ 0\;0 & \rho g \int \underline{y}_w B_w \, d\underline{x} & \begin{array}{c} \rho g \int \underline{y}_w^2 B_w \, d\underline{x} + gm \underline{z}_G \\ -\rho g \int A \underline{z}_S \, d\underline{x} \end{array} & -\rho g \int \underline{x} \underline{y}_w B_w \, d\underline{x} & 0 \\ 0\;0 & \begin{array}{c} \rho g \underline{z}_{tr} A_{tr} \\ -\rho g \int \underline{x} B_w \, d\underline{x} \end{array} & \begin{array}{c} \rho g \underline{y}_{tr} \underline{z}_{tr} A_{tr} \\ -\rho g \int \underline{x} \underline{y}_w B_w \, d\underline{x} \end{array} & \begin{array}{c} -\rho g \int A \underline{z}_S \, d\underline{x} - \rho g A_{tr} \underline{x}_{tr} \underline{z}_{tr} \\ +\rho g \int \underline{x}^2 B_w \, d\underline{x} + gm \underline{z}_G \end{array} & 0 \\ 0\;0 & -\rho g \underline{y}_{tr} A_{tr} & \begin{array}{c} \rho g \int \underline{x} A \, d\underline{x} - gm \underline{x}_G \\ -\rho g \underline{y}_{tr}^2 A_{tr} \end{array} & \begin{array}{c} -gm \underline{y}_G + \rho g \int \underline{y}_S A \, d\underline{x} \\ +\rho g \underline{x}_{tr} \underline{y}_{tr} A_{tr} \end{array} & 0 \end{pmatrix}. \tag{5.40}$$

The integrations are extended over the hull length; contributions from all hulls are added in case of multi-hull ships. A is the area of the section (below the average water surface), $B_w(\underline{x})$ is waterline breadth at $\underline{x}$, $\underline{y}_w(\underline{x})$ the mean transverse coordinate of the waterline at the section, and $(\underline{y}_S, \underline{z}_S)$ are the coordinates of the section area center; all for the transverse section at longitudinal coordinate $\underline{x}$. A_{tr} is the section area of the dry transom stern below the waterline at the side of the transom. A_{tr} is zero if the ship has no immersed transom stern, or if the transom stern is wetted (for small speed ahead). $\underline{y}_{tr}$, $\underline{z}_{tr}$ are the coordinates of the center of gravity of the dry transom area.

As an example, the element in row 2 and column 4 will be discussed. It is the transverse force per roll angle:

$$S_{24} = \rho g \int A \, d\underline{x} - gm. \tag{5.41}$$

The first term on the right-hand side is the static buoyancy force, the second one is the negative weight force. For the total body in static equilibrium both terms cancel;

thus, $S_{24} = 0$. For determining the sectional force and moment at $\underline{x}$: however, we use the same expressions to evaluate matrix $\mathbf{S}$; only the integrals are extended not over total ship length, but from $\underline{x}$ to the forward end. In this case, buoyancy and mass normally do not cancel. The resulting vertical force has, in ship-fixed coordinates, a non-zero transverse component in case of heel.

Also, for a body with high forward speed, the static buoyancy and the weight do not cancel. Whether the expressions given in (5.40) are appropriate in this case will depend on the size of the hydrodynamic effects in steady forward motion compared to weight force and moment. For planing boats, (5.40) is not applicable.

5.3.4 Radiation Forces

Radiation and diffraction forces in strip methods can be derived from a perturbation analysis; see, e.g., Salvesen et al. [7]. Such analyses are complicated, and they show inconsistencies which are inherent in strip methods, especially regarding exciting forces. Here, the relevant formulas, will not be strictly derived, but only made plausible similarly to Bertram [19].

Let $\hat{\mathbf{f}}$ denote the complex amplitude of the force and moment per length exerted by the two-dimensional water flow on the oscillating cross section at coordinate x. The vector consists of three components: the force in y- and z-directions, and the moment about the x-axis. $\hat{\mathbf{f}}$ is proportional to the three-component motion amplitude vector of the section designated here as $\hat{\mathbf{u}}_x = (\hat{u}_2, \hat{u}_3, \hat{\alpha}_1) =$ (sway, heave, roll amplitude). The proportionality matrix is called here $\omega_e^2 \bar{\mathbf{A}}$; thus

$$\hat{\mathbf{f}} = \bar{\mathbf{A}} \omega_e^2 \hat{\mathbf{u}}_x. \tag{5.42}$$

As $\omega_e^2 \hat{\mathbf{u}}_x$ is the complex amplitude of the negative section acceleration vector, $\bar{\mathbf{A}}$ constitutes the complex 3 by 3 added mass matrix of the section. Its elements $\bar{a}_{mn}$ are composed of the elements a_{mn} of the real added mass matrix, and elements d_{mn} of the real damping matrix of the section:

$$\bar{a}_{mn} = a_{mn} + \frac{d_{mn}}{i\omega_e}, \tag{5.43}$$

where m and n may be 2, 3, or 4. This is illustrated by inserting, here for the element (2, 2), (5.43) into (5.42):

$$\hat{f}_{x2} = \left(a_{22} + \frac{d_{22}}{i\omega_e} \right) \omega_e^2 \hat{u}_{x2} = -a_{22} \hat{\ddot{u}}_{x2} - d_{22} \hat{\dot{u}}_{x2}, \tag{5.44}$$

where the contributions of added mass a_{22} and damping d_{22} to the horizontal force are obvious.

One can re-write (5.42) in the form

$$\hat{\mathbf{f}} = -(i\omega_e)\bar{\mathbf{A}}(i\omega_e\hat{\mathbf{u}}_x). \tag{5.45}$$

The right-hand side can be interpreted as the complex amplitude of the negative time derivative of the momentum (mass times velocity) of the water around the section. The equation holds for the case for which $\bar{\mathbf{A}}$ was calculated: a cylindrical body, and no steady longitudinal flow along the cylinder. For a non-cylindrical body, however, $\bar{\mathbf{A}}$ is a function of x because the section shape depends on x; and for pitch motion $\hat{\mathbf{u}}_x$ depends on x. The time change of the momentum of the water around the section results, then, from a substantial instead of a partial time derivative:

$$\frac{D}{Dt} = \frac{\partial}{\partial t} - U\frac{\partial}{\partial x}, \tag{5.46}$$

where U is ship speed. The equation approximates the longitudinal flow speed of the water along the hull as the negative ship speed. (We use coordinate systems following the ship with its average speed ahead.) For complex amplitudes instead of time functions, $\partial/\partial t$ is substituted by $i\omega_e$.

Thus, for the three-dimensional case with forward speed U, we must substitute (5.45) by

$$\hat{\mathbf{f}} = \left(-i\omega_e + U\frac{d}{dx}\right)[\bar{\mathbf{A}}(x)(i\omega_e\hat{\mathbf{u}}_x(x))]. \tag{5.47}$$

The three-component cross-sectional velocity $i\omega_e\hat{\mathbf{u}}_x$ follows from the global six-component ship motion $(\hat{\mathbf{u}}, \hat{\boldsymbol{\alpha}})^T$ by left multiplication by a 3 by 6 matrix $\mathbf{W}$:

$$i\omega_e\hat{\mathbf{u}}_x = \underbrace{\begin{pmatrix} 0 & i\omega_e & 0 & -i\omega_e z_0 & 0 & i\omega_e x - U \\ 0 & 0 & i\omega_e & i\omega_e y_0 & -i\omega_e x + U & 0 \\ 0 & 0 & 0 & i\omega_e & 0 & 0 \end{pmatrix}}_{W} \begin{pmatrix} \hat{\mathbf{u}} \\ \hat{\boldsymbol{\alpha}} \end{pmatrix}. \tag{5.48}$$

In $\mathbf{W}$, the terms containing x, y_0, and z_0 take account of the fact that the origin of the section coordinate system has coordinates x, y_0, and z_0 in the ship coordinate system. For instance, in a catamaran, the section through one of the hulls will have a non-zero y_0, and the steady waterline height at a section will be non-zero if the steady wave due to forward speed is taken into account or if the origin of the ship coordinate system is not in height of the undisturbed water surface. The terms containing U take account of the steady forward motion of the ship in the inertial x-direction; in case of non-zero rotation angles α_2 and α_3, this motion contains non-zero components in the section coordinate system.

Similarly, the three-component section force/length amplitude $\hat{\mathbf{f}}$ is converted to a six-component global force/length amplitude $\hat{\mathbf{f}}_6$ by left multiplication by a 6 by 3 matrix $\mathbf{V}$:

$$\hat{\mathbf{f}}_6 = \underbrace{\begin{pmatrix} 0 & 0 & 0 \\ 1 & 0 & 0 \\ 0 & 1 & 0 \\ -z_0 & y_0 & 1 \\ 0 & -x & 0 \\ x & 0 & 0 \end{pmatrix}}_{V} \hat{\mathbf{f}}. \tag{5.49}$$

Inserting (5.48) and (5.49) into (5.47) gives

$$\hat{\mathbf{f}}_6 = \mathbf{V}\left(-i\omega_e + U\frac{d}{dx}\right)\bar{\mathbf{A}}\mathbf{W}\begin{pmatrix}\hat{\mathbf{u}}\\ \hat{\alpha}\end{pmatrix}. \tag{5.50}$$

Integration over x yields the complex amplitude of the 6-D radiation force:

$$\mathbf{B}\begin{pmatrix}\hat{\mathbf{u}}\\ \hat{\alpha}\end{pmatrix} = \int_L \mathbf{V}\left(-i\omega_e + U\frac{d}{dx}\right)\bar{\mathbf{A}}\mathbf{W}\,dx\begin{pmatrix}\hat{\mathbf{u}}\\ \hat{\alpha}\end{pmatrix}. \tag{5.51}$$

Thus, the complex matrix $\mathbf{B}$ introduced in the motion Eq. (5.39) is determined as

$$\mathbf{B} = \int_L \mathbf{V}\left(-i\omega_e + U\frac{d}{dx}\right)\bar{\mathbf{A}}\mathbf{W}\,dx. \tag{5.52}$$

The integration extends over the whole length of the body. The term $d/dx(\bar{\mathbf{A}}\mathbf{W})$ tends to infinity, where $\bar{\mathbf{A}}$ changes discontinuously over x, e.g., at a vertical trailing edge of a skeg, and it assumes large absolute values in case of substantial changes of $\bar{\mathbf{A}}$ within a small range of x values, e.g., at the stern of the body. At such locations, the steady flow due to forward motion will separate from the hull. This must be taken into account by omitting the term Ud/dx between those sections used for the numerical integration over x, where substantial flow separation is expected. In case of an immersed transom which is dry from behind, the aft integration limit must be just in front of the transom, also because the water separates from the hull at the transom. At the forward end of the hull, on the other hand, $\bar{\mathbf{A}}\mathbf{W}$ must be assumed to tend continuously to zero even if the ship has a vertical stem. If the stem is performing a transverse motion, the transverse momentum of the water in front of the stem is zero, and it will be changed to the value corresponding to $\bar{\mathbf{A}}\mathbf{W}\hat{\mathbf{u}}$ behind the stem even if this requires a large transverse force per length.

The added mass for surge motion cannot be computed by (5.52), i.e., on the basis of sway and heave added mass of sections. Instead, an approximation was derived from added masses of ellipsoids:

$$a_{11} = \rho\forall \cdot 0.774(\forall/L^3)^{0.675}, \tag{5.53}$$

where L is hull length and ∀ is buoyancy volume. Contrary to many other formulas, (5.53) is applicable to hulls of arbitrary slenderness up to that of a sphere. In case of multi-hulls, contributions from all hulls are added. If the added-mass force is assumed to act at coordinates y_B, z_B (at the buoyancy center of the hull), its contribution to $\mathbf{B}$ is

$$\Delta\mathbf{B} = \omega_e^2 a_{11}\mathbf{U}_1\mathbf{U}_1^T \tag{5.54}$$

with

$$\mathbf{U}_1 = \begin{pmatrix} 1 \\ 0 \\ 0 \\ 0 \\ z_B \\ -y_B \end{pmatrix}. \tag{5.55}$$

5.3.5 Exciting Force

As above, we assume a unit wave amplitude with a wave trough at the three-dimensional coordinate origin at $t = 0$. At the origin of a section coordinate system, i.e. at point (x, y_0, z_0) of the 3-D coordinate system, the complex amplitude of the wave is

$$\hat{\zeta}_x = e^{-ikx\cos\mu + iky_0\sin\mu}, \tag{5.56}$$

where μ is wave direction. It may be argued whether also a factor e^{-kz_0} should be applied in (5.56).

The complex amplitude of the three-component wave-excited force per length on a section at longitudinal coordinate x is designated as $\hat{\mathbf{f}}_e$. It is the sum of

- the so-called Froude-Krylov force $\hat{\mathbf{f}}_{e0}$ on the section caused by the wave pressure distribution on the section contour, and
- the diffraction force $\hat{\mathbf{f}}_{e7}$ caused by the two-dimensional diffraction potential ϕ_7.

Thus, for the case of a cylindrical body and no steady flow along the cylinder axis, we have

$$\hat{\mathbf{f}}_e = (\hat{\mathbf{f}}_{e,0} + \hat{\mathbf{f}}_{e,7})\hat{\zeta}_x. \tag{5.57}$$

For a ship with forward speed U, the diffraction force must be modified to

$$\hat{\mathbf{f}}_{e7} - \frac{U}{i\omega}\frac{d\hat{\mathbf{f}}_{e7}}{dx}, \tag{5.58}$$

as explained previously for Eq. (5.47). ($\hat{\zeta}_x$ must not be included into the x derivative. To see that, consider a cylindrical body, i.e. $\hat{\mathbf{f}}_{e7}$ being constant over x. In this case, the two-dimensional flow for which $\hat{\mathbf{f}}_{e7}$ was determined (without U) and the actual three-dimensional flow (including U) produce the same force amplitude.)

Term $\hat{\mathbf{f}}_e$ is converted into a six-component force vector in the global coordinate system by left multiplication by the transformation matrix $\mathbf{V}$ given in (5.49). The total wave excitation is then found by integrating over ship length:

$$\hat{\mathbf{F}}_e = \int_L \mathbf{V}\left(\hat{\mathbf{f}}_{e0} + \hat{\mathbf{f}}_{e7} + \frac{iU}{\omega}\frac{d\hat{\mathbf{f}}_{e7}}{dx}\right) e^{-ik(x\cos\mu - y_0\sin\mu)}dx. \tag{5.59}$$

Another representation of the same relation may be derived using (4.24):

$$\hat{\mathbf{F}}_e = \int_L \mathbf{V}\left(\hat{\mathbf{f}}_{e0} + \frac{\omega_e}{\omega}\hat{\mathbf{f}}_{e7}\right) e^{-ik(x\cos\mu - y_0\sin\mu)}dx + \frac{iU}{\omega}\int_L \mathbf{V}\frac{d}{dx}\left(\hat{\mathbf{f}}_{e7}e^{-ik(x\cos\mu - y_0\sin\mu)}\right)dx. \tag{5.60}$$

As explained for radiation forces, the term containing the x-derivative in (5.59) and (5.60) must be omitted between sections where the steady longitudinal flow separates from the body.

Like for radiation, also for wave excitation, the longitudinal force must be dealt with separately. Here, we neglect the longitudinal diffraction force as it is small compared to the longitudinal Froude-Krylov force. This fact appears plausible considering the small longitudinal added mass (formula (5.53)) compared to ship mass $\rho\nabla$; moreover, the fact was shown explicitly by Blume [20].

The complex amplitude of the longitudinal Froude-Krylov force on the ship is

$$\int_L \hat{p}\frac{dA}{dx}dx + \hat{p}_{tr}A_{tr}. \tag{5.61}$$

Here, $\hat{p}$ is the complex wave pressure amplitude at a point fixed in space and A is the (time-averaged) section area. Instead of using the average of $\hat{p}$ over a cross section, the value at the center of the cross-sectional area at (x, y_B, z_B) is used:

$$\hat{p} = -\rho g e^{-kz_B - ik(x\cos\mu - y_B\sin\mu)}. \tag{5.62}$$

This approximation appears reasonable because substantial surge motions occur only in long waves in which the pressure variation within the section area is moderate. In the last term in (5.61), the index tr indicates values at the transom. The term must be taken only if the body has an immersed transom which, due to small speed, is wetted from behind.

The contribution of the longitudinal wave force to the six-dimensional excitation vector is

$$\Delta\hat{\mathbf{F}}_e = \int_L \mathbf{U}_1\hat{p}\frac{dA}{dx}dx + \mathbf{U}_{1,tr}\hat{p}_{tr}A_{tr}, \tag{5.63}$$

where $\mathbf{U}_1$ is given in (5.55).

It remains to be discussed for which frequency the two-dimensional section forces must be determined to result in approximately correct results for the three-dimensional body. This question refers to the complex amplitudes, not to the time rate of change, which is always ω_e at fixed positions x, y, z. Obviously, $\bar{\mathbf{A}}$ in (5.52) must be computed for the encounter frequency ω_e because body motions have frequency ω_e, whereas $\hat{\mathbf{f}}_{e0}$ in (5.59) and (5.60) must be determined for the wave frequency ω because it results from the wave pressures unmodified by the body. But what about the diffraction force $\hat{\mathbf{f}}_{e7}$? Its dependence on frequency is caused both by the body boundary condition (5.16) (it contains the phase speed c and the wave number k) and by the free-surface condition (5.18) (it contains k). For purely transverse waves, the question is irrelevant because then $\omega = \omega_e$. For longitudinal and oblique waves, however, the free-surface condition is only a rough approximation because, in reality, the diffracted waves are also approximately longitudinal or oblique, respectively, whereas in the two-dimensional computation they are changed to transverse waves because of the assumed two-dimensional flow. This error is tolerable because the details of the free-surface condition are less important than those of the body boundary condition. The latter states that the diffraction flow cancels the flow through the section contour caused by the incoming wave. Thus, the correct frequency to determine $\hat{\mathbf{f}}_{e7}$ is ω, not ω_e.

By the way, it is relatively easy and computationally fast to determine the diffraction force on a partly immersed cylinder (axis in x-direction) for a flow which varies harmonically over both time and x-coordinate, thus avoiding the incorrect free-surface condition in computing the wave excitation force on a section. Essentially, only the Green function G in (5.20) must be substituted by a Hankel function of the second kind of order zero. This was noticed already in the 1970s by Grim (personal communication). However, no suitable way is known to apply this correct (in the limits of potential theory) three-dimensional diffraction force on a cylinder to non-cylindrical bodies of finite length. However, it can be used successfully in determining added masses of vibrating ship hulls from added masses of their sections.

5.3.6 Application

When $\mathbf{M}$, $\mathbf{B}$, $\mathbf{S}$, and $\hat{\mathbf{F}}_e$ have been determined using the formulas given up to here, the body motions in six degrees-of-freedom $\hat{\mathbf{u}}$ can be determined from (5.39) simply by solving a complex linear system of equations for six unknowns. This is the essence of the strip method. Chapter 15 shows that motion transfer functions computed by this method appear reasonable, but not very accurate.

Practical programs applying the strip method contain, however, much more:

- Taking account of other effects like nonlinear (e.g., viscous) forces, fin forces, motion restrictions, etc.

- Computing further results from the motions like pressures, sectional forces, and moments, drift forces, etc.
- Extensions to more complicated cases like multi-hulls, interaction of several floating bodies, etc.

Most of these extensions are independent of the method used to compute the forces due to hull pressure; therefore, they are described later on in separate chapters.

Hydrodynamic interaction between hulls in proximity can be tackled successfully by adapting the strip method to such cases, as is shown in the following section.

5.4 Hull Interaction in Multi-hull Vessels

If a floating body has several hulls, there is hydrodynamic interaction between them due to waves generated or modified by one hull and acting on another hull. Especially for low or zero speed, between hulls, there may occur resonant standing waves the amplitudes of which may far exceed those of the incoming waves, thus causing severe hydrodynamic forces on the hulls.

The method described here corresponds, in principle, to that of Söding [21]. Here, however, it is extended to an arbitrary number of hulls, and the numerical treatment is different to simplify programming at the expense of a little more computer time.

It is presupposed that the distance between hulls is large enough to take account only of the far-field waves. Near-field (standing) waves around one hull are thus neglected when calculating the force on another hull. Corresponding to the general assumption of the strip method, the propagation of radiated and diffracted waves from a cross section is assumed as if the waves were generated by a cylindrical (axis in x-direction) body of infinite length. That means: Waves due to hull motions (radiation waves) propagate in $\pm y$ direction (transversely); and incoming waves of direction μ_I generate a transmitted diffraction wave on lee side of direction μ_I, and a reflected wave on luff side of direction $-\mu_I$. These directions refer to an earth-fixed reference system.

As above, only the case of deep water will be elaborated. Shallow water is dealt with correspondingly in the program system Uthlande.

5.4.1 Hull Interaction Caused by Radiation Waves

For multi-hulls (catamarans, trimarans, ...), the hydrostatic matrix **S** and the radiation force matrix **B** are computed for each hull by the formulas given before for single hulls and are then added up for all hulls. To take account of the hydrodynamic hull interaction, a correction $\mathbf{B}_{int}$ must be added to **B**. Formulas for $\mathbf{B}_{int}$ are derived below.

In an earth-fixed system, the energy of radiation waves propagates with the group velocity

$$c_{gr,e} = \frac{1}{2}\omega_e/k_e = \frac{g}{2\omega_e} \tag{5.64}$$

in direction $\pm y$. Written in vector form, the velocity of wave energy propagation is $(0, \pm c_{gr,e}, 0)$.

In a reference system proceeding with the average ship speed U, the wave energy velocity vector has to be modified by subtracting the vector $(U, 0, 0)$ of ship speed. This has the effect that a wave generated at a ship-fixed longitudinal coordinate x_o (o for origin of the wave) hits a neighboring hull farther aft at longitudinal position x according to the relation

$$x = x_o - \frac{bU}{c_{gr,e}}, \tag{5.65}$$

where b is the transverse distance between both hulls. Practically, the distance between the hull reference planes is used. These planes contain the origins of the section coordinate systems. It might appear that this b is too large because the free water surface between neighboring hulls is smaller than b because of the hull breadths. However, the phase of the far-field wave is changed to the value at the reference plane of hull 1 (from which the wave originates) as if there were no finite hull breadth; and the force generated by a wave on the other hull 2 is determined as proportional to the complex wave amplitude (including phase) which the wave would have at reference plane 2 if there were no breadth of hull 2. Thus, the phase difference of the wave originating at x_o and arriving at x must be determined as if the hull breadths were zero. Effects of finite hull breadths on the relation between x and x_o may remain, but they are neglected.

The left-most hull (on port side) will be designated by index $j = 1$, and the right-most one by $j = N_H$. $\hat{\zeta}_{j,r}(x)$ designates the complex amplitude of an interaction wave running right (to starboard) at position x of hull j, where it produces the interaction force. The wave originates at hull $j_o < j$ at position x_o. (j_o is not necessarily $j - 1$ because hull $j - 1$ may be at a longitudinal position where it does not interfere with the wave.) Correspondingly, $\hat{\zeta}_{j,l}(x)$ is the complex amplitude of a left-running wave at hull j at coordinate x.

A right-running wave of amplitude $\hat{\zeta}_{j,r}(x)$ is the sum of up to three contributions:

1. A wave produced by the motion of a hull j_o
2. The reflection of the left-running wave $\hat{\zeta}_{j_o,l}(x_o)$ at hull j_o
3. The transmission of the right-running wave $\hat{\zeta}_{j_o,r}(x_o)$ under hull j_o

This is expressed by the formula

$$\hat{\zeta}_{j,r}(x) = \left(\mathbf{E}_{j_o,r}(x_o) \cdot \hat{\mathbf{u}}_{j_o}(x_o) + R_{j_o,r}(x_o)\hat{\zeta}_{j_o,l}(x_o) + T_{j_o,r}(x_o)\hat{\zeta}_{j_o,r}(x_o)\right) e^{-ik_e b}. \tag{5.66}$$

Here, the row vector $\mathbf{E}_{j_o,r}$ designates the wave amplitude to the right of hull j_o, which is caused by the motion of hull j_o with amplitude 1 in the three degrees of freedom of section motions (to the right, down, rotating about the x axis). The column vector $\hat{\mathbf{u}}_{j_o}(x_o)$ is the three-component actual motion of hull j_o at position x_o. $R_{j_o,r}$ is the scalar complex reflection coefficient, i.e., the amplitude ratio of the reflected wave to the right to the incoming wave from the right, both being phase changed to the reference plane of hull j_o. Correspondingly, $T_{j_o,r}$ is the scalar complex transmission coefficient, i.e., the amplitude ratio of the wave transmitted to the right under the hull to the incoming wave from left to right. The factor $e^{-ik_e b}$ takes account of the phase shift of the wave during propagation from hull j_o to hull j; k_e is the wave number of radiation waves having frequency ω_e. **E**, R, and T follow from the 2d potentials determined in the two-dimensional section flow computations. All three quantities should be determined for frequency ω_e and wave directions +90 and −90°.

An equation corresponding to (5.66) holds for the left-running waves:

$$\hat{\zeta}_{j,l}(x) = \left(\mathbf{E}_{j_o,l}(x_o)\cdot\hat{\mathbf{u}}_{j_o}(x_o) + R_{j_o,l}(x_o)\hat{\zeta}_{j_o,r}(x_o) + T_{j_o,l}(x_o)\hat{\zeta}_{j_o,l}(x_o)\right)e^{-ik_e b}. \tag{5.67}$$

The section motion $\hat{\mathbf{u}}_{j_o}(x_o)$ (3 by 1) follows from the ship motion $(\hat{\mathbf{u}},\hat{\boldsymbol{\alpha}})^T$ (6 by 1) by left multiplication by the 3 by 6 matrix $\mathbf{W}_{ja}(x_o)$:

$$\mathbf{W}_{j_o}(x_o) = \begin{pmatrix} 0 & 1 & 0 & -z_{0ja}(x_o) & 0 & x_o - U/(i\omega_e) \\ 0 & 0 & 1 & y_{0j_o} & -x_o + U/(i\omega_e) & 0 \\ 0 & 0 & 0 & 1 & 0 & 0 \end{pmatrix}. \tag{5.68}$$

Inserting $\mathbf{W}_{j_o}\hat{\mathbf{u}}$ for $\hat{\mathbf{u}}_{jo}$ into (5.66) and (5.67) gives the following relations between the interaction waves and the ship motions:

$$-\hat{\zeta}_{j,r}(x)e^{ik_e b} + R_{j_o,r}(x_o)\hat{\zeta}_{j_o,l}(x_o) + T_{j_o,r}(x_o)\hat{\zeta}_{j_o,r}(x_o) = -\mathbf{E}_{j_o,r}(x_o)\mathbf{W}_{j_o}(x_o)\begin{pmatrix}\hat{\mathbf{u}}\\ \hat{\boldsymbol{\alpha}}\end{pmatrix}; \tag{5.69}$$

$$-\hat{\zeta}_{j,l}(x)e^{ik_e b} + R_{j_o,l}(x_o)\hat{\zeta}_{j_o,r}(x_o) + T_{j_o,l}(x_o)\hat{\zeta}_{j_o,l}(x_o) = -\mathbf{E}_{j_o,l}(x_o)\mathbf{W}_{j_o}(x_o)\begin{pmatrix}\hat{\mathbf{u}}\\ \hat{\boldsymbol{\alpha}}\end{pmatrix}. \tag{5.70}$$

Because the ship motions are initially unknown, on the right-hand side $(\hat{\mathbf{u}},\hat{\boldsymbol{\alpha}})^T$ is omitted, and the unknowns $\hat{\zeta}$ are now interpreted as complex wave amplitudes per ship motion amplitudes for each of the six kinds of ship motions (surge until yaw); the 1 by 6 matrices are written as $\hat{\boldsymbol{\zeta}}$.

For all offset sections on all hulls, Eqs. (5.69) and (5.70) form a complex linear equation system for the $\hat{\zeta}_{...}$. Generally, if x is the position of an offset section, x_o is between two offset sections. Therefore, $\hat{\zeta}_{j_o,l}(x_o)$ and $\hat{\zeta}_{j_o,r}(x_o)$ are interpolated linearly between the values at the offset sections just in front of and behind x_o. Where no neighbor hull exists from which interaction waves would hit the current section, the respective line in the matrix is set to 0 except on the main diagonal which obtains the element 1. This corresponds to the equation $\hat{\zeta}_{...} = 0$. The terms on the right-hand side without $(\hat{\mathbf{u}},\hat{\boldsymbol{\alpha}})^T$ are the in-homogeneous elements of the equation system, which has six right-hand sides for the 6 degrees-of-freedom of ship motions.

Because each row of the coefficient matrix of this equation system has only between 1 and 5 non-zero elements, the system is solved in short time.

From the results, the interaction contribution $\mathbf{B}_{int}$ to the matrix $\mathbf{B}$ is obtained similarly to relations (5.59) and (5.63) for exciting forces:

$$\begin{aligned}\mathbf{B}_{int} = & \int_L \mathbf{V}\left(\hat{\mathbf{f}}_{e,x0,+90} + \hat{\mathbf{f}}_{e,x7,+90} + \frac{iU}{\omega_e}\frac{d\hat{\mathbf{f}}_{e,x7,+90}}{dx}\right)\hat{\zeta}_{j,l}(x)\, dx \\ & + \int_L \mathbf{V}\left(\hat{\mathbf{f}}_{e,x0,-90} + \hat{\mathbf{f}}_{e,x7,-90} + \frac{iU}{\omega_e}\frac{d\hat{\mathbf{f}}_{e,x7,-90}}{dx}\right)\hat{\zeta}_{j,r}(x)\, dx \\ & - \rho g \int_L \mathbf{U}_1 e^{-k_e z_B}\left(\hat{\zeta}_{j,l}(x) + \hat{\zeta}_{j,r}(x)\right)\frac{d\bar{A}_x}{dx}\, dx \\ & - \rho g \mathbf{U}_1 e^{-k_e z_B}\left(\hat{\zeta}_{j,l}(x_{tr}) + \hat{\zeta}_{j,r}(x_{tr})\right)\bar{A}_{tr}. \end{aligned} \tag{5.71}$$

The last two terms are due to the longitudinal interaction force. The last (transom) term must be taken only for immersed, wetted transoms (low speed). $\hat{\mathbf{f}}_{e,x0,+90}$ and $\hat{\mathbf{f}}_{e,x7,+90}$ are the Froude-Krylov and diffraction, respectively, wave exciting forces (three components) per length per wave amplitude for waves coming from direction $\mu = 90°$ (i.e., left-running waves); correspondingly, index $-90°$ designates forces in right-running waves. The terms containing $d\hat{\mathbf{f}}/dx$ must not be used at flow separation locations, e.g., at the aft end of a deadwood.

5.4.2 Hull Interaction Caused by Diffraction Waves

Instead of corrections, formulae will be derived here for the total wave excitation force on multi-hull vessels, taking account of the hull interaction.

In an earth-fixed system, the energy of incident and diffraction waves propagates with the group velocity

$$c_{gr} = \frac{1}{2}c = \frac{g}{2\omega} \tag{5.72}$$

in direction $\pm\mu$. (As above the real hull shape is approximated as being cylindrical.) Written in vector form, the group velocity is $(c_{gr}\cos\mu, \pm c_{gr}\sin\mu, 0)$. In a reference system proceeding with the average ship speed U, the corresponding velocity vector is $(c_{gr}\cos\mu - U, \pm c_{gr}\sin\mu, 0)$. Thus, a wave generated at a ship-fixed longitudinal coordinate x_o hits a neighboring hull at longitudinal position x according to the relation

$$x = x_o + \frac{b}{\sin|\mu|}\left(\cos\mu - \frac{U}{c_{gr}}\right), \tag{5.73}$$

where b is the transverse distance between both hull reference planes. For $\mu = 0$ or 180°, where (5.73) is not applicable, there is no hull interaction according to the simplifications made here; thus this case can be excluded.

The force on a section per wave amplitude does not depend on whether it is excited by an incident wave or a far-field diffraction wave. Therefore, if a hull j at position x is not in the incident wave 'shadow' of another hull, the complex amplitude

$$e^{-ik(x\cos\mu - y_j \sin\mu)} \tag{5.74}$$

of the incident wave at that position will be designated as $\hat{\zeta}_{j,r}(x)$ if $\mu < 0$ (right-running wave), and as $\hat{\zeta}_{j,l}(x)$ if $\mu > 0$. (Here, it is presupposed that μ is in the range -180 to $+180°$.)

The equations for determining the wave amplitudes $\hat{\zeta}$ are now:

(a) If no x_o is found in the length range of a hull to the left of hull j:

$$\hat{\zeta}_{j,r}(x) = e^{-ik(x\cos\mu - y_j \sin\mu)} \quad \text{if} \quad \mu < 0, \quad \text{otherwise} \quad 0. \tag{5.75}$$

(b) If an x_o is found in the length range of a hull j_o to the left of hull j:

$$\hat{\zeta}_{j,r}(x) = \left(R_{j_o,r}(x_o)\hat{\zeta}_{j_o,l}(x_o) + T_{j_o,r}(x_o)\hat{\zeta}_{j_o,r}(x_o)\right) e^{-ik(|(x-x_o)\cos\mu| + |b\sin\mu|)}. \tag{5.76}$$

(c) If no x_o is found in the length range of a hull to the right of hull j:

$$\hat{\zeta}_{j,l}(x) = e^{-ik(x\cos\mu - y_j \sin\mu)} \quad \text{if} \quad \mu \geq 0, \quad \text{otherwise} \quad 0. \tag{5.77}$$

(d) If an x_o is found in the length range of a hull j_o to the right of hull j:

$$\hat{\zeta}_{j,l}(x) = \left(R_{j_o,l}(x_o)\hat{\zeta}_{j_o,r}(x_o) + T_{j_o,l}(x_o)\hat{\zeta}_{j_o,l}(x_o)\right) e^{-ik(|(x-x_o)\cos\mu| + |b\sin\mu|)}. \tag{5.78}$$

These equations form a complex linear equation system for right- and left-running waves at all x-positions of each hull where section data are provided. The incident wave terms listed under (a) and (c) are the inhomogeneous terms. There is only one in homogeneous vector and thus only one set of results $\hat{\zeta}_{j,r}$ and $\hat{\zeta}_{j,l}$. The coefficient matrix is different from that for radiation interaction because the reflection and transmission coefficients R and T must be determined here for wave directions $\pm|\mu|$, whereas for radiation they refer to wave angles $\pm 90°$.

The exciting force is now found in analogy to Eqs. (5.71):

$$\hat{\mathbf{F}}_e = \int_L \mathbf{V}\left(\hat{\mathbf{f}}_{e,x0,+|\mu|} + \hat{\mathbf{f}}_{e,x7,+|\mu|} + \frac{iU}{\omega}\frac{d\hat{\mathbf{f}}_{e,x7,+|\mu|}}{dx}\right)\hat{\zeta}_{j,l}(x)\,dx \tag{5.79}$$

$$+ \int_L \mathbf{V}\left(\hat{\mathbf{f}}_{e,x0,-|\mu|} + \hat{\mathbf{f}}_{e,x7,-|\mu|} + \frac{iU}{\omega}\frac{d\hat{\mathbf{f}}_{e,x7,-|\mu|}}{dx}\right)\hat{\zeta}_{j,r}(x)\,dx \tag{5.80}$$

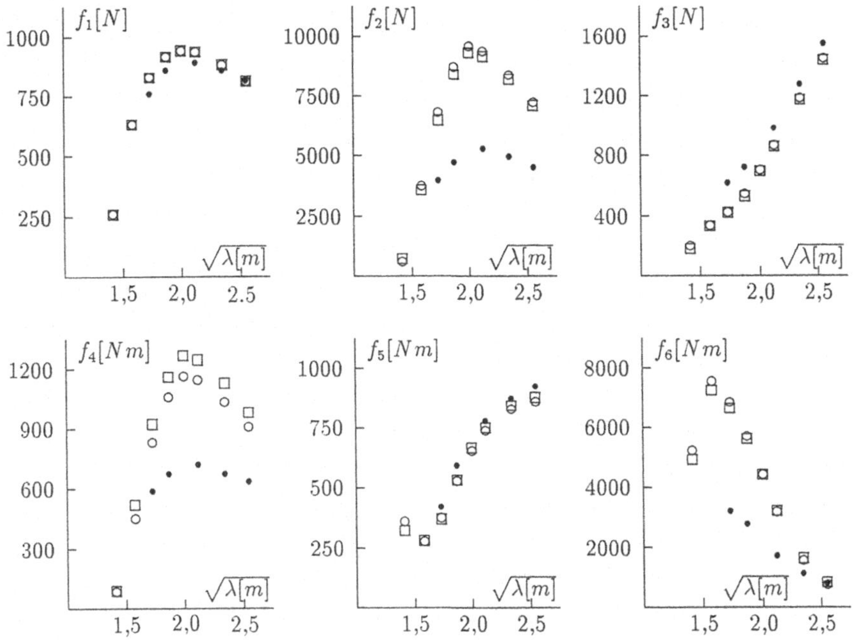

Fig. 5.6 Comparison of exciting force (f_1–f_3) and moment (f_4–f_6) determined in model experiments (dots) and computed using the above-described hull interaction method (squares: Hachmann method; circles: common method) based on [22]

$$- \rho g \int_L \mathbf{U}_1 e^{-kz_B} \left(\hat{\zeta}_{j,l}(x) + \hat{\zeta}_{j,r}(x) \right) \frac{d\bar{A}}{dx}\, dx \tag{5.81}$$

$$- \rho g \mathbf{U}_1 e^{-kz_B} \left(\hat{\zeta}_{j,l}(x_{tr}) + \hat{\zeta}_{j,r}(x_{tr}) \right) \bar{A}_{tr}. \tag{5.82}$$

As above, the terms in (5.79) and (5.80) involving x derivatives have to be omitted in regions of substantial separation of the steady longitudinal flow. The last two lines are due to longitudinal wave forces. The last line, which takes account of the transom, must be used only in case of a wetted transom.

5.4.3 Validation

Blume [22] measured wave exciting force and moment, the elements of added mass and damping matrix, and free motions of a SWATH in waves. For ship speed $U = 0$, the coincidence between measured and computed excitation, added mass, and damping was favorable. For $U = 20$ knots, on the other hand, measured and computed excitation differed very much for the 'asymmetrical' quantities sway force, roll moment, and yaw moment (see Fig. 5.6). The reason for these differences is, apparently, flow separation. Similar differences were found for the asymmetrical elements of the added mass and damping matrices. In spite of these differences, also the asymmetrical motions sway, roll, and yaw coincided nicely between experiments and computations, because errors in excitation canceled against errors in added mass and damping. (For a SWATH, the transverse added mass is many times larger than ship mass.)

References

1. B.V. Korvin-Kroukovsky, Investigation of ship motions in regular waves. Trans. Soc. Nav. Archit. Mar. Eng. (SNAME) **63**, 386–435 (1955)
2. F. Ursell, On the heaving motion of a circular cylinder on the surface of a fluid. Q. J. Mech. Appl. Math. **2**(2), 218–231 (1949)
3. O. Grim, Computing hydrodynamic forces produced by ship motions, in *Proceedings of the STG*, vol. 47 (1953) (in German)
4. O. Grim, Oscillations of floating sections (in German). Technical report, University of Hamburg, Institute for Naval Architecture
5. B.V. Korvin-Kroukovsky, W.R. Jacobs, Pitching and heaving motions of a ship in regular waves. Trans. Soc. Nav. Archit. Mar. Eng. (SNAME) **63**, 386–435 (1957)
6. H. Söding, A modification of the strip method. Ship Technol. Res. **16**, 15–18 (1969) (in German)
7. N. Salvesen, E.O. Tuck, O.M. Faltinsen, Ship motions and sea loads. Trans. Soc. Nav. Archit. Mar. Eng. **78**, 250–287 (1970)
8. V. Bertram, B. Veelo, H. Söding, K. Graf, Development of a freely available strip method for seakeeping, in *COMPIT* (2006)
9. V. Bertram, H. Söding, K. Graf, E. Mesbahi, PDSTRIP fin and sails treatment – Physics and expert knowledge (2007)
10. V. Bertram, H. Söding, Verification of the PDSTRIP 2-d radiation problem module, in *10th Numerical Towing Tank Symposium* (2007)
11. H. Söding, A method to simplify perturbation analyses of periodical flows. Eng. Marit. Environ. (2014)
12. H. Söding, Second-order seakeeping analyses using perturbators. Ship Technol. Res. **61**(1), 4–15 (2014)
13. A. Papanikolaou, On calculations of nonlinear hydrodynamic effects in ship motion. Ship Technol. Res. **31**, 91–129 (1984)
14. L. Gentaz, B. Alessandrini, G. Delhommeau, Motion simulation of a cylinder at the surface of a viscous fluid. Ship Technol. Res. **43**, 91–129 (1996)
15. F. Tasai, W. Koterayama, Nonlinear hydrodynamic forces acting on cylinders heaving on the surface of a fluid (1976)
16. K.F. Cheung, Time domain solution for second-order diffraction. Ph.D. thesis, The University of British Columbia (1991)
17. Y. Kyozuka, Non-linear hydrodynamic forces acting on two-dimensional bodies (First report, diffraction problem). J. SNAJ **148**, 45–53 (1980)
18. S. Sutulo, C.G. Soares, A boundary integral equations method for computing inertial and damping characteristics of arbitrary contours in deep fluid. Ship Technol. Res. **51**(2), 69–93 (2004)
19. V. Bertram, *Practical Ship Hydrodynamics* (Elsevier, Amsterdam, 2012)
20. P. Blume, About surge forces in following regular waves (in German). Technical report, Hamburg Model Basin, Hamburg (1976)
21. H. Söding, Computing motions and loads on SWATHs and catamarans in a seaway (in German). Technical report, University of Hamburg, Institute for Naval Architecture (1988)
22. P. Blume, H. Söding, Numerical simulation and validation for SWATH ships in waves, in *International Conference on Fast Sea Transportation (FAST'93)* (1993)

Chapter 6
Green Function Methods

Abstract This chapter 'Green function methods' and all following chapters, the three-dimensional flow around the ship is computed to determine forces and moments caused by fluid pressure, thus avoiding the approximations inherent in strip methods. In this chapter, central to computing this flow is the Green function: the potential of a pulsating source under a free surface, at which it generates waves. If the source is not fixed, but translating relative to the fluid, numerical problems in computing this Green function are large; therefore, here this flow is approximated from that of a non-moving pulsating source. The sources are continuously distributed over 'panels' covering the underwater ship hull. In each panel, the source density is determined such that there is no flow through the ship hull at all panel centers. The chapter describes also how to take account of the nonlinear effect of fluid pressure between the instantaneous and the average waterline.

6.1 Introduction

Green function panel methods discretize the hull surface up to the average waterline into a large number of small surface elements (panels). For the position of the waterline, usually, the squat (change of draft and trim due to forward speed) and the steady ship waves are neglected. For each panel, a Green function is defined. The superposition of all these Green functions, together with a parallel flow in case of forward speed, constitutes the velocity potential. Usually, the panel potentials are sources; however, if lift plays a significant role, such as for yawing or maneuvering ships, additional vortexes or dipoles must be employed to model lift effects. Contrary to the patch method described in Chap. 3, which uses point sources, usually, a distribution of sources is applied, either on the panel (hull surface) or at some distance from it within the body, to smooth flow irregularities occurring at the boundaries of the panels. All these potentials fulfill the Laplace equation, the radiation condition, the bottom or infinite depth condition, and the linearized free-surface condition. To determine the source strengths, a set of linear equations is established to satisfy the no-penetration condition at the collocation points in the middle of each panel.

B. O. el Moctar et al., *Numerical Methods for Seakeeping Problems*,
https://doi.org/10.1007/978-3-030-62561-0_6

The ship responses may be determined for regular waves of various frequencies and directions and then combined to approximate the ship behavior in a natural seaway. This so-called frequency domain method is directly applicable only to linear waves and linear responses. Nonlinear behavior can only be accounted for approximately by modifying the responses for finite wave steepness. To take account of nonlinear responses more accurately, simulations (time domain methods) must be applied, at the expense of much more computing effort.

In frequency domain methods, Green functions correspond to sources of periodically oscillating (pulsating) source strength. The Green functions may satisfy the free-surface boundary condition for the case of a superimposed parallel flow, which is appropriate for modeling the flow around a ship with speed ahead. The disturbance of the parallel flow by the ship hull cannot be taken into account in satisfying the free-surface condition. But even when neglecting the disturbance flow, the Green function is complicated and numerically difficult to determine. Therefore, usually, the flow is even more simplified by using the zero-speed frequency-domain Green function.

6.2 The Encounter Frequency Panel Method

Accurate solutions using the zero-speed free-surface Green function method are obtained for problems with linearized free-surface boundary conditions at zero forward speed, but good or reasonable approximations are attained also in case of bodies with moderate steady forward speed. The most important effect of forward speed can easily be taken into account: Ship responses oscillate not with the wave frequency ω, but with the encounter frequency

$$\omega_e = \omega - kU \cos \mu, \tag{6.1}$$

where k is the wave number $= 2\pi/\text{wave length}$, U is ship speed, and μ is the angle of wave propagation (0 for following waves).

In the following, the encounter frequency panel method GLPanel [1, 2] is described. It proved as a convenient tool to efficiently calculate motions and wave-induced loads of slender single-hull displacement ships advancing at forward speed in regular waves. The main difficulty is, as in other methods, to determine the radiation and diffraction potentials, which are required to determine fluid pressure and, following from it, the added mass, damping, and excitation force and moment.

The flow potential ϕ, which depends on location $\mathbf{x}$ and time t, is separated into a steady and an oscillatory part:

$$\phi(\mathbf{x}, t) = [-Ux + \phi_S(\mathbf{x})] + \mathrm{Re}\left(\hat{\phi}(\mathbf{x})e^{i\omega_e t}\right), \tag{6.2}$$

where $-Ux + \phi_S$ is the steady contribution to the flow potential and $\hat{\phi}$ is the complex amplitude of the oscillatory potential. As in the strip method, we use here an inertial coordinate system moving with the ship's steady ahead speed; z points downward. In the following, ϕ_S will be neglected, but the term $-Ux$ must be taken into account. $\hat{\phi}$ consists of

the amplitude $\hat{\phi}_0$ of the incoming wave potential,
the amplitude $\hat{\phi}_7$ of the disturbance (scattering) potential, which is caused by the body moving steadily (without oscillations) ahead, and
contributions generated by the oscillatory body motions in each of the six degrees of freedom:

$$\hat{\phi} = \hat{\phi}_0 + \hat{\phi}_7 + (\hat{\phi}_1, \ldots, \hat{\phi}_6) \cdot \begin{pmatrix} \hat{\mathbf{u}} \\ \hat{\boldsymbol{\alpha}} \end{pmatrix}, \tag{6.3}$$

where the dependence of all potentials on $\mathbf{x}$ has been omitted. $\hat{\mathbf{u}}$ is the complex amplitude of the translation, $\hat{\boldsymbol{\alpha}}$ that of the rotation of the body. For instance, $\hat{u}_1$ is the complex amplitude of the surge motion and $\hat{\phi}_1 = |\hat{\phi}_1|e^{i\epsilon_1}$ designates the amplitude $|\hat{\phi}_1|$ and the phase difference ϵ_1 (relative to the surge motion) of the flow potential due to a unit surge motion of the ship.

In the following, for simplification only, the case of deep water is described. The cited literature treats, however, also the case of constant finite water depth. $\hat{\phi}_0$ is the amplitude of a linear wave of amplitude 1:

$$\hat{\phi}_0 = -i\frac{\omega}{k}e^{-ikx\cos\mu + iky\sin\mu}. \tag{6.4}$$

The potentials $\hat{\phi}_j$, $j = 1$ to 7, are determined numerically by a panel method. They must satisfy the Laplace equation in the fluid domain and the zero-speed linearized free-surface boundary condition at the non-oscillating water surface $z = 0$. This condition follows from the kinematic condition (no flux through the water surface) and the dynamic condition (fluid pressure = constant air pressure): If both conditions are combined to eliminate the surface height amplitude $\hat{\zeta}$, the resulting boundary condition is

$$\left[\left(i\omega_e + U\frac{\partial}{\partial x}\right)^2 - g\frac{\partial}{\partial z}\right]\hat{\phi}_j = 0. \tag{6.5}$$

According to Wehausen and Laitone [3], the complex amplitude of the potential of a source at position $\boldsymbol{\xi} = (\xi, \eta, \zeta)$, which pulsates harmonically with frequency ω_e, under a free surface at $z = 0$ in otherwise non-moving fluid, is

$$\hat{\phi}_s = \frac{1}{r} + PV\int_0^\infty \frac{\kappa + k}{\kappa - k}e^{-\kappa(z+\zeta)}J_0(\kappa R)d\kappa + 2\pi i k e^{-k(z+\zeta)}J_0(kR). \tag{6.6}$$

Here, PV denotes the Cauchy principal value of the integral; J_0 is the Bessel function of first kind and order 0; r is the distance $|\mathbf{x} - \boldsymbol{\xi}|$ between source and actual point; and

$$R = \sqrt{(x - \xi)^2 + (y - \eta)^2} \tag{6.7}$$

is the horizontal distance between source and actual point. In [4], different methods for evaluating the principal value integral are discussed; the report covers also the determination of the required derivatives $\nabla\hat{\phi}_s$.

To apply this potential in a panel method, instead of a single source, a distribution of sources of constant density over a quadrilateral or triangular plane panel must be determined. Thus, the potential $\hat{\phi}_s(\mathbf{x}, \boldsymbol{\xi})$ is integrated to obtain a panel potential

$$\hat{\phi}_p(\mathbf{x}) = \int\int_{\text{Panel } p} \hat{\phi}_s(\mathbf{x}, \boldsymbol{\xi}) d\boldsymbol{\xi}. \tag{6.8}$$

For the possibly singular term $1/r$ in (6.6), this can be done analytically; see, e.g. [5]. For the remaining non-singular terms, it is usually (if the wave length exceeds 5 times the panel length) sufficient to use the integrand value at the panel center multiplied by the panel area.

6.3 Determination of Radiation and Diffraction Potentials

The potentials $\hat{\phi}_p$ of all panels are superimposed:

$$\hat{\phi}_j = \sum_{p=1}^{P} q_{j,p} \hat{\phi}_p, \tag{6.9}$$

where P is the number of panels. The source densities $q_{j,p}$ are determined by solving a linear equation system with seven different right-hand sides (for $j = 1$ to 7) such that the body boundary conditions (no flux through the hull) at the centers of all panels are satisfied. These conditions are
For the diffraction potential ($j = 7$):

$$\mathbf{n} \cdot \nabla\hat{\phi}_7 = -\mathbf{n} \cdot \nabla\phi_0. \tag{6.10}$$

For the radiation potentials ($j = 1$ to 6):

$$\mathbf{n} \cdot \nabla\hat{\phi}_j = i\omega_e n_j + U m_j. \tag{6.11}$$

The quantities n_j are the components of the normal vector on the hull for $j = 1$ to 3; for $j = 4$ to 6 they are defined as

$$(n_4, n_5, n_6) = \mathbf{x} \times \mathbf{n}, \tag{6.12}$$

where $\mathbf{x}$ is the position vector of the collocation point (panel center). The scalars m_j (they are different from the so-called m terms used in many other publications) are zero for $j = 1$ to 4; $m_5 = n_3$, and $m_6 = -n_2$. They take account of the fact that the panel normals are constant in a ship-fixed coordinate system, whereas the parallel flow is constant in the inertial coordinate system.

Each radiation potential ϕ_j is subdivided into two parts satisfying the two terms of the inhomogeneous boundary condition (6.11):

$$\hat{\phi}_j = \hat{\phi}_j^0 + \frac{U}{i\omega_e}\hat{\phi}_j^U, \tag{6.13}$$

where the speed-independent potentials $\hat{\phi}_j^0$ and $\hat{\phi}_j^U$ satisfy the body boundary conditions

$$\mathbf{n} \cdot \nabla\hat{\phi}_j^0 = i\omega_e n_j \tag{6.14}$$

and

$$\mathbf{n} \cdot \nabla\hat{\phi}_j^U = i\omega_e m_j. \tag{6.15}$$

The hull boundary conditions are the only inhomogeneous conditions; the other homogeneous boundary conditions are the same for all these potentials. Thus, the relations between m_j an n_j lead to

$$\hat{\phi}_j = \hat{\phi}_j^0 \quad \text{for} \quad j = 1, 2, 3, 4; \tag{6.16}$$

$$\hat{\phi}_5 = \hat{\phi}_5^0 + \frac{U}{i\omega_e}\hat{\phi}_3^0; \tag{6.17}$$

$$\hat{\phi}_6 = \hat{\phi}_6^0 - \frac{U}{i\omega_e}\hat{\phi}_2^0. \tag{6.18}$$

The potentials $\hat{\phi}_j$, $j = 1$ to 7, can now be determined: Inserting (6.9) into the boundary conditions (6.14) and (6.10) results in a complex linear equation system from which the complex source distributions $q_{j,p}$, $j = 1$ to 7, are determined. Using again (6.9) and the above three relations give then $\hat{\phi}_1$ to $\hat{\phi}_7$.

For certain (so-called irregular) frequencies, the equation system is singular. These frequencies correspond to sloshing eigenfrequencies of virtual fluid within the body. In the neighborhood of these frequencies, to determine accurate source strengths requires additional precautions. A simple remedy would be to interpolate the source strengths, or results following from them, from values at lower and higher frequencies. However, much more accurate is:

To arrange additional panels on the free surface within the body. No flow through these panels is specified as boundary condition.

Or to arrange the source distribution not on the hull (where the boundary condition is satisfied at the panel centers), but at a suitable distance inside the hull (so-called desingularization).

6.4 Pressure Force and Moment

The Bernoulli equation for unsteady flow determines the fluid pressure p (the difference to atmospheric pressure):

$$\frac{p}{\rho} = -\dot{\phi} - \frac{1}{2}|\nabla\phi|^2 + gz + \frac{1}{2}U^2. \tag{6.19}$$

Inserting for ϕ the expression (6.2) gives the complex amplitude of the pressure:

$$\frac{\hat{p}}{\rho} = -i\omega_e\hat{\phi} - U\hat{\phi}_x + \underbrace{(0,0,g)\cdot(\hat{\mathbf{u}} + \hat{\boldsymbol{\alpha}} \times \mathbf{x})}_{g(\hat{u}_3+\hat{\alpha}_1 y-\hat{\alpha}_2 x)}. \tag{6.20}$$

This equation is evaluated at each panel center, separately for the wave potential $\hat{\phi}_0$, the diffraction potential $\hat{\phi}_7$, and for the six radiation potentials $\hat{\phi}_1$ to $\hat{\phi}_6$. As (6.3) shows, the radiation potentials hold for unit amplitude motions; thus, for evaluating the pressure amplitudes for heave $\hat{p}_3$, roll $\hat{p}_4$ and pitch $\hat{p}_5$, in the hydrostatic terms, i.e. the underbraced expression in (6.20), the value 1 must be used for $\hat{u}_3$, $\hat{\alpha}_1$ and $\hat{\alpha}_2$, respectively.

In the following it is assumed that, for all panels $p = 1$ to P, the panel normal vector $\mathbf{n}_p$ is directed into the hull, and that its absolute value is equal to the area of panel p. Taking the pressure at the panel center as approximation of the average pressure over the panel, the pressure force and moment amplitudes are

$$\hat{\mathbf{F}}_j = \sum_{p=1}^{P} \hat{p}_j \mathbf{n}_p \quad \text{and} \quad \hat{\mathbf{M}}_j = \sum_{p=1}^{P} \hat{p}_j \mathbf{x} \times \mathbf{n}_p \tag{6.21}$$

for $j = 0$ to 7.

6.5 Determination of Ship Motions

An equation for linear ship motion amplitudes was given already for the strip method as (5.35). With the present notation of forces $\hat{\mathbf{F}}_j$ and moments $\hat{\mathbf{M}}_j$, the equation becomes

$$\left[-\omega_e^2\mathbf{M}-\begin{pmatrix}\hat{\mathbf{F}}_1 & \dots & \hat{\mathbf{F}}_6\\ \hat{\mathbf{M}}_1 & \dots & \hat{\mathbf{M}}_6\end{pmatrix}\right]\begin{pmatrix}\hat{\mathbf{u}}\\ \hat{\alpha}\end{pmatrix}=\begin{pmatrix}\hat{\mathbf{F}}_0+\hat{\mathbf{F}}_7\\ \hat{\mathbf{M}}_0+\hat{\mathbf{M}}_7\end{pmatrix}. \tag{6.22}$$

The mass matrix $\mathbf{M}$ is specified in (5.36).

The equation system may be complemented by corrections like additional surge and roll damping, forces on fins, etc. (Chap. 11). Solving the system of six complex scalar linear equations gives then the complex amplitude vectors of the translation $\hat{\mathbf{u}}$ and rotation $\hat{\boldsymbol{\alpha}}$.

When the motion amplitudes are known, further transfer functions like hull pressures, forces and moments in virtual cross sections, drift forces, etc., can be determined (Chaps. 12 and 14). Here, only one such additional feature is explained which is often used together with the Green function method GLPanel described above: a nonlinear correction for the effect of fluid pressure between average and actual waterline. That correction may be used, however, in combination with other linear prediction methods as well.

6.6 Nonlinear Pressure Correction

One consequence of using linear methods is, that predicted wave-induced vertical bending moments have the same magnitude in sagging (wave trough condition) and hogging (on a wave crest). Actually, however, in steep waves, absolute values of sagging moments (negative) are substantially larger than hogging moments. Most of this difference is due to the fact that flaring section shapes at both ship ends give higher upward force in sagging than downward force in hogging condition. This is a nonlinear effect; thus, it cannot be expressed by complex amplitudes for unit wave height. Instead, it requires to assume a finite wave amplitude and to simulate motions and loads during one encounter period.

To take account of the variable height of the actual waterline at a ship section, Hachmann [6] approximates the hull pressure above the waterline as

$$p(s) = p_{WL} - \rho(g\cos\alpha - \omega_e^2\zeta)s, \tag{6.23}$$

where s is a coordinate measured upward along the section contour, starting from the steady waterline (Fig. 6.1). α is the inclination of the section against the vertical (within the range of variable immersion of the section). p_{WL} is the time-dependent pressure at the waterline; it follows from the complex amplitude (using a linear method, e.g. the Green function method) of the pressure at the steady waterline of the section. ζ is the s coordinate, where $p(s) = 0$. The expression $\rho(g\cos\alpha - \omega_e^2\zeta)$ is an approximation of the pressure gradient along the section contour, caused by the static pressure $\rho g z$ modified by the girth-wise acceleration of the fluid below the surface. (This modification neglects the steady ship speed U.)

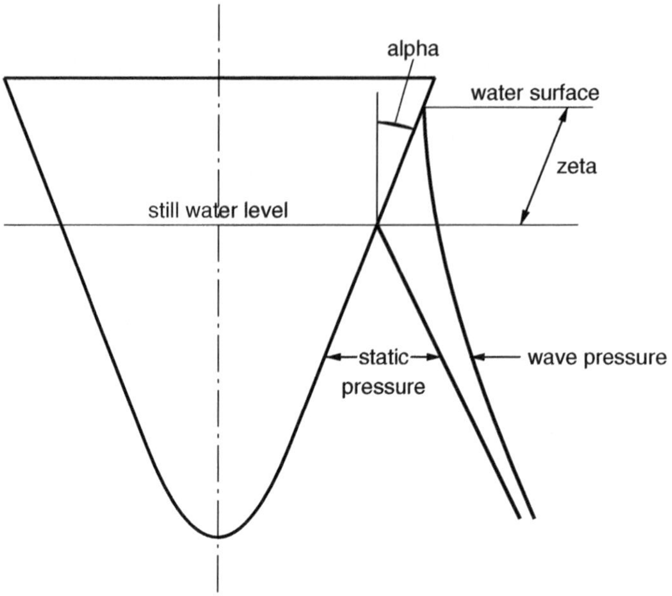

Fig. 6.1 Pressure distribution on ship cross section

Equation (6.23) is used to determine the maximum immersion coordinate ζ: For $p(s) = 0$ and $s = \zeta$, a quadratic equation for ζ is obtained. Its solution is

$$\zeta = \frac{g\cos\alpha}{2\omega_e^2}\left[1 - \sqrt{1 - \frac{p_{WL}}{\rho}\left(\frac{2\omega_e}{g\cos\alpha}\right)^2}\right]. \tag{6.24}$$

When the actual water surface is above the mean surface ($p_{WL} > 0$), between $s = 0$ and $s = \zeta$ the pressure according to (6.23) times the normal vector is integrated to obtain the nonlinear force correction. When $p_{WL} < 0$, the corresponding integral (between $s = 0$ and s for $p(s) = 0$) must be subtracted to correct the linearized negative pressure for the fact that the actual pressure near to the waterline is at least zero. This correction is performed for a sufficient number of sections to integrate its effect on the section force and moment for both sides of the waterline.

To validate this method, model experiments and corresponding computations were performed for a containership in head waves. Table 6.1 compares the results for vertical bending moments at a section in the fore-body. Because of elastic vibrations of the model consisting of separate parts connected by six-component balances, measured moments hold for low-pass filtered time histories. The table demonstrates the amount of inaccuracies, which cannot be avoided using this method. It is unknown how much of the difference is caused by experimental errors.

Table 6.1 Comparison of measured [7] and computed [8] extreme values of vertical bending moment in MNm at a section $0.8L_{pp}$ ahead of the aft perpendicular of a containership at $F_n = 0.2$ in regular head waves of length $1.1L_{pp}$

Wave height	6.9 m		13.4 m	
Bending moment	Computed	Measured	Computed	Measured
Hogging	1070	920	1780	1450
Sagging	−1130	−1260	−2750	−2500
Hogging−sagging	2200	2180	4530	3950
Hogging+sagging	−60	−340	−970	−1050

References

1. A. Papanikolaou, T.E. Schellin, A three-dimensional panel method for motions and loads with forward speed. Ship Tech. Res. **39**(4), 147–156 (1992)
2. C. Östergaard, T.E. Schellin, Development of an hydrodynamic panel method for practical analysis of ships in a seaway. Schiffbautechnische Gesellschaft **89**, 561–576 (1995) (in German)
3. J.V. Wehausen, E.V. Laitone, Surface waves. Handbuch der Physik (Springer, Berlin) **4**, 446–778 (1960)
4. A. Papanikolaou, Technical report, Technische Universität Berlin, Institut für Schiffs- und Meerestechnik (1983)
5. W.C. Webster, The flow about arbitrary three-dimensional smooth bodies. J. Ship Res. **19** (1975)
6. D. Hachmann, Determination of the wave elevation at ship sections based on pressure variations at the design waterline under the influence of the Smith effect. Technical report MTK 325 II, Germanischer Lloyd, Hamburg (in German) (1986)
7. P. Blume, Ship motion simulation in a seaway using detailed hydrodynamic force coefficients, in *3rd Conference on Stability of Ships and Ocean Vehicles (STAB'86)* (1999)
8. C. Beiersdorf, H. Rathje, Loads on the bow region of a 6700 TEU containership design (in German). Technical report FG 2000.060A, Germanischer Lloyd, Hamburg (1999)

Chapter 7
Linear Rankine Source Methods

Abstract This chapter differs from the previous chapter by using sources in the unbounded fluid as Green functions. Thus, also the free surface (including ship waves) around the hull must be covered by source panels. The described method includes the interaction of oscillatory and stationary potentials; the latter is assumed stationary not in inertial, but in ship-fixed coordinates for higher accuracy. Also, the method to satisfy the radiation conditions (no disturbance from outside, no wave reflections) is important for accuracy: For the stationary flow, a back-shift of sources by 1 panel length is used, whereas for the oscillatory flow damping is applied everywhere on the free surface.

7.1 Introduction

A Rankine source flow is the potential flow caused by a point source in unbounded ideal fluid. Its flow potential is, in three-dimensional space:

$$\phi(\mathbf{x}) = \frac{-q}{|\mathbf{x} - \boldsymbol{\xi}|} . \tag{7.1}$$

Here, $\boldsymbol{\xi}$ is the source location, and q will be called source strength. The ejected fluid volume per time is $4\pi q$.

The source flow satisfies the continuity equation

$$\Delta\phi = \phi_{xx} + \phi_{yy} + \phi_{zz} = 0 \tag{7.2}$$

everywhere except at the source location $\boldsymbol{\xi}$. The continuity equation is also satisfied by arbitrary superpositions of point sources, and by distributions of sources over arbitrary surfaces, except at the source locations or the source surfaces, respectively. Thus, to describe the flow of an ideal fluid, the point sources, or the source surfaces, must be outside of or along the boundary of the fluid region. Numerical methods which approximate a flow by Rankine source potentials are called Rankine source methods.

B. O. el Moctar et al., *Numerical Methods for Seakeeping Problems*,
https://doi.org/10.1007/978-3-030-62561-0_7

For modeling time-periodic free-surface flows, Rankine source methods have been used first, to our knowledge, by Bai and Yeung [1] in 2 space dimensions, and by Bertram [2, 3] in 3 space dimensions. For steady free-surface flows in three-dimensional, Rankine source methods seem to have been used first by Gadd [4] and Dawson [5].

The body boundary condition and the free surface boundary condition in a certain region around the body must be satisfied approximately in Rankine source methods by choosing appropriate source strengths. This requires sources, or source distributions, arranged both along the wetted body surface and along the discretized part of the free surface. Thus, the number of sources, or source panels, required to approximate the flow accurately is higher than when using Green functions that satisfy the free-surface condition. However, compared to the latter methods, Rankine source methods have important advantages:

- The computation of Green functions is much easier and faster.
- They must not substitute the steady flow around bodies with forward speed (or in a current) by a parallel flow.

Thus, Rankine source methods can model the real flow in all details except for effects of viscosity, flow separation, and wave breaking. To take account of these influences requires empirical corrections.

This chapter deals only with responses to the incoming waves that depend linearly on wave amplitude. Bernoulli's equation for fluid pressure contains a term $\frac{1}{2}\rho v^2$, where v is fluid speed. If v is the sum of a stationary velocity v_0 and a time-harmonic velocity v_1, the above pressure term is

$$\frac{1}{2}\rho(v_0^2 + 2v_0v_1 + v_1^2). \tag{7.3}$$

The middle term between parentheses is linear in periodical speed; thus, it contributes to the first-order seakeeping force. Because it contains also v_0, to solve the linearized seakeeping problem requires to determine also the stationary flow field.

The stationary potential flow around a steadily moving body at the free surface was, for a long time, of major interest for computing the *wave* resistance of ships. Today, this application is less interesting because RANS solvers can compute the *total* ship resistance with moderate effort and, apparently, with better accuracy than potential flow methods with empirical viscous corrections. Theoretically, one could combine a steady RANS computation with a time-harmonic potential flow calculation for seakeeping. Because of the very different computational meshes for RANS and for potential methods, and because of the smaller accuracy required for the stationary flow if only seakeeping is of interest, this combination appears less favorable than computing both the stationary and the time-harmonic flow by potential methods.

7.2 Rankine Source Method for the Stationary Flow Around a Ship with Forward Speed

The method described here is, essentially, a further development of the methods by Jensen et al. [6] and Söding and Conrad [7]. It determines the flow around a body in forward motion at the surface of an otherwise undisturbed ideal fluid. Whereas the program applied by Söding and Conrad [7] can deal also with lifting and slightly instationary flows, in the following description a steady, a non-lifting flow is presupposed.

With present personal computers, the stationary flow can be computed using Rankine source methods down to Froude numbers F_n of about 0.13; smaller F_n would require to solve excessively large equation systems. Therefore, for small F_n the free surface is approximated by a horizontal plane at which the symmetry condition is satisfied. The resulting so-called double-body flow field is used as a steady basis flow interacting with the time-harmonic flow. From the following description of computing the steady free-surface flow, computing the double-body flow can be derived simply by omitting all details relating to the free surface, and adding the symmetry condition at the plane water surface as described in Chap. 3.

7.2.1 Basic Boundary Value Problem

A body-fixed coordinate system is used. Its origin is at the undisturbed water surface. The coordinate axes x, y, z point forward, to starboard and downward, respectively. The flow potential is

$$\Phi = -Ux + \phi, \tag{7.4}$$

where ϕ is the disturbance potential and U the speed ahead (in $+x$ direction) of the body. The gradient of Φ is the fluid velocity relative to the body. ϕ should satisfy the continuity equation for incompressible flow, $\Delta\phi = 0$, and the following boundary conditions:

1. No flow through the wetted body surface:

$$\nabla\Phi \cdot \mathbf{n} = \nabla\phi \cdot \mathbf{n} - Un_1 = 0, \tag{7.5}$$

   where n_1 is the x component of the normal vector $\mathbf{n}$.
2. No flow through the free surface: The upward normal vector on the free surface is

$$\mathbf{n} = (\zeta_x, \zeta_y, -1), \tag{7.6}$$

   where $\zeta(x, y, t)$ is the vertical coordinate of the free surface, and indices x, y designate partial derivatives. Thus, the so-called kinematic (Greek kinema = motion) free-surface condition is, just like the body boundary condition:

$$\nabla\phi \cdot \mathbf{n} - Un_1 = 0. \tag{7.7}$$

3. The pressure at the water surface is equal to the air pressure. The latter is assumed constant; static air pressure variations between wave trough and crest as well as dynamical pressure variations due to wind are neglected. Then Bernoulli's equation gives the so-called dynamical (force) free-surface condition:

$$g\zeta = -U\phi_x + \frac{1}{2}(\nabla\phi)^2. \tag{7.8}$$

4. In case of shallow water of constant depth d, there must be no flow through the bottom at $z = d$:

$$\phi_z(x, y, d) = 0. \tag{7.9}$$

   This condition is satisfied by using, instead of sources or source panels, pairs of sources or source panels of equal source strength lying symmetrically on both sides of (above and below) the bottom.
5. In case of deep water, the disturbance flow must tend to zero for $z \to \infty$. This condition is satisfied automatically if sources are arranged only within a finite depth range. However, the resulting decrease of $\nabla\phi$ with increasing z is sometimes too slow. To avoid this, it may be recommended to add an additional source relatively high (e.g., 15% of the length of the body) above the water surface, and to satisfy the additional condition that the sum of all source strengths is zero. Then, far below the body, the disturbance potential will decrease like that of a dipole, whereas it will decrease like that of a source if the sum of source strengths is non-zero. Whether this additional condition appears necessary or not (the sum of source strengths may be approximately zero also without such a condition) depends on the case (e.g., deep or shallow water) and on details of the discretization (e.g., on the size of the region in which the free-surface condition is satisfied). In any case, such a condition does not decrease the approximation quality, and it has nearly no influence on computing time.
6. If the body is symmetrical to the plane $y = 0$, the mirror principle (see Sect. 3.7) is applied.
7. We want to handle the case of a body proceeding into undisturbed water. In that case, waves appear beside and behind the body, not in front of it. This so-called radiation condition will be satisfied by a numerical trick explained later. Ahead of the ship, there are near-field disturbances the amplitudes of which decrease rapidly with distance from the body. The fact that no far-field waves appear in front of the body is explained by the fact that the group velocity of the waves is lower than the phase velocity. A stationary wave field requires that the component of ship speed in wave propagation direction is equal to the wave phase speed. As the group speed is less, the far-field waves remain, in a ship-fixed reference system, behind the point where they are generated. Nonetheless, the condition is necessary because there exist solutions to the boundary value problem of conditions 1–6 which violate condition 7. This is illustrated by the

linearized free-surface condition (7.14) below: If the condition is satisfied for a potential field ϕ and speed U, it is also satisfied for $-\phi$ and $-U$. Also the body boundary condition remains satisfied after the change of signs, however for a ship moving in opposite direction. The waves, which appeared behind the stern in the original solution, remain there also in case of backward ship motion according to the solution $-\phi$, $-U$, which is contrary to reality.

7.2.2 Wave Breaking

Even slender ships with moderate forward speed generate a track of foam generated by breaking waves and splashes, at least at the bow. The free-surface conditions given above cannot describe these phenomena. Because we are not interested in them, we have only to make sure that splashes and wave breaking do not preclude computing the large-scale waves in which we are interested because of their influence on wave resistance, squat, and seakeeping.

Splashes are, normally, not a problem if the free-surface grid is not finer than required to resolve the interesting waves. In our programs, it proved sufficient to use a grid spacing of $0.75U^2/g$ in front of the body, and $0.6U^2/g$ besides and behind it, to resolve the waves accurately. This corresponds to $\lambda/8.4$ and $\lambda/10.5$, respectively, where λ is the length of the transverse waves behind the ship. (This free-surface grid length is much larger than that used by most other authors. Here, it is feasible because of our numerical method satisfying the radiation condition.)

Breaking waves, on the other hand, often require a modification of the free-surface condition, especially for blunt ships and low forward speed. In principle, a potential method is not suitable to describe a breaking wave accurately, because even the large-scale flow (neglecting turbulence) in a breaking wave has substantial values of rot v near to the surface behind the breaking region; thus the flow field does not have a potential. However, because the wave amplitude decreases behind the breaking region, wave damping can be used as an approximation to wave breaking.

Wave damping is attained by modifying the kinematic free-surface condition (7.7):

$$\nabla\phi(\mathbf{x}-\mathbf{d})\cdot\mathbf{n}(\mathbf{x}) - Un_1 = 0, \tag{7.10}$$

where $\mathbf{n}$ is given by (7.6). $\mathbf{d}$ is a vector in the direction of longitudinal grid lines, i.e., approximately in flow direction. $|\mathbf{d}|$ determines the strength of damping.

To understand the effect of $\mathbf{d}$, we consider a parallel flow of speed U and the disturbance potential

$$\phi = \text{Re}(ae^{-kz-ikx}). \tag{7.11}$$

It describes a wave which is stationary in our body-fixed coordinate system moving forward with speed $U = \sqrt{g/k}$. We linearize (only for explaining the effect of $\mathbf{d}$) the modified kinematic boundary condition (7.10) and the dynamic free-surface condition (7.8):

$$U\zeta_x + \phi_z(x-d) = 0; \tag{7.12}$$

$$\zeta = -\frac{U}{g}\phi_x(x). \tag{7.13}$$

Both conditions can be combined to eliminate ζ:

$$\frac{U^2}{g}\phi_{xx}(x) = \phi_z(x-d). \tag{7.14}$$

Inserting (7.11) results in

$$\text{Re}\left(\frac{U^2}{g}k^2 e^{-kz-ikx}\right) = \text{Re}\left(ke^{-kz-ik(x-d)}\right), \tag{7.15}$$

For arbitrary x, this condition is satisfied only if

$$k = \frac{g}{U^2}e^{ikd} = \frac{g}{U^2}[\cos(kd) + i\sin(kd)]. \tag{7.16}$$

Inserting the complex $k = k_r + ik_i$ into (7.11) results in

$$\phi = \text{Re}[ae^{k_i x}e^{-k_z - ik_r x}]. \tag{7.17}$$

Thus, the imaginary part of k produces a decrease of the stationary wave amplitude for decreasing x, i.e., within the region where a positive d is specified.

Non-zero **d** is applied where the slope of the free surface behind a wave crest exceeds a limit value. This increases the robustness of the iteration to satisfy the boundary conditions very much. Additionally to the localized damping in regions of strong downward slope in flow direction, sometimes an overall slight wave damping is required to attain convergence. To reduce or eliminate its influence on the final solution, the overall damping is reduced or omitted after the first few iteration steps.

7.2.3 Representation of the Disturbance Potential by Source Potentials

The disturbance potential will be superimposed from point sources. Thus, the disturbance potential is approximated as

$$\phi(\mathbf{x}) = \sum_{j=1}^{n} q_j G(\mathbf{x}, \boldsymbol{\xi}_j), \tag{7.18}$$

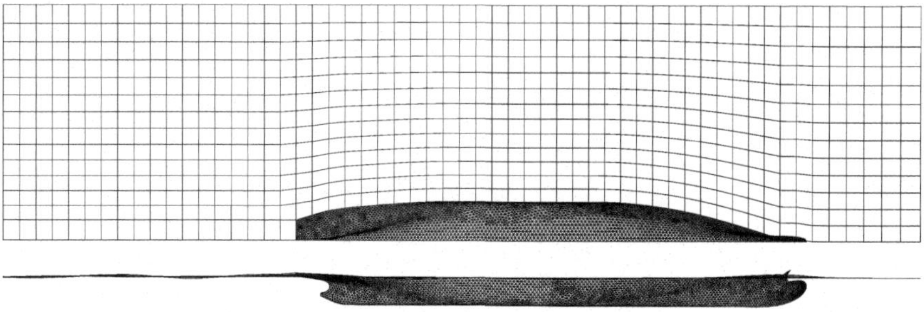

Fig. 7.1 Body and free-surface grid for steady flow computation around a container-ship at $F_n = 0.24$

where the sum is extended over all sources. The Green function G is, in the simplest case,

$$G(\mathbf{x}, \boldsymbol{\xi}_j) = \frac{-1}{|\mathbf{x} - \boldsymbol{\xi}_j|} . \tag{7.19}$$

Depending on the case (boundary conditions 4, 6), associations of several unit source potentials are used sometimes instead of the single source in (7.19).

The source strengths q_j are determined from the condition that there is no net flux through all body and free-surface patches. The wetted body surface is covered by triangular patches, and a region of the free surface around the body is covered by quadrilaterals (Fig. 7.1). Source points $\boldsymbol{\xi}_j$ are arranged within the body a short normal distance behind each body triangle midpoint and about 1 cell breadth above the free surface cell midpoints. The aforementioned numerical trick used to satisfy the radiation condition is [6]: The free-surface panels of the foremost transverse panel row lack a source above their midpoint; the same applies to the free-surface panels immediately behind a transom. Instead, an additional transverse row of sources is arranged half a cell length behind the aft end of the free-surface grid. All this refers to the 'original' sources; depending on case, the required mirror sources are added. Figure 7.2 illustrates the effect of this backshift of sources.

7.2.4 Patch Method and Body Boundary Condition

The patch method [9] is used in our stationary Rankine source program to increase the accuracy of computed body forces and to simplify the formulae for determining the potential and its patch-averaged gradient. For two space dimensions, it was described already in Sects. 3.8 and 5.2.3; here, it is used in three-dimensional flow.

A triangular patch on the body surface may be defined by corner points $\mathbf{x}_A$, $\mathbf{x}_B$, and $\mathbf{x}_C$ arranged clockwise if seen from outside of the body (Fig. 7.3). Patch side vectors are defined by

$$\mathbf{a} = \mathbf{x}_C - \mathbf{x}_B; \quad \mathbf{b} = \mathbf{x}_A - \mathbf{x}_C; \quad \mathbf{c} = \mathbf{x}_B - \mathbf{x}_A . \tag{7.20}$$

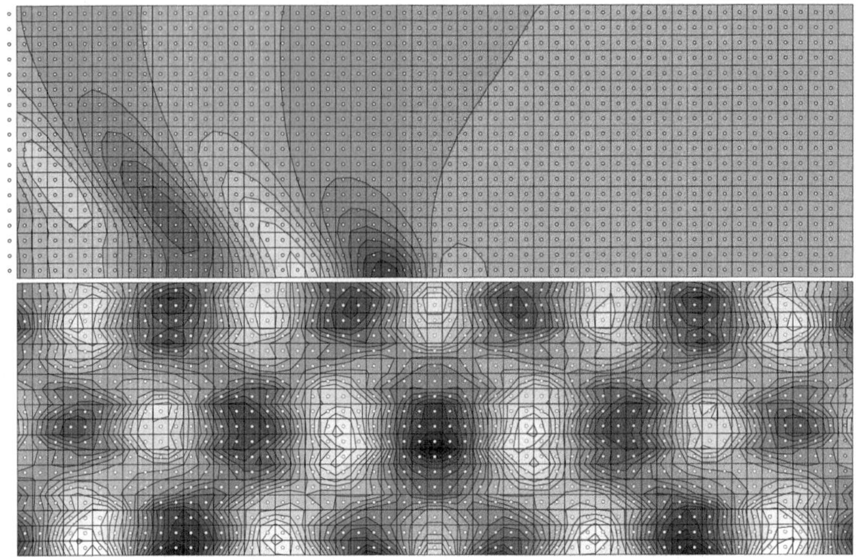

Fig. 7.2 Wave pattern of a fully submerged ellipsoid with free surface panel grid and free surface source points. Top: with backshifted sources; bottom: with all sources above the panel centers [8]

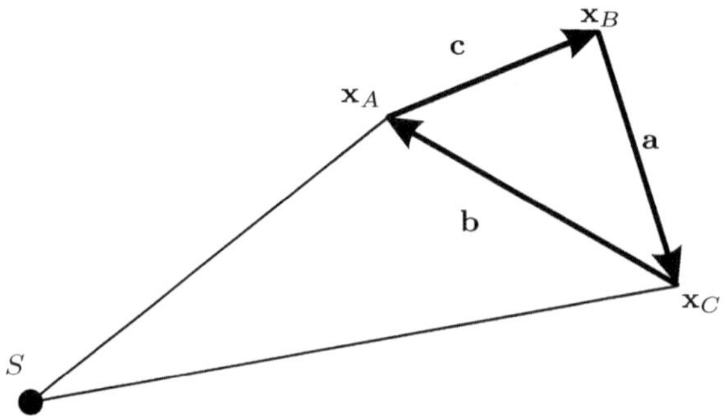

Fig. 7.3 Source point S and patch ABC

Two further vectors are defined as

$$\mathbf{n}_{AB} = \mathbf{b} - \frac{\mathbf{c} \cdot \mathbf{b}}{\mathbf{c}^2}\,\mathbf{c}; \quad \mathbf{n}_{AC} = \mathbf{c} - \frac{\mathbf{b} \cdot \mathbf{c}}{\mathbf{b}^2}\,\mathbf{b}. \tag{7.21}$$

The potential at the corners A, B, C due to a unit source is determined using (7.19). That equation is used also to determine the potential at the midpoints of the panel sides (these are supports in the three-point Gauß integration over a triangular area).

$$\phi_a = \phi([\mathbf{x}_B + \mathbf{x}_C]/2); \tag{7.22}$$

correspondingly for ϕ_b and ϕ_c. Then the flow potential averaged over the patch is approximated as

$$\bar{\phi} = (\phi_a + \phi_b + \phi_c)/3. \tag{7.23}$$

As an alternative, an analytical formula for the average of the Green function over the panel may be used.

The velocity due to a unit source, averaged over the panel, is approximated as

$$\overline{\nabla\phi} = \frac{\phi_A - \phi_C}{\mathbf{b} \cdot \mathbf{n}_{AB}} \mathbf{n}_{AB} + \frac{\phi_B - \phi_A}{\mathbf{c} \cdot \mathbf{n}_{AC}} \mathbf{n}_{AC} - \frac{\alpha}{A_0} \mathbf{n}, \tag{7.24}$$

where $\mathbf{n}$ is the unit normal vector on the patch pointing into the body,

$$\mathbf{n} = \frac{\mathbf{a} \times \mathbf{b}}{|\mathbf{a} \times \mathbf{b}|}, \tag{7.25}$$

and A_0 is the panel area:

$$A_0 = \frac{1}{2} |\mathbf{a} \times \mathbf{b}|. \tag{7.26}$$

The first two terms on the right-hand side of (7.24) give the patch-averaged tangential velocity. The terms follow from assuming a bi-linear variation of ϕ within the patch. The last term in (7.24) gives the average normal velocity; α is the solid angle under which the patch appears as seen from the source. In the present application, the normal component of the fluid velocity is zero on the patch, but the angles α are required to determine the matrix of the equation system for determining the source strengths.

The solid angle α may be determined by the formula for the area of a triangle in spherical geometry:

$$\alpha = \beta_{SAB,SBC} + \beta_{SBC,SCA} + \beta_{SCA,SAB} - \pi. \tag{7.27}$$

Term $\beta_{SAB,SBC}$ is the angle between (a) the plane containing the source and patch corners A and B, and (b) the plane containing the source and corners B and C; correspondingly for the other β values. In the following, $\mathbf{A}$, $\mathbf{B}$, $\mathbf{C}$ designate the vectors pointing from the source to corners A, B, and C, respectively. Then $\beta_{SAB,SBC}$ is determined as

$$\beta_{SAB,SBC} = \arctan \left| \frac{[(\mathbf{A} \times \mathbf{B}) \times (\mathbf{B} \times \mathbf{C})] \cdot \mathbf{B}}{(\mathbf{A} \times \mathbf{B}) \cdot (\mathbf{B} \times \mathbf{C}) |\mathbf{B}|} \right| . \tag{7.28}$$

correspondingly for the other angles β.

Alternatively to (7.27) and (7.28), the formula of Oosterom and Strackee may be used [8]:

$$\alpha = 2 \arctan \left| \frac{\det(\mathbf{A}, \mathbf{B}, \mathbf{C})}{|\mathbf{A}||\mathbf{B}||\mathbf{C}| + \mathbf{A} \cdot \mathbf{B}|\mathbf{C}| + \mathbf{A} \cdot \mathbf{C}|\mathbf{B}| + \mathbf{B} \cdot \mathbf{C}|\mathbf{A}|} \right| . \tag{7.29}$$

In both formulas for α, the sign of α has to be changed if, seen from the source, the points A, B, C turn anti-clockwise, i.e. if

$$(\mathbf{a} \times \mathbf{b}) \cdot \mathbf{d} < 0, \tag{7.30}$$

where $\mathbf{d} = \frac{1}{3}(\mathbf{A} + \mathbf{B} + \mathbf{C})$.

For large (compared to the triangle sides) distance of the source from the patch, the approximation

$$\alpha = \frac{(\mathbf{a} \times \mathbf{b}) \cdot \mathbf{d}}{2|\mathbf{d}|^3} \tag{7.31}$$

may be used to reduce the complexity and rounding errors, which may be large for large $|\mathbf{d}|$.

The body boundary condition (7.5) requires that the normal component of $\overline{\nabla\phi}$, added from all sources with their actual source strengths, and that of the parallel flow cancel. Using the last term in (7.24) for the normal component, this results in

$$\sum_j (\Delta q_j + q_j^{prev})\alpha_j + A_0 U n_1 = 0 \tag{7.32}$$

for each body panel. Here, q_j^{prev} is the strength of source j of the previous iteration step (0 at the beginning), and Δq_j is the required change.

7.2.5 Numerical Treatment of Free-Surface Conditions

An iteration is required to satisfy the nonlinear free-surface condition (7.8), to correct the ship's floating position (squat), to shift the free-surface grid to the actual height of the free surface, and to modify the body grid to reach up to the actual, deformed waterline.

For any step within this iteration, it is supposed that the current free-surface panel grid points have vertical coordinates satisfying the dynamic free-surface condition (7.8) using the current solution for ϕ, but that the kinematic condition (7.7) is not yet satisfied. Thus, the source strengths q_j must be changed to cancel the net flow through each free-surface quadrilateral. For two reasons, the method used for the body boundary condition would not be successful here:

1. A change of free-surface source strengths has influence on ϕ and, because of the dynamic boundary condition, also on ζ, and further, because of (7.6), on $\mathbf{n}$ and thus on the kinematic condition itself. To postpone this, influence to the next iteration step would result in divergence of the iteration.
2. There would be a decoupling between the ζ values at neighboring longitudinal free-surface grid lines, leading to alternating errors of ζ along transverse grid lines.

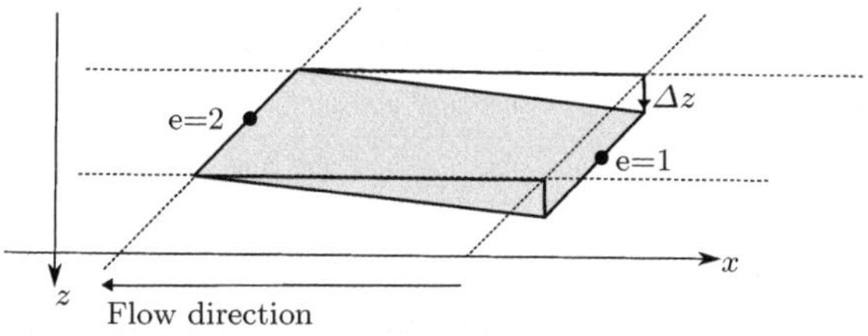

Fig. 7.4 Change of panel position is approximated by vertical shift of midpoints (e = 1 and e = 2)

Therefore, the following method is used to correct the net flux through free-surface panels: The residual (error) r_i of (7.7) for panel i is

$$r_i = \int_i \nabla\phi \cdot \mathbf{n}\, da - Un_1 = \sum_j q_j \int_i \nabla G(\mathbf{x}, \boldsymbol{\xi}_j)\, da \cdot \mathbf{n} - Un_1, \tag{7.33}$$

where $\int_i \ldots da$ means integration over the area of panel i.

Neglecting nonlinear effects, the residuum is reduced to zero by determining the required change of source strengths Δq_j according to the equation

$$r_i + \sum_j \frac{dr_i}{dq_j} \Delta q_j = 0. \tag{7.34}$$

The derivative dr_i/dq_j is composed of the direct effect of q_j on the right-hand side of (7.33) for constant panel position and of the indirect effect (as described under 1 above) due to the change of the panel height and inclination (Fig. 7.4):

$$\frac{dr_i}{dq_j} = \int_i \nabla G(\mathbf{x}, \boldsymbol{\xi}_j) \cdot \mathbf{n}\; da - \sum_{e=1}^{2} \frac{\partial}{\partial \zeta_e} \int_i [\nabla\phi - (U, 0, 0)] \cdot \mathbf{n}\, da \frac{d\zeta_e}{dq_j}\ . \tag{7.35}$$

Here, the change of panel position is approximated by the vertical shift of the midpoints of the forward (e = 1) and aft (e = 2) side of the panel.

The first integral in (7.35) (it occurs also in (7.33)) is simply the solid angle α_j of panel i seen from source j. The second integral in (7.35) is approximated by assuming that the flow is in the direction $\mathbf{x}_1 - \mathbf{x}_2$, where $\mathbf{x}_1$ and $\mathbf{x}_2$ are the midpoints of the front and aft panel side. This gives

$$\frac{\partial}{\partial \zeta_e} \int_i [\nabla\phi - (U, 0, 0)]\mathbf{n}\, da \approx \pm[\nabla\phi - (U, 0, 0)] \frac{\mathbf{x}_1 - \mathbf{x}_2}{|\mathbf{x}_1 - \mathbf{x}_2|} b, \tag{7.36}$$

where b is panel breadth. The plus sign is used for $e = 1$, the minus for $e = 2$. Finally, $d\zeta_e/dq_j$ follows from (7.8) and (7.18):

$$\frac{d\zeta_e}{dq_j} = \frac{1}{g}[\nabla\phi - (U, 0, 0)]\nabla G(\mathbf{x}_e, \boldsymbol{\xi}_j)\ . \tag{7.37}$$

$\nabla G(\mathbf{x}_e, \boldsymbol{\xi}_j)$ follows from (7.19):

$$\nabla G(\mathbf{x}_e, \boldsymbol{\xi}_j) = \frac{\mathbf{x}_e - \boldsymbol{\xi}_j}{|\mathbf{x}_e - \boldsymbol{\xi}_j|^3} . \tag{7.38}$$

(This is the only case in which the analytical derivative of G is used.) Together, these formulae allow to evaluate the matrix elements dr_i/dq_j in (7.34).

7.2.6 *Solution of the Equation System*

Equations (7.32) (for each body triangle) and (7.34) (for each free-surface quadrilateral) constitute the linear equation system for determining the changes of source strength Δq_j. As discussed above, possibly the condition

$$\sum_j q_j = \sum_j (\Delta q_j + q_j^{prev}) = \sum_j \Delta q_j = 0 \tag{7.39}$$

should be added. In solving the equation system, the free `LAPACK` package proved to be much faster than the simple Gauß algorithm given in Chap. 3. Iterative solutions proved to be not robust even after preconditioning of the matrix. The reason is the high condition number of the system matrix, caused in parts by the backshifting of sources relative to the free-surface panel grid.

Because of the various nonlinearities, under-relaxation is required, i.e., only a (user-specified) fraction of the computed change Δq_j is applied. Further, the approximations in (7.37) preclude a convergence of $|\Delta q_j|$ to zero; however, a reduction of the maximum residual by two to three orders of magnitude is attained and sufficient.

7.2.7 *Further Computations*

The following is required in each iteration step when the new source strengths have been determined:

1. Computing the potential at body grid points
2. Computing average velocity and pressure at body triangles
3. Computing pressure force and moment
4. Correcting the floating condition (if squat is not suppressed)
5. Correcting the waterline along the body from the condition $p = 0$. Deviations from the condition $p = 0$ must be allowed, mostly at both ship ends, to limit values of $|dy/ds|$ and $|dz/ds|$ of the waterline, where s is waterline arc length. If the waterline intersects a nearly horizontal part of the hull surface in the afterbody, it may be necessary to approximate the waterline in this range by a straight

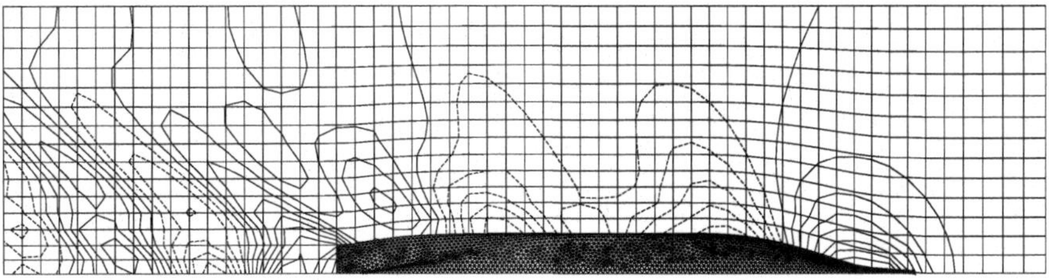

Fig. 7.5 Free-surface height contours around another container-ship at $F_n = 0.24$. Vertical distance 0.2 m; broken lines below the undisturbed surface

line, using the least-squares method, to avoid a rugged or even disconnected shape. A smooth waterline is required to obtain a smooth panel mesh on the water surface.

6. Correcting the body grid to reach up to the new waterline. The much simpler method to let the body grid reach above the waterline is neither accurate nor robust because it may lead to body sources located too near to free-surface sources, or to free-surface grid points or midpoints of the front and aft side of free-surface quadrilaterals lying within the body.
7. Improving the body grid for more equilateral triangles
8. Generating a new free-surface panel grid around the new waterline
9. Adapting the height of the interior free-surface grid points using (7.8) at the midpoints of the forward and aft panel sides
10. Writing and storing results in a file for possible later use
11. Test if iteration should end
12. Computing new source locations
13. Computing the potential at free-surface grid points and the average velocity in free-surface quadrilaterals
14. Computing the wave damping vector **d** for all free-surface panels

Additionally, means of importing and evaluating the body shape, mesh generation on body and free surface, plotting results, etc., must be provided. As an example, Fig. 7.5 shows free-surface height contours around another container-ship at $F_n = 0.24$.

7.3 Rankine Source Method for the Time-Harmonic Flow

Here, we deal with determining transfer functions of ship motions in regular waves. Specifically, the method of Söding et al. [10] is described. It is confined to determine responses (motions, hull pressure, and loads in transverse ship sections) depending linearly on wave amplitude. Based on these, stationary second-order effects like added resistance are also determined. The method takes account of the interaction between the time-harmonic and the stationary free-surface flow computed as described before. Besides the inertial coordinate system x, y, z moving ahead with the average ship speed, which was used above, a body-fixed coordinate system $\underline{x}$, $\underline{y}$, $\underline{z}$ will be used. Averaged over time, both systems coincide with respect to origin and axis directions.

7.3.1 Superposition of Potentials

Conventionally, the total flow potential ϕ^t is composed of the parallel flow (backward in the moving coordinate system used here), the steady flow disturbance $\phi^0(\mathbf{x})$, and the time-harmonic flow potential described by its complex amplitude $\hat{\phi}(\mathbf{x})$:

$$\phi^t(\mathbf{x}, t) = -Ux + \phi^0(\mathbf{x}) + \text{Re}(\hat{\phi}(\mathbf{x})e^{i\omega_e t}). \tag{7.40}$$

Here, t is time, and ω_e is the encounter angular frequency of the wave. U is ship speed.

For reasons explained below, it appears better to use another superposition of potentials:

$$\phi^t(\mathbf{x}, t) = -Ux + \phi^0(\underline{\mathbf{x}}) + \text{Re}(\hat{\phi}e^{i\omega_e t}). \tag{7.41}$$

The difference is with respect to the steady disturbance potential ϕ^0. In (7.40), it is independent of time in the inertial reference system. In (7.41), on the other hand, it is time-independent in the body-fixed reference frame; thus, in the inertial system the steady potential field moves together with the body. Equation (7.41) was applied first (as far as known) by Hachmann [11] to improve the determination of hull pressures in connection with a strip method.

Both (7.40) and (7.41) are correct and describe the same total potential ϕ^t, but $\hat{\phi}$ is different in both cases. When using (7.40), the ship is oscillating in a stationary pressure field generated by ϕ^0 and is feeling corresponding restoring forces and moments. These restoring forces must be canceled largely by opposite terms in the periodical potential. This canceling cannot be exact, e.g., because different panel grids are used on the free surface for computing ϕ^0 and $\hat{\phi}$. This difficulty becomes obvious in long waves. Here the ship should follow the orbital motion of the water particles, but due to the above-mentioned restoring forces, which are not exactly canceled, computed motions are somewhat different. When using (7.41), on the other hand, these false restoring forces and moments do not occur, because the pressure field due to ϕ^0 is moving with the body. Less obvious, but practically more important are differences at wavelengths comparable to ship length. These will be discussed below.

In the following, $\hat{\phi}$ is always meant to represent the function occurring in (7.41), not that in (7.40). Furthermore, quantities depending nonlinearly on wave or motion amplitude are omitted.

The relation between the position vectors $\underline{\mathbf{x}}$ and $\mathbf{x}$ of the same point, expressed in the body-fixed or inertial coordinate systems, respectively, is

$$\mathbf{x} = \underline{\mathbf{x}} + \mathbf{v}, \tag{7.42}$$

where

$$\mathbf{v} = \mathrm{Re}(\hat{\mathbf{v}} e^{i\omega_e t}) \tag{7.43}$$

is the shift of a point fixed to the body relative to its mean position in the inertial system. $\mathbf{v}$ follows from the body motion at the coordinate origin, $\mathbf{u}$, and the rotation vector $\boldsymbol{\alpha}$ of the body:

$$\mathbf{v} = \mathbf{u} + \boldsymbol{\alpha} \times \mathbf{x}. \tag{7.44}$$

Using a first-order Taylor expansion of ϕ^0 around $\underline{\mathbf{x}}$, Eqs. (7.41)–(7.43) result in

$$\phi^t(\mathbf{x}, t) = -Ux + \phi^0(\mathbf{x}) + \mathrm{Re}[(\hat{\phi} - \hat{\mathbf{v}} \cdot \nabla\phi^0) e^{i\omega_e t}]. \tag{7.45}$$

A comparison with (7.40) shows that $\hat{\phi}$ in (7.40) is equal to $\hat{\phi} - \hat{\mathbf{v}} \cdot \nabla\phi^0$ in the Hachmann method used here.

Later, the gradient of ϕ^t is required. It follows from (7.45) as

$$\nabla\phi^t(\mathbf{x}, t) = \mathbf{U} + \nabla\phi^0(\mathbf{x}) + \mathrm{Re}[(\nabla\hat{\phi} - \nabla\phi^0 \times \hat{\boldsymbol{\alpha}} - \hat{\mathbf{v}}\nabla\nabla\phi^0) e^{i\omega_e t}], \tag{7.46}$$

where

$$\mathbf{U} = (-U, 0, 0), \tag{7.47}$$

and $\nabla\nabla\phi^0$ is the Hesse matrix (matrix of second partial derivatives) of the stationary potential ϕ^0. In deriving (7.46), the following relation was used:

$$(\nabla\hat{\mathbf{v}})\nabla\phi^0 = \nabla(\hat{\boldsymbol{\alpha}} \times \mathbf{x})\nabla\phi^0 = \nabla\phi^0 \times \hat{\boldsymbol{\alpha}}. \tag{7.48}$$

In analogy to ϕ^t in (7.45), the vertical coordinate of the free surface is

$$\zeta^t(\mathbf{x}, t) = \zeta^0(\mathbf{x}) + \mathrm{Re}[(\hat{\zeta} - \hat{\mathbf{v}} \cdot \nabla\zeta^0) e^{i\omega_e t}]. \tag{7.49}$$

Here, ζ^0 describes the steady wave field due to forward speed, and $\hat{\zeta}$ is the complex amplitude of the time-harmonic wave field.

The periodic potential will be superimposed from the Airy wave potential ϕ_w and from the potentials of point sources:

$$\hat{\phi} = \hat{\phi}_w + \sum_{j=1}^{J} \hat{q}_j G_j. \tag{7.50}$$

Here, $\hat{q}_j$ are the complex amplitudes of the source strengths (in units of 4π). The Green function G is the same as that used in the stationary potential. $\hat{\phi}_w$ is the complex amplitude of the Airy wave potential for wave amplitude of unity

$$\hat{\phi}_w = \frac{\omega}{ik} e^{\boldsymbol{\nu} \cdot \mathbf{x}} \tag{7.51}$$

with

$$\nu = ik(-\cos\mu, \sin\mu, i), \tag{7.52}$$

where μ is the wave angle. The complex amplitude of the corresponding undisturbed wave surface height is

$$\hat{\zeta}_w = e^{\boldsymbol{\nu} \cdot \mathbf{x}}. \tag{7.53}$$

7.3.2 Body Boundary Condition

The body boundary condition states that there is no net flow through the body patches. The same body patch grid is used in determining the stationary and the time-harmonic potentials. Contrary to the stationary case, however, now the body patches perform an oscillatory motion relative to the inertial system.

The fluid velocity relative to the inertial coordinate system and expressed in inertial coordinate directions is, per definition of the potential,

$$\mathbf{w}(\mathbf{x}) = \nabla\phi^t(\mathbf{x}). \tag{7.54}$$

To express the condition of no flow through the body patches, $\mathbf{w}$ is transformed to the body-fixed coordinate system and expressed in body-fixed coordinate directions. As shown by Bertram [3], this gives

$$\underline{\mathbf{w}}(\underline{\mathbf{x}}) = \nabla\phi^t(\mathbf{x}) - \boldsymbol{\alpha} \times \nabla\phi^t(\mathbf{x}) - \dot{\mathbf{v}}. \tag{7.55}$$

On the right-hand side, the term containing the body rotation $\boldsymbol{\alpha}$ takes account of the different coordinate directions of both systems and the last term is due to the relative motion $\mathbf{v}$ between both systems at point $\mathbf{x}$.

Using (7.55), the condition of no flux through the body surface is easily expressed:

$$\underline{\mathbf{n}}(\underline{\mathbf{x}}) \cdot \underline{\mathbf{w}}(\underline{\mathbf{x}}) = \underline{\mathbf{n}}(\underline{\mathbf{x}}) \cdot [\nabla\phi^t(\mathbf{x}) - \boldsymbol{\alpha} \times \nabla\phi^t(\mathbf{x}) - \dot{\mathbf{v}}] = 0. \tag{7.56}$$

Inserting the expression (7.46) for $\nabla\phi^t$ gives

$$\underline{\mathbf{n}}(\underline{\mathbf{x}}) \cdot \left[\mathbf{U} + \nabla\phi^0(\mathbf{x}) + \mathrm{Re}\left(\nabla\hat{\phi} - \nabla\phi^0 \times \hat{\boldsymbol{\alpha}} - \hat{\mathbf{v}}\nabla\nabla\phi^0 - \hat{\boldsymbol{\alpha}} \times (\mathbf{U} + \nabla\phi^0) - i\omega_e\hat{\mathbf{v}}\right) e^{i\omega_e t}\right] = 0. \tag{7.57}$$

Here, the term $\nabla\phi^0(\mathbf{x})$ is constant over t at fixed $\mathbf{x}$, but we need it at body-fixed points $\underline{\mathbf{x}}$, where it contains also an oscillatory part. To separate both parts, we use the general relation (see, e.g., [3]) between the partial derivatives in the inertial system (∇) and those in the body-fixed system ($\underline{\nabla}$):

$$\nabla = \underline{\nabla} + \alpha \times \nabla. \tag{7.58}$$

From this, the following relation may be derived for any differentiable zeroth-order function $f(\mathbf{x})$ and the first-order shift $\mathbf{v}$:

$$\nabla f(\mathbf{x}) = \underline{\nabla} f(\underline{\mathbf{x}}) + \mathbf{v}\nabla\nabla f. \tag{7.59}$$

The derivation involves a first-order Taylor expansion of f around $\underline{\mathbf{x}}$ and Eqs. (7.42) and (7.44).

Equation (7.59) is used to substitute $\nabla\phi^0(\mathbf{x})$ in (7.57). Then the terms containing $\nabla\nabla\phi^0$ cancel; this might have been expected because both the body and the field ϕ^0 do not move relative to each other. Further, two terms cancel. The remaining equation contains constant and oscillatory terms. The constant terms add up to zero because ϕ^0 satisfies the body boundary condition. Omitting these terms, the body boundary condition for $\nabla\hat{\phi}$ results in the following simple formula:

$$\underline{\mathbf{n}}(\underline{\mathbf{x}}) \cdot (\nabla\hat{\phi} - \hat{\alpha} \times \mathbf{U} - i\omega_e\hat{\mathbf{v}}) = 0. \tag{7.60}$$

The fact that (7.60) does not contain second derivatives of ϕ^0 is the main reason for preferring (7.41)–(7.40), because determining these derivatives involves much larger discretization errors than determining first derivatives, even if the best-known approximation methods are used. If (7.40) is used, second derivatives of ϕ^0 occur in the body boundary condition. They cause inaccuracies predominantly along the body surface; thus, they have a large influence on body pressures and on the forces following from them. On the other hand, when using (7.41), second derivatives of ϕ^0 occur not in the body, but instead in the kinematic free-surface boundary condition. Thus, the accompanied inaccuracies are concentrated more along the free surface where they have less influence on the pressure field on the body surface. This has been demonstrated in numerical experiments showing that for large Froude number (thus large ϕ^0) the Hachmann method is producing much more consistent results than the conventional method based on (7.40).

Inserting (7.50) into the boundary condition (7.60) gives a condition for the complex amplitudes $\hat{q}_j$ of the source strengths. Like for the steady potential, the condition is satisfied in the average over each patch by integrating over the patch area A. This gives

$$\int \underline{\mathbf{n}}(\underline{\mathbf{x}}) \cdot \nabla\hat{\phi}_w \, dA + \sum_j \hat{q}_j \int \underline{\mathbf{n}}(\underline{\mathbf{x}}) \cdot \nabla G_j \, dA - \underline{\mathbf{n}} \cdot \left(\hat{\alpha} \times \mathbf{U} + i\omega_e\hat{\mathbf{v}}\right) A = 0. \tag{7.61}$$

The first integral over the normal component of the wave orbital velocity is evaluated analytically. The second integral is, simply, the solid angle under which the patch is seen from the source point $\boldsymbol{\xi}_j$. In the last term, $\hat{\mathbf{v}}$ is evaluated at the midpoint of the patch.

The complex amplitudes of the body motions $\hat{\mathbf{u}}$ and $\hat{\alpha}$ are still unknown. Therefore, one assumes, initially, unit motions in the six degrees of freedom in evaluating the last term in (7.61); this gives six right-hand sides for determining the six source distributions for the so-called radiation potentials; they describe the waves radiated from the oscillating ship. Another right-hand side is established by the first (wave) term in (7.61); it gives the source strengths generating the so-called diffraction potential, which describes the diffraction of the incoming wave at the non-oscillating body. Later, the seven potentials are superimposed so that the body motion equations are satisfied.

Instead of only one wave case, one can evaluate the first (wave) term in (7.61) for various wave angles. Because the computer time for solving the equation system is only slightly increased with the number of right-hand sides, this can reduce the computer time by factors of 10 or more if waves of various directions and frequencies are of interest. However, like for the steady flow, the equation system (7.61) must be combined with that for the free-surface conditions. The coefficient matrix of the latter equation system depends on encounter frequency ω_e. Thus, only combinations of wave frequency and wave direction which lead to the same encounter frequency can be combined to a single equation system.

7.3.3 Dynamic Free-Surface Boundary Condition

The condition of constant pressure at the water surface gives the boundary condition to be satisfied at $z = \zeta^t$:

$$\frac{1}{2}(\nabla\phi^t)^2 + \dot{\phi}^t - g\zeta^t = \frac{1}{2}U^2. \tag{7.62}$$

Here, the partial time derivative $\dot{\phi}$ designates the time change for fixed $\mathbf{x}$, not for fixed $\underline{\mathbf{x}}$. Inserting the above-given expressions for ϕ^t, $\nabla\phi^t$ and ζ^t results in

$$\frac{1}{2}\left(\mathbf{U} + \nabla\phi^0(\mathbf{x}) + \mathrm{Re}[(\nabla\hat{\phi} + \hat{\alpha} \times \nabla\phi^0 - \hat{\mathbf{v}}\nabla\nabla\phi^0)e^{i\omega_e t}]\right)^2$$

$$+ \mathrm{Re}[i\omega_e(\hat{\phi} - \hat{\mathbf{v}} \cdot \nabla\phi^0)e^{i\omega_e t}] - g\zeta^0(\mathbf{x}) - \mathrm{Re}[g(\hat{\zeta} - \hat{\mathbf{v}} \cdot \nabla\zeta^0)e^{i\omega_e t}] = \frac{1}{2}U^2. \tag{7.63}$$

If the squared expression is expanded, the resulting equation contains terms which—for fixed $\mathbf{x}$—are constant over time. These terms cancel if the steady quantities ϕ^0 and ζ^0 satisfy the free-surface boundary conditions. The time-harmonic terms add up to zero at all times only if their complex amplitudes add up to zero. Thus, the dynamic boundary condition for the complex amplitudes $\hat{\phi}$ and $\hat{\zeta}$ becomes

$$(\mathbf{U}+\nabla\phi^0)\cdot(\nabla\hat{\phi}+\hat{\alpha}\times\nabla\phi^0-\hat{\mathbf{v}}\nabla\nabla\phi^0)+i\omega_e(\hat{\phi}-\hat{\mathbf{v}}\cdot\nabla\phi^0)-g(\hat{\zeta}-\hat{\mathbf{v}}\cdot\nabla\zeta^0)=0. \tag{7.64}$$

Compared to the dynamic free-surface boundary condition when using (7.40) instead of (7.41), additional terms containing $\hat{\mathbf{v}}$ occur here. Because $\hat{\mathbf{v}}$ depends on the body motions, these terms are handled as described for the body boundary condition by assuming a unit body motions in the six degrees of freedom.

7.3.4 *Kinematic Free-Surface Boundary Condition*

The condition states that there is no flow through the free surface. Thus, a fluid particle with vertical coordinate z_F, which is at the free surface $z=\zeta^t$, will remain at the free surface. This is expressed by

$$\frac{D}{Dt}(z_F-\zeta^t)=0. \tag{7.65}$$

Here, D/Dt designates the substantial time derivative, i.e. the change for a point moving with the fluid:

$$\frac{D}{Dt}=\frac{\partial}{\partial t}+\nabla\phi^t\cdot\nabla. \tag{7.66}$$

The substantial time derivative of z_F is the vertical component of the fluid velocity ϕ^t_z. Thus, the kinematic condition becomes

$$\left(\frac{\partial}{\partial t}+\nabla\phi^t\cdot\nabla\right)\zeta^t=\frac{\partial}{\partial z}\phi^t. \tag{7.67}$$

Also here, partial time derivatives are meant to imply fixed $\mathbf{x}$, not fixed $\underline{\mathbf{x}}$.

Inserting the expressions (7.45), (7.46) and (7.49) for ϕ^t, $\nabla\phi^t$ and ζ^t gives a complicated expression, from which—like for the dynamic boundary condition—the stationary terms can be deleted because they cancel if the stationary solution satisfies its kinematic free-surface condition. After a number of non-trivial transformations one obtains, finally, the kinematic free-surface boundary condition in the form

$$\hat{R}=i\omega_e(\hat{\zeta}-\hat{\mathbf{v}}\cdot\nabla\zeta^0)+\nabla\zeta^0\cdot\nabla\hat{\phi}+(\mathbf{U}+\nabla\phi^0)\cdot\nabla\hat{\zeta}-\hat{\phi}_z+(-\phi^0_y,\phi^0_x,U\zeta^0_y)\cdot\hat{\alpha}+\hat{A}=0, \tag{7.68}$$

where $\hat{R}$ is the complex residual, and $\hat{A}$ is an expression containing the Hesse matrices of ϕ^0 and ζ^0:

$$\hat{A}=\frac{-\hat{\mathbf{v}}\cdot\mathbf{n}^0}{|\mathbf{n}^0|^2}\left[\mathbf{n}^0(\nabla\nabla\phi^0)\mathbf{n}^0+\mathbf{n}^0(\nabla\nabla\zeta^0)(\mathbf{U}+\nabla\phi^0)\right]. \tag{7.69}$$

$\mathbf{n}^0$ is the upward normal vector on the stationary free surface:

$$\mathbf{n}^0 = (\zeta_x^0, \zeta_y^0, -1). \tag{7.70}$$

As explained above, $\nabla\phi^0$ must not be determined from the derivatives of the Green functions; instead, the average value over a free-surface panel is determined from (a) the values of the potential at the corners of the patch (these give the tangential components) and (b) the flux through the panel (it gives the normal component of $\nabla\phi^0$).

This average value is used as an approximation of $\nabla\phi^0$ at the panel midpoint.

More difficult is a correct approximation of $\mathbf{n}^0(\nabla\nabla\phi^0)\mathbf{n}^0$ occurring in A. From the Laplace equation $\Delta\phi^0 = 0$ follows

$$\mathbf{n}^0(\nabla\nabla\phi^0)\mathbf{n}^0 = -\mathbf{s}^0 B\mathbf{s}^0 - \mathbf{t}^0 B\mathbf{t}^0, \tag{7.71}$$

where $\mathbf{s}^0$ and $\mathbf{t}^0$ are vectors which are tangential to the steady free surface and normal to each other, and B is defined as

$$B = \nabla\nabla\phi^0 - (\mathbf{n}^0\nabla\nabla\phi^0) \circ \mathbf{n}^0. \tag{7.72}$$

Here, $\circ$ designates the outer product; in tensor notation, the last term in (7.72) is $n_k^0\phi_{,ki}^0 n_j^0$. B may be interpreted as the Hesse matrix of ϕ^0 minus its normal component. Thus, B can be determined from the values of $\nabla\phi^0$ at the midpoints of the four panels surrounding the actual panel, assuming a bi-linear variation of $\nabla\phi^0$ over the region of the four panels. Along the boundary of the free-surface grid, where one or two neighboring panels are missing, the actual panel must be substituted for one or two missing neighbor panels.

Another second-order derivative occurring in A is $\nabla\nabla\zeta^0$. Its computation differs from that of $\nabla\nabla\phi^0$ because ζ^0 does not depend on z, and it does not satisfy a Laplace equation. $\nabla\nabla\zeta^0$ can be approximated directly from the ζ^0 values at the corners of the actual panel and its (usually four) neighbors.

7.3.5 Numerical Treatment of Free-Surface Conditions

To establish a condition for $\hat{q}_j$, Eqs. (7.68) and (7.50) are combined:

$$\underbrace{i\omega_e(\hat{\zeta}_w - \hat{\mathbf{v}}\cdot\nabla\zeta^0) + \nabla\hat{\phi}_w\cdot\nabla\zeta^0 + (\mathbf{U}+\nabla\phi^0)\cdot\nabla\hat{\zeta}_w - \hat{\phi}_{w,z} + (-\phi_y^0, \phi_x^0, U\zeta_y^0)\cdot\hat{\alpha} + \hat{A}}_{\hat{R}} + \sum_j \frac{d\hat{R}}{d\hat{q}_j}\hat{q}_j = 0. \tag{7.73}$$

The derivative $d\hat{R}/d\hat{q}_j$ follows from (7.68):

$$\frac{d\hat{R}}{d\hat{q}_j} = i\omega_e\frac{d\hat{\zeta}}{d\hat{q}_j} + (\mathbf{U}+\nabla\phi^0)\cdot\nabla\frac{d\hat{\zeta}}{d\hat{q}_j} + \underbrace{[\nabla\zeta^0 - (0,0,1)]}_{\mathbf{n}^0}\cdot\nabla G_j, \tag{7.74}$$

and $d\hat{\zeta}/d\hat{q}_j$ follows from the dynamic free-surface condition (7.64):

$$\frac{d\hat{\zeta}}{d\hat{q}_j} = \frac{1}{g}\left((\mathbf{U} + \nabla\phi^0)\cdot\nabla G_j + i\omega_e G_j\right). \tag{7.75}$$

Also here, the boundary condition is integrated over a free-surface panel to satisfy it in the average over each free-surface panel. After integration, the second and fourth terms in (7.73) constitute together the flux (positive upward) through the (stationary) panel caused by the undisturbed wave. It is determined analytically by integrating the wave-induced velocity. The last term in (7.74), integrated over a panel, is the flux through the panel caused by a unit source at $\boldsymbol{\xi}_j$; it is simply the solid angle of the panel seen from $\boldsymbol{\xi}_j$. In (7.75), only the component of ∇G_j in the direction of the stationary flow is required. It is approximated from the difference of G_j values at the midpoints of the forward and aft sides of the panel.

The quantity $\nabla\hat{\phi}_w$ in (7.73) may cause problems in very short waves. On a crest ($z < 0$) of the stationary wave field, the wave potential (7.51) and its velocity assume large absolute values due to the factor e^{-kz}. In principle, these large values are canceled by a correspondingly large diffraction potential $\hat{\phi}$. However, this involves also large discretization errors. To avoid these, the z dependence of the wave potential can be modified to $e^{-k(z-\zeta^0)}$. Because ζ^0 depends on x and y, this modification violates the Laplace equation. Nonetheless, the modification seems to improve results in very short waves.

7.3.6 Boundary Condition at a Transom

Except for small forward speed, a transom remains dry even if the waterline intersects the transom contour. This holds both for the stationary and for the time-harmonic flow. In determining the stationary flow, the no-flux condition of the panels immediately behind the transom determines the longitudinal inclination of these panels. Their forward height is given by the transom contour, not by the dynamic free-surface condition. The latter is used only at the other free-surface grid points.

Because of the linearization with respect to wave and motion amplitude, the free-surface panels are not moving with the oscillatory flow, but remain stationary. To ensure that the free surface remains at the immersed transom contour requires, thus, an additional transom condition. It substitutes the kinematic free-surface condition for the panels immediately behind the transom.

The transom condition for periodical flow follows from the dynamic free-surface condition (7.64) with the additional statement

$$\hat{\zeta} = \hat{v}_3, \tag{7.76}$$

where $\hat{v}_3$ is the z component of the shift (in the inertial system) $\hat{\mathbf{v}}$ of body-fixed points at the transom. This gives the following condition:

$$(\mathbf{U} + \nabla\phi^0) \cdot \nabla\hat{\phi} - \mathbf{U} \cdot (\nabla\phi^0 \times \hat{\alpha}) - (\mathbf{U} + \nabla\phi^0) \cdot (\hat{\mathbf{v}}\nabla\nabla\phi^0) + i\omega_e(\hat{\phi} - \hat{\mathbf{v}} \cdot \nabla\phi^0) + g(\zeta_x^0, \zeta_y^0, -1) \cdot \hat{\mathbf{v}} = 0. \quad (7.77)$$

The term containing $\nabla\nabla\phi^0$ is difficult to evaluate accurately and has only small influence on the results; in our program it is neglected.

In the first term, the longitudinal component of $\nabla\hat{\phi}$ is approximated as above by the values of $\hat{\phi}$ at the midpoints of the forward and aft side of the free-surface panels immediately behind the transom. Inserting (7.50) for $\hat{\phi}$ gives then a condition for the source strengths $\hat{q}_j$.

7.3.7 Other Conditions

For the present case of a floating body in a regular wave, the radiation condition ensures that the flow around the body is influenced alone by the incoming wave, not by any other disturbances from outside the discretized fluid region. Further, the outer boundary of the discretized free surface should not reflect waves. In contrast to the stationary flow, radiation and diffraction waves can occur also in front of the body if

$$\tau = \frac{U\omega_e}{g} < 0.25. \quad (7.78)$$

The backward shifting of sources relative to the panel centers, which is an elegant way of satisfying the radiation condition in stationary flow, proved to give no robust method in the time-harmonic flow problem. Even in combination with damping and conditions allowing only out-going waves, backshifting of sources did not always give accurate results. Thus, in our programs, all free-surface sources are located above the free-surface panel midpoints; and the radiation condition is satisfied alone by damping the radiation and diffraction waves (but not the incoming wave). This is easily attained by assuming a complex instead of a real gravity acceleration g in the dynamic free-surface condition. The amount of damping, i.e., the positive imaginary part of g, is increased quadratically with distance from the ship's waterline. A small positive damping is assumed already at the waterline; presumably, it is required to eliminate waves trapped in certain regions near to the waterline by the stationary flow field.

If the ship is symmetric, this is utilized to reduce the computing time as described in Sect. 3.7.

The case of shallow water is handled like in steady flow by using pairs of original sources and mirror sources below the bottom as Green functions. Naturally, the wave potential $\hat{\phi}_w$ given in (7.51) has to be substituted by its shallow-water equivalent. For a symmetrical body in shallow water, both symmetry conditions can be combined using quadruples of sources.

Like in the steady flow case, it may be advisable to add the condition that the source strength amplitudes add up to zero. To retain an equal number of sources and conditions, an additional source must then be arranged at a suitable height (e.g., $0.15L$), preferably above the center of the free-surface grid.

7.3.8 Free-Surface Panel Grid

The free-surface grid used for time-harmonic flow is, in our programs, similar to that used in determining the steady flow (Fig. 7.1). However, the mesh length and the extent of the mesh are adapted to the wave. The mesh length is determined as a fraction (typically 1/15) of the minimum of the lengths of incoming and radiated waves. However, there are exceptions. The mesh length must not exceed a certain fraction of the body length, and in short incoming waves, in which the ship does not move substantially, the length of radiation waves is not considered.

The quantities ϕ^0 and ζ^0 are interpolated from the grid used for the steady problem to that for time-harmonic flow.

7.3.9 Solving the Equation System

Contrary to the steady flow problem, here no iteration is required, because only a linear set of boundary conditions must be satisfied. Because of the several inhomogeneous vectors relating to the 6 motions and at least one wave, iterative solutions are not recommended. Instead, the LR decomposition for complex linear systems contained in the LAPACK package is used in our programs.

7.3.10 Calculation of Hull Pressure

Bernoulli's equation gives the fluid pressure minus air pressure

$$\frac{p(\mathbf{x})}{\rho} = \frac{1}{2}U^2 - \frac{1}{2}\left(\nabla\phi^t(\mathbf{x})\right)^2 - \dot{\phi}^t(\mathbf{x}) + gz. \tag{7.79}$$

The time derivative $\dot{\phi}^t$ refers to fixed $\mathbf{x}$, not fixed $\underline{\mathbf{x}}$. Inserting ϕ^t from (7.45) and (7.46) results in

$$\frac{p(\mathbf{x})}{\rho} = \frac{1}{2}U^2 - \frac{1}{2}\left(\mathbf{U} + \nabla\phi^0(\mathbf{x}) + \mathrm{Re}[\nabla\hat{\phi} - \nabla\phi^0 \times \hat{\alpha} - \hat{\mathbf{v}}\nabla\nabla\phi^0]e^{i\omega_e t}\right)^2$$

$$-\operatorname{Re}[i\omega_e(\hat{\phi} - \hat{\mathbf{v}} \cdot \nabla\phi^0)e^{i\omega_e t}] + gz. \tag{7.80}$$

To compute the pressure force and moment on the body, the pressure is required at body-fixed points. Pressures $p(\mathbf{x})$ and $\underline{p}(\underline{\mathbf{x}})$ are equal if $\mathbf{x}$ and $\underline{\mathbf{x}}$ designate the same point expressed in inertial and body-fixed coordinates, respectively. But the function p is different from $\underline{p}$; thus, two different symbols are used here. Also derivatives ∇ in the inertial directions must be distinguished from derivatives $\underline{\nabla}$ in the body-fixed directions; the relation was given already in (7.58). Using this relation transforms (7.80) into

$$\frac{\underline{p}(\underline{\mathbf{x}})}{\rho} = \frac{1}{2}U^2 - \frac{1}{2}\left(\mathbf{U} + \underline{\nabla}\phi^0(\underline{\mathbf{x}}) + \operatorname{Re}[\nabla\hat{\phi} - \nabla\phi^0 \times \hat{\alpha}]e^{i\omega_e t}\right)^2$$

$$-\operatorname{Re}[i\omega_e(\hat{\phi} - \hat{\mathbf{v}} \cdot \nabla\phi^0)e^{i\omega_e t}] + g(\underline{z} + v_3). \tag{7.81}$$

Also here, the terms containing second derivatives of the stationary potential ϕ^0 cancel because the stationary potential field is assumed to follow the body motions, thus allowing a more accurate determination of fluid pressure on the hull.

The complex amplitude of the pressure $\underline{p}$ follows from (7.81) by omitting stationary terms and substituting time-dependent terms by their complex amplitudes:

$$\frac{\hat{\underline{p}}}{\rho} = -\left(\mathbf{U} + \nabla\phi^0(\mathbf{x})\right) \cdot \left(\nabla\hat{\phi} - \nabla\phi^0 \times \hat{\alpha}\right) - i\omega_e(\hat{\phi} - \hat{\mathbf{v}} \cdot \nabla\phi^0) + g\hat{v}_3. \tag{7.82}$$

As above, the average pressure over a panel is determined by using panel averages of $\nabla\hat{\phi}$ and $\nabla\phi^0$. These averages are determined from the values of the potential at the panel corners by applying (7.24).

7.3.11 Pressure Force and Moment

The amplitudes of the pressure force and moment follow from the amplitudes of the average pressure on the body patches:

$$\hat{\mathbf{F}} = \sum_{Patches} \hat{\underline{p}}\,\underline{\mathbf{a}}, \quad \hat{\mathbf{M}} = \sum_{Patches} \hat{\underline{p}}\,\underline{\mathbf{x}} \times \underline{\mathbf{a}}, \tag{7.83}$$

where $\underline{\mathbf{a}}$ is the area vector of he patch: normal on the patch, directed into the body, magnitude = patch area. $\underline{\mathbf{x}}$ designates the patch midpoint. Here, we do not need to distinguish between the inertial and the body-fixed system, because the difference of first-order quantities like $\hat{\underline{p}}$ or $\hat{\mathbf{F}}$ in both systems is of second order.

Normally, further contributions to the force and moment must be added; for details see Chap. 11

7.3.12 *Motion Equation*

The quantities given up here are used to determine the body motion. The motion equation was given already as (5.39). Slightly adapted to the present notation, it becomes

$$(-\omega_e^2\mathbf{M} - \mathbf{B} + \mathbf{S})\begin{pmatrix}\hat{\mathbf{u}}\\ \hat{\alpha}\end{pmatrix} = \begin{pmatrix}\hat{\mathbf{F}}_e\\ \hat{\mathbf{M}}_e\end{pmatrix}. \tag{7.84}$$

Here, $\mathbf{M}$ is the 6 by 6 mass matrix as given by formula (5.36) of chapter 'Strip Methods'. The 6 by 6 restoring force matrix $\mathbf{S}$ is, essentially, the same as (5.40); however, the integrals over x have to be substituted by corresponding formulae based on the hull patches, as elaborated for S_{44} (roll restoring moment per roll angle) in Sect. 11.2. The exciting force and moment amplitudes $\hat{\mathbf{F}}_e$ and $\hat{\mathbf{M}}_e$ are determined from (7.82) and (7.83) using only pressure contributions which are independent of body motions. Correspondingly, the six columns of the complex matrix $\mathbf{B}$ (it is the complex added mass matrix times ω_e^2) are the six-component radiation force vectors determined likewise, but using the radiation pressure distributions for surge, sway, heave, roll, pitch, and yaw motion for columns 1–6, respectively.

Solving the small linear complex equation system (7.84) gives the complex amplitudes of the translation vector $\hat{\mathbf{u}}$ and the rotation vector $\hat{\alpha}$.

References

1. K.J. Bai, R.W. Yeung, Numerical solutions to free-surface flow problems, in *10th Symposium on Naval Hydrodynamics* (1974), pp. 609–647
2. V. Bertram, A Rankine source approach to forward speed diffraction problems, in *Symposium on Water Waves and Floating Bodies, Manchester* (1990)
3. V. Bertram, Ship motions by a Rankine source method. Ship Technol. Res. **37** (1990)
4. G.E. Gadd, National Maritime Institute (Feltham), *A Method of Computing the Flow and Surface Wave Pattern Around Full Forms*, vol. 118 (1976)
5. C.W. Dawson, A practical computer method for solving ship-wave problems, in *Proceedings of Second International Conference on Numerical Ship Hydrodynamics, Berkeley* (1977), pp. 30–38
6. G. Jensen, H. Söding, Z. Mi, Rankine source methods for numerical solutions of the steady wave resistance problem, in *16th Symposium on Naval Hydrodynamics* (1986)
7. H. Söding, F. Conrad, Analysis of overtaking manoeuvres in a narrow waterway. Ship Technol. Res. **52** (2005)
8. A. von Graefe, A Rankine source method for ship-ship interaction and shallow water problems. Ph.D. thesis, University of Duisburg-Essen (2014)
9. H. Söding, A method for accurate force calculations in potential flow. Ship Technol. Res. **40** (1993)
10. H. Söding, A. von Graefe, O. el Moctar, V. Shigunov, Rankine source method for seakeeping prediction, in *Proceedings of the ASME 2012 31st International Conference on Ocean, Offshore and Arctic Engineering, OMAE* (2012), p. 2012
11. D. Hachmann, Calculation of pressures on a ship's hull. Ship Technol. Res. **38**, 111–133 (1991)

Chapter 8
Nonlinear Rankine Panel Methods

Abstract This chapter describes a new, fully nonlinear simulation method for ship motions and loads based on potential theory. On the hull surface, it uses a panel mesh fixed to the body, whereas on the free surface the mesh is generated a new for each time step. Special account is taken of partly submerged hull panels. To improve the accuracy, the panel method determines not only the flow potential, but also its time derivative from derivatives of the boundary conditions. The radiation conditions for the disturbance potential are satisfied using a combination of the Dawson operator for lengthwise derivatives and wave damping, leaving the potential of the incoming waves unaffected. The chapter comprises also a comparison of motions and loads in steep head and quartering waves computed by this method with results of model experiments and RANS calculations.

8.1 Introduction

The success of linear strip and panel methods for motion and load predictions motivated many attempts to further improve these methods by adding nonlinear terms. For instance, the hull pressure was integrated up to the actual waterline (instead of up to the mean waterline in linear methods) to derive force, moment, and load corrections. Often this resulted in improvements of predicted bending moment amplitudes. Absolute values of the extreme midship bending moments are higher in sagging (wave trough) condition than in hogging (on a wave crest). But predictions of other nonlinear responses, at least in certain frequency regions, may become worse if not all relevant nonlinearities are taken into account, because often various nonlinear contributions nearly cancel each other. A floating body in long waves, for instance, largely follows the motion of the free surface without substantial pressure amplitudes at ship-fixed points. Thus, taking into account nonlinear contributions either of radiation (due to the body motion in still water) or of diffraction (due to the wave acting on a fixed body), but not of both, will result in increased errors. The same effect occurs in short waves for the pressure on the ship's bottom. Here amplitudes of Froude-Krylov (undisturbed wave) pressure and diffraction pressure (change by the body) largely cancel.

B. O. el Moctar et al., *Numerical Methods for Seakeeping Problems*,
https://doi.org/10.1007/978-3-030-62561-0_8

Roll motions and the accompanied torsional moments are most influenced by nonlinear effects, followed by vertical bending moments and shear forces. For simulating nonlinear roll motions, many specialized methods have been developed, e.g., [1]. Normally, other responses (especially motions) are only slightly nonlinear. Thus, it makes sense to use the most accurate linear potential method, i.e., the Rankine source panel method, for extending it to a generally applicable fully nonlinear procedure. A comprehensive older review of such methods is given by Beck [2]. He states that fully nonlinear potential seakeeping methods for real ship shapes (contrary to Wigley hulls, which have parabolical waterlines, etc.) were not available. According to the most recent (2017) review of the International Towing Tank Conference [3], that statement seems to be true even now.

Special difficulties with nonlinear Rankine panel codes are

- Large motions between water surface and body cause corresponding motions between the body panel mesh and the free-surface mesh. A free-surface mesh extending into the body would give inaccurate results. Thus, the free-surface panel mesh must follow the waterline. For typical ship shapes, this involves severe, even topological mesh changes, especially if a transom stern changes from immersed to emerged.
- Surface irregularities like breaking waves and spray form at the waterline, especially in upward relative motion of the water along flaring sections. Because wave breaking and spray cannot be resolved by a reasonable three-dimensional panel method, they must be 'skipped' somehow, without causing a failure of the simulation.

In the following, the Rankine panel method [4], which overcomes these and other difficulties, is described. It is suitable for rigid single-hull ships in deep water with or without forward speed.

8.2 Coordinate Systems

Like in most methods for this purpose, three different coordinate systems are used:

- An inertial system having its origin at the mean water surface. In most methods, it follows the mean forward speed of the vessel; however, if maneuvering in waves should be included, there is no mean forward speed. Thus, here the system is earth-fixed or—in case of a steady homogeneous current—it follows the current flow. Inertial coordinates x and y are horizontal, z is directed downward. The system is used to describe the motion of the ship, the water surface, and the motion of water particles by the gradient of a flow potential.
- A ship-fixed coordinate system with coordinates $\underline{x}$ (longitudinal to the bow), $\underline{y}$ (to starboard), $\underline{z}$ (in ship-fixed downward direction). It is used to describe the ship shape and its mass distribution.

- An intermediate system $\tilde{x}, \tilde{y}, \tilde{z}$, which participates in the horizontal translations and the yaw motion of the ship, but not in heave, roll, and pitch motions. It is used to generate the free-surface panel mesh.

The relation between a vector $\mathbf{x} = (x, y, z)$ expressed in inertial coordinates and the same vector $\underline{\mathbf{x}} = (\underline{x}, \underline{y}, \underline{z})$ expressed in ship-fixed coordinates, is

$$\mathbf{x} = \mathbf{u} + \mathbf{T}\underline{\mathbf{x}}, \tag{8.1}$$

where $\mathbf{u} = (u_x, u_y, u_z)$ is the translation vector (position vector of the ship-fixed coordinate origin), and $\mathbf{T}$ is the matrix representing the combined rotations φ (heel), θ (pitch), and ψ (yaw) given in (2.33). The corresponding relations involving vectors expressed in the intermediate system are

$$\tilde{\mathbf{x}} = (0, 0, u_z)^T + \mathbf{T}_\theta \mathbf{T}_\varphi\, \underline{\mathbf{x}} \tag{8.2}$$

and

$$\mathbf{x} = (u_x, u_y, 0)^T + \mathbf{T}_\psi \tilde{\mathbf{x}}. \tag{8.3}$$

The transformation matrices $\mathbf{T}_\varphi$, $\mathbf{T}_\theta$, and $\mathbf{T}_\psi$ are given in (2.31) and (2.32).

8.3 Subdivision of the Flow Potential

Usually, incident waves are generated in simulations by 'numerical wave-makers', mostly by wave generated boundary conditions. Here, another method is preferred. The potential ϕ of the water flow is combined from

ϕ_s, the potential of the exciting regular wave or irregular seaway, and
ϕ_d, the potential of the disturbance caused by the body.

Potential ϕ_s is assumed to be known; Chap. 4 treats potential methods to determine the potential of nonlinear regular waves and of nonlinear natural seaways. Thus, in the following, a numerical method to compute the disturbance potential is described.

The subdivision of ϕ is essential for solving a common problem of numerical computations of free surface flows: wave reflections at the outer boundary of the computational domain. To model the case of a body in unbounded water, the outer boundaries of the computational domain must be either transparent to the waves, or the waves must be damped. Both measures involve numerical errors. If we know the flow due to the exciting waves, we must avoid only reflections of the disturbance waves. These decrease anyway with increasing distance from the body. Thus, inaccuracies caused by damping the disturbance waves are small compared to those by damping of all waves.

8.4 Panel Meshes and Time Derivatives

For a Rankine panel method, we need panels both on the hull and on the free water surface. To allow possible extreme variations of the waterline with time, the free surface panel mesh is generated a new at each time step. To generate also a body panel mesh up to the actual waterline reliably at each time step appears difficult and time-consuming. Therefore, a body panel mesh moving with the body appears preferable. It has time-invariable grid points $\underline{\mathbf{x}}$ and extends at least up to the line of maximum immersion. (If non-smooth obstacles above the weather deck like hatches, containers, and superstructures are immersed, the applicability of a panel method for ships with forward speed appears questionable.)

The fluid pressure depends on the partial time derivative (i.e., at points fixed in the inertial system) of the flow potential. To determine this derivative from finite differences between ϕ values at successive time steps would be inaccurate, especially at the water surface where the panel mesh may change extremely between time steps. It appears much more accurate to directly compute $\dot{\phi}$ from its boundary conditions, as was proposed by Bandyk and Beck [5]. As also $\dot{\phi}$ satisfies the Laplace equation, the same panel method can be used to determine ϕ and $\dot{\phi}$, and even the same panel mesh and influence matrix are applicable to determine both ϕ and $\dot{\phi}$.

In the following, partial time derivatives $\partial/\partial t$ (at constant points $\mathbf{x}$) are designated by dots like in $\dot{\phi}$, whereas time derivatives at points moving with the mesh (within a single time step) are designated by d/dt. Both operators are related by

$$\frac{d}{dt} = \frac{\partial}{\partial t} + \mathbf{v} \cdot \nabla, \tag{8.4}$$

where $\mathbf{v}$ is the velocity of the mesh point in the inertial system.

8.5 Body Boundary Conditions

To suppress any flow through the hull, the condition

$$(\nabla\phi - \mathbf{v}) \cdot \mathbf{n} = 0 \tag{8.5}$$

must be satisfied for points $\mathbf{x}$ on the wetted ship surface. $\phi(\mathbf{x}, t)$ is the flow (velocity) potential, $\mathbf{v}$ the body velocity at $\mathbf{x}$, and $\mathbf{n}$ a normal vector on the hull at $\mathbf{x}$.

The time derivative of (8.5) at body-fixed points is the boundary condition for $d\phi/dt$:

$$\left(\frac{d\nabla\phi}{dt} - \frac{d\mathbf{v}}{dt}\right) \cdot \mathbf{n} + (\nabla\phi - \mathbf{v}) \cdot \frac{d\mathbf{n}}{dt} = 0. \tag{8.6}$$

The time derivatives in this equation are

$$\frac{d\nabla\phi}{dt} = \nabla\dot{\phi} + (\nabla\nabla\phi)\mathbf{v}. \tag{8.7}$$

Here $\nabla\nabla\phi$ is the Hesse matrix (matrix of second partial derivatives) of ϕ. The time derivative of (8.1) results in

$$d\mathbf{x}/dt = \mathbf{v} = \dot{\mathbf{u}} + \dot{T}\underline{\mathbf{x}} = \dot{\mathbf{u}} + \omega \times T\underline{\mathbf{x}}, \tag{8.8}$$

where ω is the angular velocity of the body. The time derivative of this equation gives

$$d\mathbf{v}/dt = \ddot{\mathbf{u}} + \omega \times (\omega \times T\underline{\mathbf{x}}) + \dot{\omega} \times T\underline{\mathbf{x}}. \tag{8.9}$$

The last derivative in (8.6) is

$$d\mathbf{n}/dt = \omega \times \mathbf{n}. \tag{8.10}$$

Inserting $\phi = \phi_s + \phi_d$ into the body boundary conditions (8.5) and (8.6) gives the required boundary conditions for the potentials ϕ_d and $d\phi_d/dt$:

$$\nabla\phi_d \cdot \mathbf{n} = (\mathbf{v} - \nabla\phi_s) \cdot \mathbf{n}; \tag{8.11}$$

$$\nabla\frac{d\phi_d}{dt} \cdot \mathbf{n} = -\nabla\dot{\phi}_s \cdot \mathbf{n} - \mathbf{v}(\nabla\nabla\phi_s)\mathbf{n} + (\mathbf{v} - \nabla\phi_d - \nabla\phi_s) \cdot \frac{d\mathbf{n}}{dt} + \frac{d\mathbf{v}}{dt} \cdot \mathbf{n}. \tag{8.12}$$

In many panel methods, computing the Hesse matrix $\nabla\nabla\phi$ of the flow potential causes problems because the method may converge (for decreasing panel size) to correct first derivatives of the potential, but—due to insufficient smoothness—not to correct second derivatives. In (8.12), however, there are no such problems, because $\nabla\nabla\phi_s$ is accurately computed from the formula for ϕ_s, whereas the Hesse matrix of ϕ_d is not required.

The last term in (8.12) involves $d\mathbf{v}/dt$, which as shown by (8.9) contains the translational and rotational body accelerations $\ddot{\mathbf{u}}$ and $\dot{\omega}$. As these are not known initially when dealing with a new time step, $d\phi_d/dt$ is subdivided into seven contributions:

$$\frac{d\phi_d}{dt} = \phi_1\ddot{u}_x + \phi_2\ddot{u}_y + \phi_3\ddot{u}_z + \phi_4\dot{\omega}_x + \phi_5\dot{\omega}_y + \phi_6\dot{\omega}_z + \phi_7. \tag{8.13}$$

The potentials ϕ_1 until ϕ_6 will be used to determine the added mass and inertia matrix, whereas ϕ_7 comprises all contributions to $d\phi_d/dt$ which are not proportional to the body accelerations, e.g., diffraction and effects of body angular velocity. Corresponding to this subdivision, the following body boundary conditions are prescribed for ϕ_1 to ϕ_6:

$$\nabla\phi_1 \cdot \mathbf{n} = n_x; \quad \nabla\phi_2 \cdot \mathbf{n} = n_y; \quad \nabla\phi_3 \cdot \mathbf{n} = n_z; \tag{8.14}$$

$$\nabla\phi_4 \cdot \mathbf{n} = (T\underline{\mathbf{x}} \times \mathbf{n})_x; \quad \nabla\phi_5 \cdot \mathbf{n} = (T\underline{\mathbf{x}} \times \mathbf{n})_y; \quad \nabla\phi_6 \cdot \mathbf{n} = (T\underline{\mathbf{x}} \times \mathbf{n})_z. \tag{8.15}$$

When the subdivision (8.13) is inserted into the body boundary condition (8.12), many terms cancel because of the above boundary conditions for ϕ_1 to ϕ_6. The remaining rest is the body boundary condition for the 'rest potential' ϕ_7:

$$\nabla\phi_7 \cdot \mathbf{n} = -\nabla\dot{\phi}_s \cdot \mathbf{n} - \mathbf{n}(\nabla\nabla\phi_s)\mathbf{v} + (\mathbf{v} - \nabla\phi_d - \nabla\phi_s) \cdot \frac{d\mathbf{n}}{dt} + [\omega \times (\omega \times T\underline{\mathbf{x}})] \cdot \mathbf{n}. \tag{8.16}$$

8.6 Free-Surface Boundary Conditions

If p is the difference between fluid and air pressure (the latter is approximated as constant), at the water surface $z = \zeta(\mathbf{x}, t)$ Bernoulli's equation requires

$$\frac{p}{\rho} = g\zeta - \dot{\phi} - \frac{1}{2}(\nabla\phi)^2 = 0. \tag{8.17}$$

Thus, we have a Dirichlet-type free surface boundary condition for $\dot{\phi}$:

$$\dot{\phi} = g\zeta - \frac{1}{2}(\nabla\phi)^2. \tag{8.18}$$

The kinematic condition states that a fluid particle at the free surface will stay at the free surface:

$$\frac{D}{Dt}(z - \zeta) = \left(\frac{\partial}{\partial t} + \nabla\phi \cdot \nabla\right)(z - \zeta) = -\dot{\zeta} + \phi_z - \nabla\phi \cdot \nabla\zeta = 0, \tag{8.19}$$

from which follows:

$$\dot{\zeta} = \phi_z - \nabla\zeta \cdot \nabla\phi. \tag{8.20}$$

Like ϕ, also ζ is subdivided into a known surface deformation by the incident waves ζ_s, and an initially unknown surface deformation ζ_d caused by the body. Inserting $\zeta = \zeta_s + \zeta_d$ into the boundary conditions (7.8) and (8.20) gives

$$\dot{\phi}_s + \dot{\phi}_d = g(\zeta_s + \zeta_d) - \frac{1}{2}(\nabla\phi_s + \nabla\phi_d)^2; \tag{8.21}$$

$$\dot{\zeta}_s + \dot{\zeta}_d = \phi_{s,z} + \phi_{q,z} - (\nabla\zeta_s + \nabla\zeta_d) \cdot (\nabla\phi_s + \nabla\phi_d). \tag{8.22}$$

The incident waves ϕ_s, ζ_s alone should satisfy both free surface conditions. But because we will use only approximations for ϕ_s and ζ_s (in the most simple case, a superposition of Airy waves), there remain errors which, when using the conditions

(8.21) and (8.22), constitute forcing terms for ϕ_d, ζ_d also far from the body. This contradicts the decrease of the disturbance waves ϕ_d, ζ_d to zero at the outer boundary of the free surface panel mesh. We need this decrease to avoid wave reflections there. Thus, to avoid these forcing terms, the dynamic condition for ϕ_s alone is subtracted from (8.21), and the kinematic condition for ζ_s alone from (8.22). The effect of this is: Errors of the assumed incident waves regarding their free surface boundary conditions are not canceled by ϕ_d and ζ_d, but remain in the final solution. Whether these errors are relevant for the responses of interest can be tested by using more accurate formulae for the incident waves (see Chap. 4).

The conditions for ϕ_s and ζ_s hold at ζ_s, whereas the conditions for ϕ_d and ζ_d hold at $\zeta_s + \zeta_d$. To take account of this difference, a first-order Taylor expansion for ϕ_s over z is applied to convert the free surface conditions for ϕ_s and ζ_s to the position $\zeta_s + \zeta_d$ before subtracting these conditions from (8.21) and (8.22). This results in the following free surface boundary conditions for $\dot{\phi}_d$ and $\dot{\zeta}_d$:

$$\dot{\phi}_d = \zeta_d(g - \dot{\phi}_{s,z} - \nabla\phi_s \cdot \nabla\phi_{s,z}) - \nabla\phi_s \cdot \nabla\phi_d - \frac{1}{2}(\nabla\phi_d)^2 + \frac{1}{2}(\nabla\phi_{s,z}\zeta_d)^2. \tag{8.23}$$

$$\dot{\zeta}_d = \phi_{q,z} + \phi_{s,zz}\zeta_d - \nabla\zeta_s \cdot \nabla\phi_d - \nabla\zeta_d \cdot (\nabla\phi_s + \nabla\phi_d) - \nabla\zeta_s \cdot \nabla\phi_{s,z}\zeta_d. \tag{8.24}$$

At a point moving with the free surface panel mesh (local speed $\dot{\mathbf{x}}$), we have

$$\frac{d\phi_d}{dt} = \dot{\phi}_d + \nabla\phi_d \cdot \dot{\mathbf{x}} = \phi_1\ddot{u}_x + \phi_2\ddot{u}_y + \phi_3\ddot{u}_z + \phi_4\dot{\omega}_x + \phi_5\dot{\omega}_y + \phi_6\dot{\omega}_z + \phi_7, \tag{8.25}$$

where the last expression is taken from (8.13). The condition is satisfied by choosing on the free surface

$$\phi_i = 0 \quad \text{for} \quad i = 1 \quad \text{to} \quad 6, \tag{8.26}$$

and

$$\phi_7 = \dot{\phi}_d + \nabla\phi_d \cdot \dot{\mathbf{x}}. \tag{8.27}$$

The conditions for ϕ_1 to ϕ_6 correspond to an added mass matrix for infinite frequency. This is no approximation, because the condition (8.25) is exactly satisfied by the appropriate rest potential ϕ_7.

Like for the acceleration potentials ϕ_1 to ϕ_7, on the free surface, we have a Dirichlet boundary condition also for ϕ_d. Its values follow from the time integration (within the current time step) of ϕ_d using $d\phi_d/dt$. The value at the beginning of each time step must be interpolated from the end value of the previous time step, using interpolation between the previous and the current meshes.

The surface deformation ζ_d is determined correspondingly by time integration of $d\zeta_d/dt$. The latter is determined as

$$d\zeta_d/dt = \dot{\zeta}_d + \nabla\zeta_d \cdot \dot{\mathbf{x}}, \tag{8.28}$$

where $\dot{\zeta}_d$ is given by (8.24).

8.7 Transom Condition

If a transom is immersed to a depth exceeding U^2/g, the transom is assumed to be wetted and treated like other parts of the hull. For less immersion, the transom is assumed dry, and the pressure at the hull panels immediately in front of the transom is forced to air pressure.

If U-shaped transverse sections at the aft end of the hull cause a very blunt waterline, the case is treated like a dry transom of submergence zero, and the free-surface panels immediately after the end of the waterline are assumed to be horizontal.

8.8 Radiation Condition

Usually, one wants to model the response of a body due to a given incident wave and the accompanied disturbance by the body. The body and free surface boundary conditions, however, do not exclude the presence of other external waves. They could stem from the starting process of the simulation, or they may be accumulating from numerical errors during long simulations. To avoid such unwanted waves, damping must be added. Here, wave damping is produced by assuming a higher or lower pressure at the free surface where it moves upward or downward, respectively, in the intermediate coordinate system. This results in an additional term

$$\delta_w c \, d\zeta_d/dt \tag{8.29}$$

on the right-hand side of (8.23), where c is the phase velocity of the wave or the seaway at its significant frequency, and δ_w is a non-dimensional parameter. Usually, wave damping is applied in a 'damping zone' near the outer boundary of the computational domain. Numerical tests showed, however, that a constant damping parameter δ_w (for instance $\delta_w = 0.2$) gives superior results. This may be understood from the fact that waves are reflected not only at the outer boundary, but partially also at isolines of a variable damping parameter.

Waves due to forward speed of a ship, which travel along with it, are not damped by the term (8.29). These waves should have a far-field only behind, not in front of the disturbance causing them. That is attained here by using the Dawson [6] operator: When determining $\nabla\phi_d$ or $\nabla\zeta_d$ at a point P, the derivative in the ship's longitudinal direction $\tilde{x}$ is estimated as that of an interpolating fourth-order polynomial passing through the function values at P and at the three next mesh points upstream of P. All together four points suffice because the interpolating polynomial used by Dawson lacks the third-order term:

$$f(\tilde{x}) = a_0 + a_1(\tilde{x} - \tilde{x}_P) + a_2(\tilde{x} - \tilde{x}_P)^2 + a_4(\tilde{x} - \tilde{x}_P)^4. \tag{8.30}$$

To use only upstream points avoids that disturbances generate far-field waves upstream.

In other directions and for bodies without forward speed, derivatives of ϕ_d and ζ_d are determined as described in the next section.

8.9 Numerical Method to Satisfy the Body Boundary Condition

Also here the patch method (Sects. 3.8 and 7.2.4) is applied. On the body, it uses a mesh of triangular panels (Fig. 8.1). For partly submerged panels, the ratio r of the submerged area to the total panel area is determined and used to decrease the strength of the source next to the panel by factor r. To accomplish this, the element of the matrix corresponding to that panel and that source is divided by r. Fully emerged body panels ($r = 0$) are eliminated from the equation system. Within each time step, the change of r is under-relaxed to avoid oscillations of r. To take account of the time dependence of r in a partly submerged panel i, the term

$$\nabla \frac{1}{|\mathbf{x} - \mathbf{x}_{qi}|} \frac{-q_i}{r_i^2} \frac{dr_i}{dt} \cdot \mathbf{n} \tag{8.31}$$

must be added on the right-hand side of the discretized form of the boundary condition (8.16).

8.10 Numerical Method to Satisfy the Free Surface Boundary Conditions

At the beginning of each time step, the panel mesh on the free surface is generated around the waterline. As the nearly horizontal hull in the stern region of most modern ships may cause highly curved or even discontinuous waterline shapes, the aft part of the waterline is approximated by a straight line to avoid irregular panel meshes. During each time step, waterline changes are taken into account by appropriate mesh deformations in the intermediate coordinate system. In the inertial system, the mesh is moving also according to the change from intermediate to inertial coordinate system.

For each quadrilateral free surface panel, a point source is arranged approximately one panel length above the free surface. A single additional source is arranged higher above the midship section. It is required to take account of another condition for instationary flows, because the sum of all source strengths must be zero to attain the correct far-field behavior of the fluid pressure.

The kinematic boundary condition gives the (panel-averaged) vertical speed of the free surface. For each mesh point, the average value of the (mostly four) surrounding

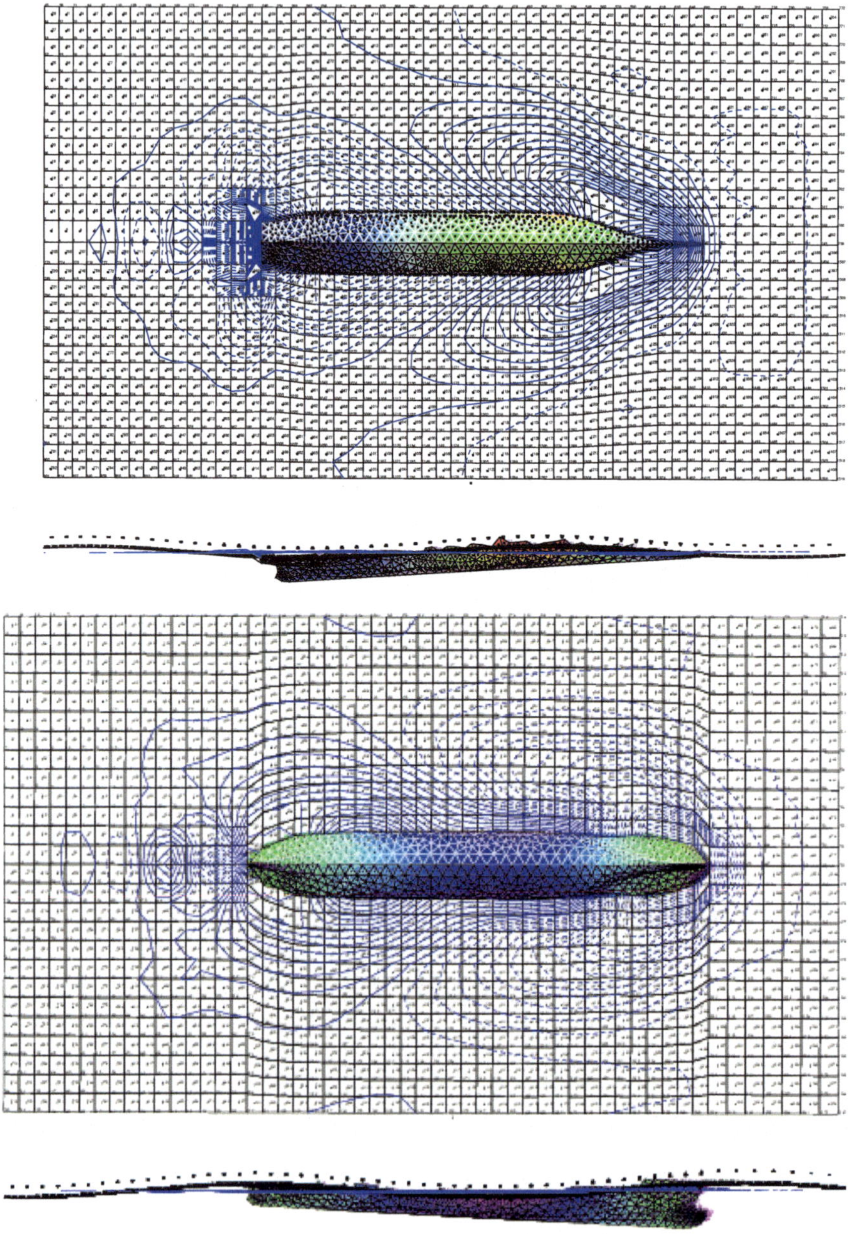

Fig. 8.1 Panel mesh of the WILS containership at two instants. Colors on hull panels code the dynamic pressure. Contour lines show the surface deformation ζ_d, excluding the exciting waves ζ_s

panels is used to update the vertical coordinate of the mesh point using a fourth-order Runge-Kutta method. Exception are the points on the waterline. In reality, the physically correct boundary conditions generate often spray and breaking waves along the waterline. To avoid these irregularities, which we cannot resolve by a reasonably fine panel mesh, the surface height at the waterline is extrapolated from the next two mesh points besides the waterline.

The dynamic free-surface condition, on the other hand, gives the time derivative of the (panel-averaged) potentials ϕ_d. Again, the fourth-order Runge-Kutta method is used to determine ϕ_d itself.

8.11 Determination of Source Strengths

The velocity potential ϕ_d is combined from that of point sources:

$$\phi_d(\mathbf{x}, t) = \sum_j \frac{-q_j(t)}{|\mathbf{x} - \mathbf{x}_{qj}|}, \tag{8.32}$$

where the sum comprises body sources, free-surface sources, and the additional source above the midship section. Corresponding formulas are used for the acceleration potentials ϕ_i, $i = 1$ to 7; however, with different sets of source strengths $q_j^{(i)}$. The boundary conditions for each panel form a linear equation system having the same coefficient matrix for all eight potentials; however, with eight different inhomogeneous vectors. The inhomogeneous vector for ϕ_7 can only be determined when the source strengths of ϕ_d are known. Therefore, at first, an LU factorization of the coefficient matrix is made; then the source strengths for ϕ_d and ϕ_1 to ϕ_6 are determined; the q_j allow to compute the inhomogeneous vector for ϕ_7; and finally, the source strengths of ϕ_7 are determined. This is done using subroutines of the Linear Algebra Package (LAPACK), which proved to be efficient. Iterative solvers are not applicable here, because of the bad condition number of the system, and because we must solve for eight different inhomogeneous vectors.

8.12 Determination of Body Force and Moment

The pressure force F_p on the body is added from contributions of all fully or partly submerged body panels:

$$\mathbf{F}_p = \sum_i p_i \mathbf{n}_i r_i. \tag{8.33}$$

Here, the normal vector $\mathbf{n}_i$ of panel i is scaled such that it points into the body, and that its absolute value is the panel area. r_i is the submergence ratio of panel i (1 for fully submerged panels). The pressure difference to the air pressure p is

$$p = \rho\left(gz - \dot{\phi} - \frac{1}{2}(\nabla\phi)^2\right) = \rho\left(gz - \phi_1\ddot{u}_x \ldots - \phi_6\dot{\omega}_z - \phi_7 + \nabla\phi_d \cdot \mathbf{v} - \dot{\phi}_s - \frac{1}{2}(\nabla\phi)^2\right). \tag{8.34}$$

Term F_p is subdivided into contributions independent and dependent from the accelerations:

$$\mathbf{F}_p = \mathbf{F}_0 - A_u\begin{pmatrix}\ddot{\mathbf{u}}\\ \dot{\omega}\end{pmatrix}. \tag{8.35}$$

Here,

$$\mathbf{F}_0 = \sum_i \rho\left(gz_i - \phi_{7i} + \nabla\phi_{qi} \cdot \mathbf{v}_i - \dot{\phi}_{si} - \frac{1}{2}(\nabla\phi_i)^2\right)\mathbf{n}_i r_i, \tag{8.36}$$

and A_u is the upper half of the 6 by 6 added mass matrix. Its elements in rows 1–of column j are

$$A_{1:3,j} = \rho\sum_i \phi_j \mathbf{n}_i r_i. \tag{8.37}$$

Correspondingly, the pressure moment about G is subdivided:

$$\mathbf{M}_p = \mathbf{M}_0 - A_l\begin{pmatrix}\ddot{\mathbf{u}}\\ \dot{\omega}\end{pmatrix}, \tag{8.38}$$

where A_l is the lower half of the added mass matrix. $\mathbf{M}_0$ and the elements of A_l are determined by equations analogous to (8.36) and (8.37), however containing $(\mathbf{x}_i - \mathbf{x}_G) \times \mathbf{n}_i$ instead of $\mathbf{n}_i$.

To obtain the total force $\mathbf{F}$ and the moment $\mathbf{M}$ on the body, we have to add a residual force $\mathbf{F}_r$ and its moment $\mathbf{M}_r$. These stem from the following contributions:

- The weight force
- Viscous and other corrections, e.g. due to flow separation especially at the stern and the rudder. Section 11.6 gives a simple approximation. For ship maneuvering, more involved approximations must be used.
- Control forces. They can be used to keep, approximately, the planned speed and path, and are quantified by parameters of a PID control loop.
- Possibly wind and other external forces and moments.

8.13 Motion Equations

If $\underline{\mathbf{x}}_G$ is the position of the center of gravity of (rigid) masses in ship-fixed coordinates, its acceleration follows from (8.9) as

$$\ddot{\mathbf{u}}_G = \ddot{\mathbf{u}} + \omega \times (\omega \times \mathbf{T}\underline{\mathbf{x}}_G) + \dot{\omega} \times \mathbf{T}\underline{\mathbf{x}}_G. \tag{8.39}$$

From Newton's law

$$m\ddot{\mathbf{u}}_G = \mathbf{F} \tag{8.40}$$

together with (8.39) and (8.35) follows:

$$m[\ddot{\mathbf{u}} + \omega \times (\omega \times \mathbf{T}\underline{\mathbf{x}}_G) + \dot{\omega} \times \mathbf{T}\underline{\mathbf{x}}_G] = \mathbf{F}_0 - A_{1:3,1:6}\begin{pmatrix}\ddot{\mathbf{u}}\\ \dot{\omega}\end{pmatrix} + \mathbf{F}_r. \tag{8.41}$$

Body rotations are governed by the equation

$$\mathbf{I}\dot{\omega} + \omega \times \mathbf{I}\omega = \mathbf{M} = \mathbf{M}_0 - A_l\begin{pmatrix}\ddot{\mathbf{u}}\\ \dot{\omega}\end{pmatrix} + M_r, \tag{8.42}$$

where the right-hand expression follows from the subdivision of $\mathbf{M}$ into pressure (8.38) and residual moment. $\mathbf{I}$ is the inertia matrix of rigid body masses referring to G in inertial coordinates. It is determined from the inertia matrix in ship-fixed coordinates using (2.43).

Equations (8.41) and (8.42) for translational and rotational motions will be combined to one matrix equation for the six accelerations. Therefore, the vector product involving $\dot{\omega}$ in (8.41) is written in index notation. For arbitrary three-dimensional vectors $\mathbf{a} = (a_1, a_2, a_3)$ and $\mathbf{b} = (b_1, b_2, b_3)$ their vector product can be written as

$$\mathbf{a} \times \mathbf{b} = \epsilon_{ijk} a_j b_k, \tag{8.43}$$

where ϵ_{ijk} are the elements of the Levi-Civita tensor: ϵ_{ijk} is 1 if ijk constitute an even permutation of (1, 2, 3); -1 for an odd permutation; and 0 if i, j, k are not all different. Using this notation, (8.41) and (8.42) can be combined into the matrix equation

$$\left[\begin{pmatrix} m\mathbf{E} & m\epsilon_{ijk}(\mathbf{T}\underline{\mathbf{x}}_G)_k \\ \mathbf{0} & \mathbf{I}\end{pmatrix} + \mathbf{A}\right]\begin{pmatrix}\ddot{\mathbf{u}}\\ \dot{\omega}\end{pmatrix} = \begin{pmatrix}\mathbf{F}_0 + \mathbf{F}_r - m\omega \times (\omega \times \mathbf{T}\underline{\mathbf{x}}_G) \\ \mathbf{M}_0 + \mathbf{M}_r - \omega \times \mathbf{I}\omega\end{pmatrix}. \tag{8.44}$$

Here, $\mathbf{E}$ is the 3 by 3 unit matrix, $\mathbf{0}$ the 3 by 3 zero matrix, and $\mathbf{A}$ the 6 by 6 added mass matrix. The rigid mass matrix on the left-hand side (in parentheses) appears unusual, for instance, in comparison with (5.36). The reason is that, here, moments and moments of inertia refer to the center of gravity of ship mass G, but the translation

u refers to the origin of the body-fixed coordinate system, which may be different from G.

The 6 by 6 linear equation system (8.44) is solved for the accelerations $\ddot{\mathbf{u}}$ and $\dot{\omega}$. Then $\dot{\mathbf{u}}$, $\mathbf{u}$ and ω follow by Runge-Kutta time integration. The time derivatives of the rotation angles follow from ω as

$$\begin{pmatrix} \dot{\varphi} \\ \dot{\theta} \\ \dot{\psi} \end{pmatrix} = \mathbf{R}^{-1}\omega, \tag{8.45}$$

where $\mathbf{R}^{-1}$ is given in (2.39). Finally, the rotation angles follow from $(\dot{\varphi}, \dot{\theta}, \dot{\psi})$ by Runge-Kutta integration.

8.14 Verification

Results of this method are compared with those published by Sigmund and El Moctar [7] for the containership Duisburg Test Case (DTC; Table 8.1) in steep regular head waves. The wave data are given in Table 8.2. In some cases, the wave height used in the reference had to be reduced to obtain a solution using the nonlinear Rankine method.

Figure 8.2 shows nearly full coincidence of heave and pitch motion amplitudes between model experiments, RANS calculations, and calculations using the above nonlinear Rankine panel method for both panel meshes used. The wiggle in the heave response near 0.4 rad/s is caused by the fact that, near to this frequency, there is

(a) a minimum of the heave exciting force integrated over ship length because upward forces at the wave crest are canceled by downward forces at the wave trough, and
(b) the heave (and nearby the pitch) resonance, where heave restoring and mass effects cancel each other.

Fact (a), together with the strong damping of heave and pitch motions by wave radiation, is the reason that no marked resonance peak occurs in the heave (and pitch) response curve. In this frequency region, the higher sensitivity due to canceling effects causes a larger deviation between linear non-dimensional responses (curve) and the nonlinear responses (markers).

In spite of the excellent coincidence of both computational methods with experimental results for heave and pitch motions in longitudinal waves, other responses, and other wave conditions show substantial deviations. The right-most diagram of Fig. 8.2 shows this for the added resistance R_a; displayed is the non-dimensional coefficient

$$C_{AR} = \frac{R_a}{\rho g B^2 A^2 / L_{pp}}, \tag{8.46}$$

Table 8.1 Main data of the DTC container ship

L_{pp}	355.0 m
B_{wl}	51.0 m
T_m	14.5 m
GM	1.5 m
Displacement mass	173467 m^3
r_{Gy}	88.77 m
No. of body panels, coarse submerged on average	1520 ≈1000
No. of body panels, fine submerged on average	3976 ≈2750

Table 8.2 Wave data used by [7] and own calculations

λ/L_{pp}	H_w/λ [%], Sigmund and El Moctar	H_w/λ [%], coarse mesh	H_w/λ [%], fine mesh
0.36	6.47	3.23	
0.44	7.74	3.25	4.48
0.60	3.53	3.53	3.53
0.80	3.50	3.50	3.50
0.91	2.80	2.36	2.36
1.00	2.87	2.24	2.24
1.09	2.90	2.63	2.63
1.20	2.39	2.39	
1.40	2.56	2.56	2.56
1.80	1.99	1.99	1.99
2.50	1.43	1.43	1.43

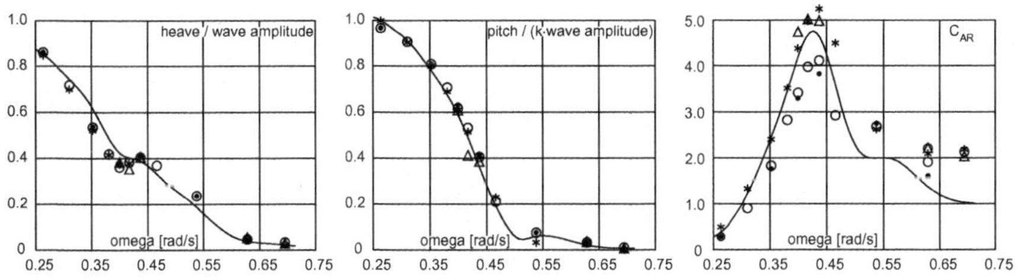

Fig. 8.2 Non-dimensional heave and pitch motion amplitudes and added resistance coefficient of the DTC in steep head waves at $F_n = 0.139$. $*$: RANS calculations; Δ: model experiments (both taken from [7]); ∘: present method using coarse body panels; •: same using fine body panels. Curves: results of the linear program GLRankine

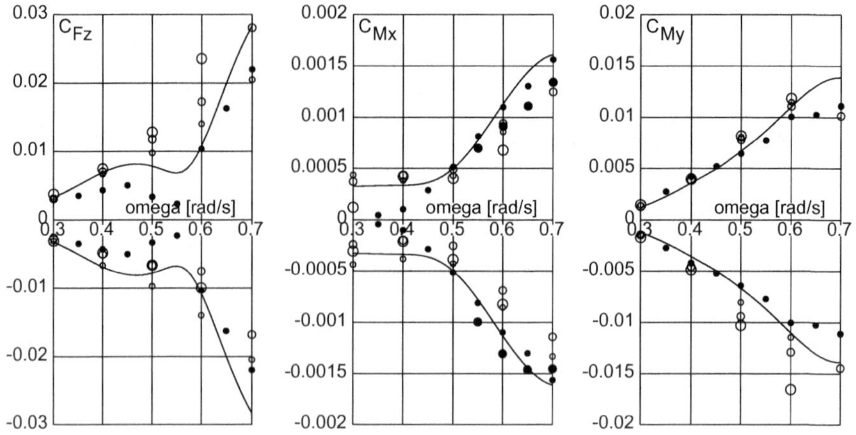

Fig. 8.3 Maximum and minimum non-dimensional vertical shear forces, torsional moment, and vertical bending moment at the midship section of the WILS containership at a speed of 5km in quartering ($\mu = 60°$) waves. Curves: results of the linear Rankine panel method GLRankine; small and big dots: experimental results by MOERI for small and 3.5 m, respectively, wave amplitude; small, medium, and big circles: results of the nonlinear Rankine panel method for small, 3.5 and 5.0 m wave amplitude, respectively

where A is wave amplitude. The coefficient is very sensitive because the added resistance is, even in steep waves, extremely small compared to, for instance, the heave force (but not compared to the ship resistance in calm water or propeller thrust). Whereas heave and pitch amplitudes become constant after a few encounter periods, added resistance requires simulations or measurements for, typically, more than 25 encounter periods.

Also, because of canceling effects, accurate predictions of loads in transverse ship sections are more difficult than motion predictions. The difficulties increase further in beam and oblique waves. This is illustrated by Fig. 8.3, which shows loads in the midship section of another containership (WILS; for details see Table 1 of chapter 'Comparison Study') in quartering ($\mu = 60°$) regular waves. The figure shows nondimensional load coefficients for the vertical shear force F_z,

$$C_{Fz} = \frac{F_z}{\rho g L_{pp} B A}, \tag{8.47}$$

while moment components M are characterized by coefficients

$$C_M = \frac{M}{\rho g L_{pp}^2 B A}. \tag{8.48}$$

Curves (results of the linear Rankine panel method GLRankine), small dots (model experiments by MOERI), and small circles (results of the nonlinear Rankine panel method SIS), both in waves of small amplitude, should coincide. On the other hand, big dots and medium circles are for wave height 7 m, and large circles for 10 m wave height. Practically important is the vertical bending moment M_y (right diagram). Here, the coincidence of responses in low waves is good. In steep waves, very substantial nonlinear effects are shown for negative (sagging) moments. For containerships with their large hatch openings, torsional moments M_x (middle diagram) are also of much practical interest. The larger differences between measured and predicted values are, at least to a large part, caused by different roll angles in experiments and predictions; roll angles are extremely sensitive to small errors both in experiments and calculations. The very large deviations of vertical shear force between experiments and both predictions (left diagram) occur because the shear force is very small at the midship section; its amplitude is much larger at sections in the fore- and afterbody.

References

1. H. Söding, V. Shigunov, T. Zorn, P. Soukup, Method rolls for simulating roll motions of ships. Ship Technol. Res. **60**(4), 70–84 (2013)
2. R. Beck, Modern computational methods for ships in a seaway. Trans. Soc. Nav. Archit. Mar. Eng. **109**, 1–51 (2001)
3. ITTC, in *ITTC Recommended Procedures and Guidelines* (2017)
4. H. Söding, A potential flow method for fully non-linear wave responses of ships, in *International Workshop on Ship Hydrodynamics, Hamburg* (2019)
5. P.J. Bandyk, R.F. Beck, The acceleration potential in fluid-body interaction problems. J. Eng. Math. **70**, 147–163 (2011)
6. C.W. Dawson, A practical computer method for solving ship-wave problems, in *Proceedings of Second International Conference on Numerical Ship Hydrodynamics, Berkeley* (1977), pp. 30–38
7. S. Sigmund, O. el Moctar, Numerical and experimental investigation of added resistance of different ship types in short and long waves. Ocean Eng. **147**, 11–67 (2018)

Chapter 9
Viscous Field Methods

Abstract In this chapter, we briefly describe numerical approaches, based on computational fluid dynamics (CFD), developed to compute wave-induced ship motions. We do not claim completeness; instead, this chapter describes examples of numerical techniques commonly employed in this field. The listed references may be of help to the reader wishing to obtain additional details. Further on, we briefly discuss numerical errors and, specifically, procedures available to estimate discretization errors. Finally, we present an example of applying CFD to predict ship motions and loads in extreme seas.

9.1 Introduction

Numerical methods for computing fluid flows, which solve the mass conservation and Navier-Stokes equations, so-called field methods, have their merits in applications where effects depending nonlinearly on wave elevation are of major concern. Such nonlinearities comprise wave-induced extreme ship motions and loads as well as ship responses affected by viscosity. In the literature, corresponding applications of field methods have mostly been restricted to relatively short duration simulations because of the required high computational effort. Procedures combining the strengths of potential flow solvers (efficiency) and field methods (accuracy for nonlinear problems) may be advantageous. el Moctar et al. [1, 2], Seng and Jensen [3], Oberhagemann [4], Oberhagemann et al. [5], Ley et al. [6, 7], Jiang et al. [8], for example, discuss and apply procedures concerning field methods to predict long-term extreme wave-induced loads and motions. A broad range of numerical methods and mathematical models to solve mass conservation and Navier-Stokes equations are available. The choice between these methods should depend on numerous flow characteristics that have to be dealt with, e.g. compressible or incompressible fluids, pressure- or viscosity-dominated flows, transient responses or stationary conditions, etc. There are numerous textbooks and publications that describe various numerical techniques in detail, e.g. Ferziger and Perić [9], Ferziger et al. [10], Wendt [11], Lomax et al. [12], Versteeg and Malalasekera [13].

B. O. el Moctar et al., *Numerical Methods for Seakeeping Problems*,
https://doi.org/10.1007/978-3-030-62561-0_9

In this chapter, we briefly describe numerical approaches used to compute wave-induced ship motions and loads. We do not claim completeness; instead, this chapter describes examples of numerical techniques that are commonly used in this field. The given references may help the reader with more details. Part of the following description is based on [14, 15].

In Chap. 2, the mass (continuity) and momentum (Navier-Stokes) conservation equations were derived already:

$$\frac{\partial}{\partial t}\int_V \rho\, dV + \int_S \rho\mathbf{v}\cdot\mathbf{n}\, dS = 0, \tag{9.1}$$

$$\frac{\partial}{\partial t}\int_V \rho\mathbf{v}\, dV + \int_S \rho(\mathbf{v}\mathbf{v})\cdot\mathbf{n}\, dS = \int_S \mathbf{T}\cdot\mathbf{n} dS + \int_V \mathbf{f}\, dV, \tag{9.2}$$

where ρ denotes the fluid density, V is the volume, S is the bounding surface of a control volume (CV), $\mathbf{n}$ is the outward normal vector on S, and $\mathbf{v}$ is the velocity vector of the fluid. $\mathbf{f}$ denotes a volumetric force and t the time. $\mathbf{T}$ is the *stress tensor*; it represents the molecular rate of momentum transport. For Newtonian fluids, the stress tensor is a linear function of the velocity gradient:

$$\mathbf{T} = -\mathbf{E}\, p + 2\,\mu\mathbf{D} + \frac{2}{3}\,\mu\,(\nabla\cdot\mathbf{v})\mathbf{E} \quad \text{with} \quad \mathbf{D} = \frac{1}{2}\left[\nabla\mathbf{v} + (\nabla\mathbf{v})^{\mathrm{T}}\right],$$

where $\mathbf{D}$ denotes the rate of strain tensor, $\mathbf{E}$ the unit tensor, p the pressure, and μ the dynamic viscosity.

The conservation equations for mass and momentum form a closed system of equations to compute velocity and pressure. These equations are valid for both laminar and turbulent flows. Turbulent flows are unsteady and chaotic on small time and space scales. They are three-dimensional even if the flow geometry is one- or two-dimensional. Pressure and velocity vary stochastically in space and time. Turbulent flows are diffusive: their fluctuations lead to an increased exchange of momentum, energy, and mass, also normal to the main flow direction. Although the transport process is convective, it acts like an increased diffusion and is often modeled in this way. Turbulent flows are also dissipative: the kinetic energy of the fluid is converted into heat as a result of the turbulence. This is because the turbulence leads to an increase in $|\nabla\mathbf{v}|$, thus increasing friction compared to the corresponding laminar flow.

In principle, turbulent flows could be determined by solving the above conservation equations. However, for the relevant Reynolds numbers $>10^5$, resolving the smallest vortices would require unreasonably small control volumes and time steps. Thus, either the effect of viscosity and turbulence must be neglected completely, or a turbulence model must be applied. In the following, two groups of turbulence models are sketched.

9.2 Reynolds-Averaged Navier-Stokes Equations

According to Reynolds, velocity and pressure may be decomposed into time-averaged values $\overline{\mathbf{v}}$, $\overline{p}$, and turbulent fluctuations $\mathbf{v}'$, p', the time average of which is zero:

$$\mathbf{v} = \overline{\mathbf{v}} + \mathbf{v}' \; ; \; p = \overline{p} + p'. \tag{9.3}$$

In case of wave-induced responses of ships, averaging is performed for times which are small compared to the encounter period between ship and wave, but large compared to turbulent fluctuations. Inserting (9.3) into the Navier-Stokes equations and taking the time average results in the Reynolds-averaged Navier-Stokes equations (RANSE):

$$\frac{\partial}{\partial t}\int_V \rho \, dV + \int_S \rho\overline{\mathbf{v}} \cdot \mathbf{n} \, dS = 0, \tag{9.4}$$

$$\frac{\partial}{\partial t}\int_V \rho\overline{\mathbf{v}} \, dV + \int_S \rho(\overline{\mathbf{v}}\overline{\mathbf{v}}) \cdot \mathbf{n} \, dS = \int_S (\mathbf{T} - \rho\overline{\mathbf{v}'\mathbf{v}'}) \cdot \mathbf{n} \, dS + \int_V \mathbf{f} \, dV. \tag{9.5}$$

Here, the average of the convective term $\rho\mathbf{v}\mathbf{v}$ resulted in $\rho(\overline{\mathbf{v}}\overline{\mathbf{v}} + \overline{\mathbf{v}'\mathbf{v}'})$. The additional term $\rho\overline{\mathbf{v}'\mathbf{v}'}$ contains six scalar unknowns $v_i'v_j'$, $i, j = 1, 2, 3$. In analogy to the viscous stress tensor, $\rho\overline{\mathbf{v}'\mathbf{v}'}$ is called the Reynolds stress tensor although it describes a transport process. To determine the additional unknowns, we must have additional equations, which are called collectively turbulence model, and which express the components of the Reynolds stress tensor as functions of the averaged flow quantities.

Launder and Spalding [16] introduced such a model in 1974, which is called today standard k-ϵ turbulence model. It is based on the Boussinesq hypothesis that expresses the effect of turbulence as added diffusion; in Cartesian coordinates:

$$\rho\overline{v_i'v_j'} = \mu_t \left(\frac{\partial \overline{v}_{\mathrm{i}}}{\partial x_{\mathrm{j}}} + \frac{\partial \overline{v}_{\mathrm{j}}}{\partial x_{\mathrm{i}}} \right) - \frac{2}{3}\rho\delta_{ij}k \quad \text{with} \quad \mu_t = C_\mu \rho \frac{k^2}{\epsilon}, \tag{9.6}$$

where i, j denote the coordinate directions 1, 2, 3; μ_t denotes the turbulent eddy viscosity, which is a function of the local turbulence; C_μ is an empirical constant; and δ_{ij} denotes the components of the unit tensor. k is the turbulent kinetic energy per unit mass ($k = \frac{1}{2}\overline{\mathbf{v}' \cdot \mathbf{v}'}$), and ϵ is its dissipation rate ($\epsilon = \frac{\mu}{\rho}\overline{\nabla\mathbf{v}' : (\nabla\mathbf{v}')^T}$). The symbol : means a scalar product of two tensors.

Then transport equations are solved for k and ϵ:

$$\frac{\partial}{\partial t}\int_V \rho k \mathrm{d}V + \int_S \rho k\overline{\mathbf{v}} \cdot \mathbf{n}\mathrm{d}S = \int_S \mathbf{q}_k \cdot \mathbf{n}\mathrm{d}S + \int_V (\mathrm{P} - \rho\epsilon)\mathrm{d}V, \tag{9.7}$$

$$\frac{\partial}{\partial t}\int_V \rho\epsilon \mathrm{d}V + \int_S \rho\epsilon\overline{\mathbf{v}} \cdot \mathbf{n}\mathrm{d}S = \int_S \mathbf{q}_\epsilon \cdot \mathbf{n}\mathrm{d}S + \int_V \left(C_1 P\frac{\epsilon}{k} - C_2\rho\frac{\epsilon^2}{k} - C_4\rho\epsilon\,(\nabla \cdot \overline{\mathbf{v}}) \right) \mathrm{d}V.$$

Table 9.1 Empirical constants of the standard k-ϵ model

C_μ	C_1	C_2	C_4	σ_k	σ_ϵ
0.09	1.44	1.92	−0.33	1.0	1.3

where $\mathbf{q}_k$ and $\mathbf{q}_\epsilon$ are the diffusive fluxes of k and ϵ. They are defined as

$$\mathbf{q}_k = \left(\mu + \frac{\mu_t}{\sigma_k}\right)\nabla k \quad \text{and} \quad \mathbf{q}_\epsilon = \left(\mu + \frac{\mu_t}{\sigma_\epsilon}\right)\nabla \epsilon, \tag{9.8}$$

and P is the production of kinetic energy. It is modeled as

$$P = -\rho\overline{\mathbf{v}'\mathbf{v}'} : \nabla\,\overline{\mathbf{v}}. \tag{9.9}$$

In Table 9.1 C_1, C_2, C_4, σ_k, σ_μ, are typical empirical constants.

For rotating bodies, e.g., in turbomachinery, the boundary layer is influenced by the circumferential shear stress. Therefore, turbulence models are sometimes modified [17]. Such an extension is usually omitted for simulations of the flow around propellers or maneuvering ships because the rotational speeds are low. For simulating ship motions in waves, turbulence modeling is of secondary importance because the pressure is dominant in hydrodynamic forces, which depend mainly on the convective terms, especially for high Reynolds numbers.

In marine applications, the Shear Stress Transport turbulence model ($SST\ k-\omega$) by Menter [18] is widely used.

9.3 Large Eddy Simulation and Hybrid Models

The Large Eddy Simulation (LES) technique is another group of turbulence models. LES turbulence models average only over small-scale turbulent fluctuations, whereas larger ones must be resolved by the numerical grid. Depending on the Reynolds number, typically 100 million grid points or more, and very small time steps (microseconds) are required. Thus, the computational effort is too high for marine applications.

Hybrid methods that combine RANS and LES models have been developed for flows of high Reynolds numbers. They use the RANS model to simulate the flow in close proximity to a wall so that the small vortices there do not have to be resolved, whereas LES models are used to resolve the larger turbulent vortex structures in the rest of the computational domain. Details may be found in Fureby et al. [19] and Shevchuk [20].

With the exception of resonant roll motions and damping of hull vibrations, effects of viscosity and, thus, of turbulence on wave-induced ship motions and loads are rather small. Therefore, both RANS and Euler equations can often be used for such applications. For vortex-induced motions and vibrations, we recommend the use of hybrid models.

9.4 Discretization

The finite volume method is commonly employed to discretize the Navier-Stokes equations. For this method, the computational domain surrounding the ship is filled with a numerical grid consisting of cells or control volumes (CVs) of arbitrary shape. Most of the finite volume methods use cell-centered variables. The area and volume integrals in the conservation equations are approximated numerically for each CV. The approximations use variables at the respective CV and some of its neighbors. Linearization of nonlinear terms may be done, e.g., by the so-called Picard iteration [10]. For each CV, in three-dimensional flows, there are at least four scalar equations. Together they yield a system of coupled equations for the entire fluid domain, which must be solved to determine the flow variables. The following is required:

- Areal and volume integrals must be approximated.
- Values at the CV faces must be interpolated using values at the CV centers.
- The gradients in the diffusive fluxes must be approximated.

9.4.1 Approximation of Area and Volume Integrals

The area and volume integrals may be computed by the midpoint rule, which is a second-order approximation. Higher order methods lead to large computational and memory requirements [21]. Such methods are usually employed only for structured grids, where they are easier implemented. First-order approximations yield more stable and smooth solutions; however, they cause excessive numerical diffusion. Hence, in most cases, second-order methods are the option with the best cost-benefit ratio, especially for locally refined grids.

The volume integral approximated by the midpoint rule is

$$\int_v f \mathrm{d}V \approx f_{P_0} \Delta V_{P_0}, \tag{9.10}$$

where ΔV_{P_0} denotes the volume of the CV_0, and f_{P_0} is the value of f at the center of CV_0 (see Fig. 9.1).

Correspondingly, the area integrals are approximated as

$$\int_{S_j} f \mathrm{d}S \approx f_j S_j, \tag{9.11}$$

where the function f stands for the convective or the diffusive flux and f_j for its value at the center of cell face j. S_j is the area of face j.

9.4.2 Convective Fluxes

The convective flux of a variable ϕ per mass through the control volume face j is approximated as

$$\int_{S_j} \rho\phi\mathbf{v}\cdot\mathbf{n}\mathrm{d}S \approx \phi_j \int_{S_j} \rho\mathbf{v}\cdot\mathbf{n}\mathrm{d}S \approx \phi_j\rho\mathbf{v}_j\cdot\mathbf{S}_j = \phi_j\dot{m}_j. \tag{9.12}$$

Here, $\mathbf{S}_j = S_j\mathbf{n}$. The mass flux $\dot{m}_j$ is taken from the previous iteration (Picard iteration). Then the only remaining unknown is the value of ϕ at the center of the control volume face j. The nonlinear equations are satisfied when the iterative changes become sufficiently small. The interpolation scheme for the variables at the face centers is important to obtain convergence of the iteration and accurate solutions. Different interpolation methods may be used. Examples of a first-order and a second-order schemes are described below. Here, a first-order scheme means one in which the interpolation error is proportional to the interpolation distance. To compute wave-induced motions and loads, second-order approximations such as the central difference scheme (see below) or the linear-upwind scheme [22] are recommended.

9.4.2.1 Central Difference Scheme (CDS)

The value of the variable ϕ at a control volume face j is determined by linear interpolation (see Fig. 9.1).

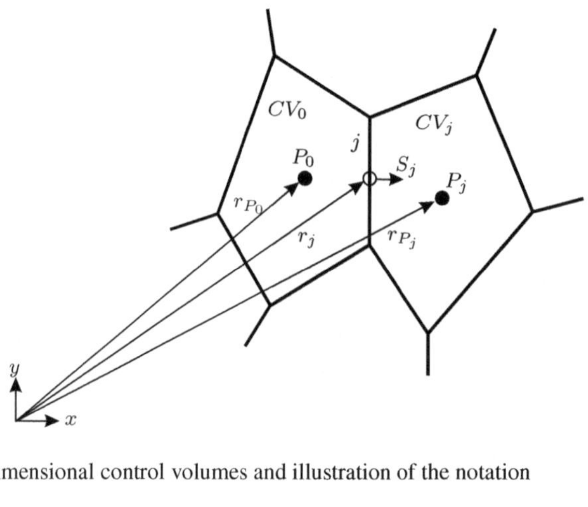

Fig. 9.1 Two-dimensional control volumes and illustration of the notation

$$\phi_j = e_j \phi_{P_0} + (1 - e_j)\phi_{P_j} \ , \quad \text{with} \ \ e_j = \frac{|\mathbf{r}_j - \mathbf{r}_{P_j}|}{|\mathbf{r}_{P_j} - \mathbf{r}_{P_0}|} \ . \tag{9.13}$$

Here $\mathbf{r}_j$ is the position vector of the center of control volume face j, $\mathbf{r}_{P_j}$ is the position vector of the center of the control volume CV_j (with center P_j), $\mathbf{r}_{P_0}$ is the position vector of the center of the considered control volume CV_0 (with center P_0), and e_j is the geometric interpolation factor.

The approximation is second-order accurate only if the straight line between CV centers P_0 and P_j passes through the shared face center j, as it is in Cartesian grids. Otherwise, a correction to improve the accuracy may be used; see, e.g., [23, 24].

9.4.2.2 Upwind Difference Scheme (UDS)

$$\phi_j = \phi_{P_0} \ \text{ if } \ \mathbf{v}_j \cdot \mathbf{S}_j \geq 0, \quad \text{and} \quad \phi_j = \phi_{P_j} \ \text{ if } \ \mathbf{v}_j \cdot \mathbf{S}_j < 0. \tag{9.14}$$

The value of ϕ_j is set equal to the value of the next upwind node. This approximation is first-order accurate.

9.4.2.3 Combination of UDS and CDS

Second-order discretization schemes may lead to physically unrealistic oscillations on coarse grids, especially if the Peclet number (a Reynolds number based on grid spacing) is greater than two in regions of large flow gradients. The first-order UDS, on the other hand, yields stable and smooth, but less accurate solutions. To avoid unrealistic oscillations of the numerical solution, both methods may be blended:

$$\phi_j = \phi_j^{UDS} + \lambda_j(\phi_j^{CDS} - \phi_j^{UDS})^{old} \ \text{ with } \ 0 \leq \lambda_j \leq 1. \tag{9.15}$$

The blending factor λ_j can be prescribed differently on each control volume face j. For regions of large flow gradients, one may choose higher blending factors (e.g., 0.95); near to the outer boundaries of the fluid domain, a smaller factor (e.g., 0.8) may be more appropriate. This approach has led to consistent and reliable results, especially for three-dimensional flows of high Reynolds numbers. For these one uses a finer grid near to the body surface, and a coarser one farther away, to keep the cell number moderate. For details, the reader may refer to Kohsla and Rubin [25].

9.4.3 Diffusive Fluxes

The diffusive flux D_j of variable ϕ through face j may be approximated using various methods [26]; besides others, by the midpoint rule and the Gauss' algorithm.

According to the midpoint rule,

$$D_j = \int_{S_j} \Gamma \nabla \phi \cdot \mathbf{n} \mathrm{d}S \approx \Gamma_j (\nabla \phi)_j \cdot \mathbf{S}_j. \tag{9.16}$$

grad ϕ can be approximated by a polynomial function of $\mathbf{x}$ in the vicinity of CV_0 and its neighboring cells CV_j. A linear expression yields:

$$\mathbf{d}_j \cdot (\nabla \phi)_{P_0} = \phi_{P_j} - \phi_{P_0} \quad (j = 1, n), \quad \text{with} \quad \mathbf{d}_j = \mathbf{r}_{P_j} - \mathbf{r}_{P_0}. \tag{9.17}$$

This equation may be solved using the least-squares method:

$$(\text{grad } \phi)_{P_0} = \mathbf{D}^{-1} \sum_j \mathbf{d}_j^T (\phi_{P_j} - \phi_{P_0}), \quad \text{with} \quad \mathbf{D} = \sum_j \mathbf{d}_j \mathbf{d}_j^T. \tag{9.18}$$

Alternatively, the gradient can be approximated up to second-order using the Gauss theorem. If the grid is unstructured or skewed, this approach may be more robust in terms of convergence than the midpoint rule. The method requires determining the gradients of ϕ at the midpoints of all cell faces. If the gradients are stored at the CV centers, they can be interpolated at the midpoints of the faces in the same way as described for the convective terms.

9.4.4 Computation of Source Terms

The second-order accurate midpoint rule is often used to approximate the integrals over the CV for computing the source terms.

9.4.5 Time Marching Methods

Time can be thought of as a fourth directional coordinate. The main difference between spatial and time coordinates is that a change in flow at a particular location and time may affect the flow everywhere (for elliptical problems), but not at earlier times. Hence, flow simulations proceed from earlier to later times. Several schemes for this 'marching in time' exist. They can be distinguished according to their order of accuracy, and whether they are implicit (i.e., require an iteration within each time step) or explicit. The step size of temporal and spatial discretizations of a flow are related to each other by the *Courant number*, also called *Courant-Friedrichs-Lewy number*. For a flow problem with a characteristic velocity v_{ref}, it is defined as follows:

$$\mathrm{CFL} = v_{\mathrm{ref}} \frac{\Delta t}{\Delta x_{\mathrm{ref}}} . \tag{9.19}$$

Term Δx_{ref} is the reference length of the control volume and Δt is the time step. While explicit schemes are stable only if CFL is smaller than a critical value, many implicit methods are stable for arbitrarily large time steps. Different temporal discretization methods are discussed in [10]; some example schemes are presented below.

Schemes for time integration can be derived from Taylor series expansions. For a multiple differentiable variable $\phi(t)$, the Taylor expansion about time $t = t_n$ and $\phi(t_n) = \phi^n$, is

$$\phi(t) = \phi^n + (t - t_n)\left(\frac{\partial\phi}{\partial t}\right)_n + \frac{(t - t_n)^2}{2!}\left(\frac{\partial^2\phi}{\partial t^2}\right)_n + \frac{(t - t_n)^3}{3!}\left(\frac{\partial^3\phi}{\partial t^3}\right)_n + \mathrm{HOT}, \tag{9.20}$$

where HOT are terms of higher order. If we delete the second-order and all higher order terms, take as t the next time instant $t = t_{n+1} = t_n + \Delta t$, and if $(\partial\phi/\partial t)_n = f(t_n, \phi^n)$ can be determined from the conservation equation for ϕ, we have for the next time instant $t = t_{n+1}$:

$$\phi^{n+1} = \phi^n + \Delta t f(t_n, \phi^n)\ . \tag{9.21}$$

Here, all terms on the right-hand side have been determined before. Equation (9.21) is a first-order approximation called explicit Euler scheme.

Analogously, the Euler backward difference scheme, also called the implicit Euler method, can be derived from a Taylor series expansion at time t_{n+1}, using $t = t_n$.

$$\phi^{n+1} = \phi^n + \Delta t f(t_{n+1}, \phi^{n+1})\ . \tag{9.22}$$

This equation is called Euler backward scheme or implicit Euler scheme. The equation cannot be solved directly, because ϕ^{n+1} occurs also on the right-hand side, where it is coupled to ϕ^{n+1} at other locations. Thus, an iteration is required to determine ϕ^{n+1}.

Both Euler methods can be combined to improve accuracy. If both enter with equal weight, one obtains the Crank-Nicolson scheme

$$\phi^{n+1} = \phi^n + \frac{1}{2}\Delta t\left[f(t_n, \phi^n) + f(t_{n+1}, \phi^{n+1})\right]. \tag{9.23}$$

This method is implicit and of second order; therefore, it may lead to oscillating solutions.

Another second-order implicit method is the three-time-level scheme. It follows from the second-order approximation

$$\phi(t) = \phi^{n+1} + f(t_{n+1}, \phi^{n+1})(t - t_{n+1}) + a(t - t_{n+1})^2, \tag{9.24}$$

which contains a coefficient a. Using this formula for $t = t_n$ to give ϕ^n, and $t = t_{n-1}$ to give ϕ^{n-1}, one obtains two equations. They are combined to eliminate a. The resulting equation is

$$\phi^{n+1} = \frac{4}{3}\phi^n - \frac{1}{3}\phi^{n-1} + \frac{2}{3}\Delta t f(t_{n+1}, \phi^{n+1})\ . \tag{9.25}$$

It is implicit and of second order.

Implicit second-order time integration methods, e.g., the three-time level and the Crank-Nicolson method, are recommended for typical marine applications because they allow relatively large time steps, for which explicit methods could become instable, and first-order methods become inaccurate. The disadvantage of implicit methods, i.e., that they require iterations within each time step, is not important in field methods, because the large equation systems are solved iteratively anyway.

9.5 Moving Grids

For flows around moving bodies, such as ships in waves or rotating propellers, the grid may need to be moved. Here, the volume conservation equation must be satisfied:

$$\frac{\mathrm{d}}{\mathrm{d}t}\int_V \mathrm{d}V - \int_S \mathbf{v}_b \cdot \mathbf{n}\mathrm{d}S = 0, \tag{9.26}$$

Term $\mathbf{v}_b$ denotes the velocity of the control volume face. Equation (9.26) describes the volume conservation when the control volume changes its form or position in time. The mass conservation yields:

$$\frac{\mathrm{d}}{\mathrm{d}t}\int_V \rho \mathrm{d}V + \int_S \rho(\mathbf{v} - \mathbf{v}_b) \cdot \mathbf{n}\mathrm{d}S = 0. \tag{9.27}$$

Combining both Eqs. (9.26) and (9.27) and assuming the fluid density to be constant yields

$$\int_S \mathbf{v} \cdot \mathbf{n}\mathrm{d}S = 0. \tag{9.28}$$

The numerical scheme needs to ensure that, at each time step, the sum of volume fluxes through a control volume face equals the change in volume. Otherwise, (9.28) is violated. The volume fluxes originate from the face motions of the control volume.

Details may be found in Ferziger et al. [10] and Demirdžić and Perić [23].

9.6 Algebraic System of Equations

Once all terms in the conservation equation for a CV are approximated, we obtain a linear algebraic equation of the form

$$A_{P_0}\phi_{P_0} + \sum_j A_j \phi_{P_j} = b_{P_0}. \tag{9.29}$$

Term ϕ is the variable to be determined from the equation. ϕ_{P_0} is its value at the center P_0 of CV_0, and ϕ_{P_j} are the corresponding desired variables at the centers P_j of the adjacent CV_j. A_j arise from contributions of the area integrals of shared control volume faces of the CV_0 and the CV_j. A_{P_0} contains also contributions from time derivatives of the integrals. b_{p_0} contains known quantities. For the entire computational domain, we have then an equation system

$$\mathbf{A}\phi = \mathbf{b}. \tag{9.30}$$

Term $\mathbf{A}$ is an N by N matrix, where N is the number of control volumes. ϕ denotes the vector of unknowns and $\mathbf{b}$ that of the source terms.

The coefficient matrix $\mathbf{A}$ is sparse. For unstructured grids, $\mathbf{A}$ has an irregular structure. Besides others, a preconditioned conjugate gradient method known as incomplete Cholesky decomposition with inner iterations can be used to solve the system of Eqs. (9.30), [10].

9.6.1 Under-Relaxation

To obtain a converged solution, equations may need to be underrelaxed. Except for the pressure correction equations, under-relaxed equations are generated by transforming (9.30) to

$$\left(\mathbf{A} + \frac{1-\alpha_e}{\alpha_e}\mathbf{D}\right)\phi = b + \frac{1-\alpha_e}{\alpha_e}\mathbf{D}\phi^{m-1}. \tag{9.31}$$

Here, α_e denotes the under-relaxation factor, which may range between zero and one. $\mathbf{D}$ is the diagonal matrix. Its elements are equal to the diagonal elements of $\mathbf{A}$. ϕ^{m-1} is the vector of variables known from the previous iteration. Obviously, in case of convergence (9.30) and (9.31) are equivalent. The modified coefficient matrix is diagonally dominant for sufficiently small α_e. This improves the range of convergence of the iterative solver; on the other hand, smaller α_e requires more iterations to obtain a given reduction factor of the residuals. The required value of α_e to attain convergence depends on grid quality and time step size. Different values of the under-relaxation factor are applied to the momentum conservation equations and the pressure correction equation (see section pressure velocity coupling).

9.7 Initial Values and Boundary Conditions

Initial values for velocity and pressure must be given in the centers of all CVs. For steady flows, well-estimated initial conditions decrease the required number of iterations. In case of parameter studies, initial values may be taken from the solution for the previous parameter value.

At the boundary surfaces of the computational domain, either variable must be given, or conditions must be specified from which the variables can be determined.

At the inlet (i.e., where fluid enters the computational domain). Generally, all variables necessary for calculating the convective flux are given. Along impermeable walls and on symmetry planes, convective fluxes are prescribed as zero. At the outlet (where fluid leaves the computational domain), convective fluxes are usually calculated by extrapolation from the interior of the fluid domain. At symmetry planes, diffusive fluxes are zero for all scalar quantities and for the components of velocity parallel to the plane. At the inlet and at walls, normal derivatives occurring in the diffusive flux are approximated as one-sided difference quotients.

Usually, the following boundary conditions are specified for incompressible flow:

- At the inlet boundary, velocities are given.
- On a symmetry plane, zero gradients in normal direction are specified for all scalar quantities and for the parallel velocity components.
- At walls, all velocity components are equal to those of the wall due to the no-slip condition.
- At the outlet boundary, gradients of all variables are often prescribed as zero. Thus, velocities are extrapolated from the neighboring control volume center to the outlet boundary. The pressure is prescribed at the outlet boundary.
- Periodic boundary conditions may be applied for propellers under open water axial flow conditions. The flow region is then confined to a sector containing only one propeller blade. Flow quantities at a radial cut on one side of the sector are equal to those on the other radial cut.

For ships in waves, initial values of the velocity, the pressure, and the free surface elevation at CV centers may be computed using the applied wave theory (see Chap. 4). The same may be done at later times along the inlet boundary. If waves are eliminated at the outlet, a static pressure distribution can be given there. Section 9.12 describes how wave reflections can be avoided at the boundaries.

9.7.1 Wall Functions

Wall functions may be used to calculate the wall shear stress. They presume the logarithmic velocity profile of boundary layer flows, which is valid in a region (without flow separation) of thin, fully turbulent boundary layers in the equilibrium of turbulence production and dissipation. Wall functions allow to bridge the near-wall region by a single cell layer and are often used in marine applications. Details may be found in [9, 27].

9.8 Pressure-Velocity Coupling

For incompressible flows, there is no explicit equation for the pressure. When the momentum equations are used to determine the three velocity components, the only equation left is the continuity equation. However, this equation does not contain the pressure. The equation to determine the pressure is derived by taking the divergence of the momentum conservation equation for incompressible flows considering the continuity equation. This yields the Poisson equation for the pressure (9.32):

$$\nabla \cdot (\nabla p) = \nabla \cdot \left(-\rho \frac{\partial \mathbf{v}}{\partial t} - \rho \nabla \cdot (\mathbf{v}\mathbf{v}) + \mu \nabla \cdot (\nabla \mathbf{v}) + \rho \mathbf{g} \right). \tag{9.32}$$

There are different methods for pressure-velocity coupling. A solution algorithm based on so-called projection methods comprises the following steps:

1. Initialize all variables (velocities, pressures, turbulence quantities, etc.) at initial time $t = t_0$.
2. Set $t_1 = t_0 + \Delta t$
3. Start the outer iteration loop (in which the nonlinear and coupling terms are updated) with outer iteration counter m.
4. Establish the system of Eqs. (9.30) for the discretized and linearized momentum equations for velocity components.The associated pressures are taken from the previous iteration or time step.
5. Solve this system of Eqs. (9.30) for velocity components iteratively to yield v_i^{m*}. These iterations are called inner iterations. An exact solution is unnecessary, because the coefficient matrix and the source terms will be corrected in the next outer iteration. Equation (9.30) is solved successively for each velocity component, whereby the same storage may be used for the coefficient matrix **A** and the same source term vector **b**. Interpolate the computed velocity components v_i^{m*} to the cell faces and use these to compute the mass fluxes, which are needed to determine the source terms in the pressure Eq. (9.32) (or pressure correction Eq. (9.43)). The velocities v_i^{m*} do not satisfy the mass conservation equation.
6. Establish the pressure (or pressure correction) equation using the velocity components v_i^{m*} to obtain the source term of the Poisson equation (9.32).
7. Solve the pressure (or pressure correction) equation.
8. Correct the pressure and velocity components at cell centers and the mass fluxes through cell faces. The corrected velocity field now satisfies the mass conservation equation.
9. Solve the transport equations for scalar values (e.g. k and ϵ, volume of fraction).
10. Repeat steps (3)–(9) using the corrected velocities and pressures from step 8 until the defined residuals from all equations are sufficiently small.
11. Repeat steps (2)–(10) until reaching the prescribed number of time steps.

The discretized momentum equation for a control volume is then written as follows:

$$A_{P_0} v_{i,P_0}^{n+1} + \sum_j A_{P_j} v_{i,P_j}^{n+1} = b_{P_0}^{n+1} - \left(\frac{\delta p^{n+1}}{\delta x_i} \right)_{P_0}, \tag{9.33}$$

where P_0 refers to the center of control volume CV_0 and j the index to the centers of its neighboring control volumes CV_j. A_{P_j} embody the contributions from convective and diffusive terms in the coefficient matrix (9.30) and A_{P_0} contains, in addition, contributions from time derivatives of the integrals. b_{P_0} represents the known quantities.

Superscript $n+1$ denotes the new time step and m the outer iteration counter. Symbol δ in the pressure term in Eq. (9.33) denotes its derivative, indicating that its discretization does not depend on the discretization of the other spatial derivatives. The momentum equation is solved within the new time step, taking the pressure field from the previous outer iteration $(m-1)$ or the previous time step:

$$A_{P_0} v_{i,P_0}^{m*} + \sum_j A_{P_j} v_{i,P_j}^{m*} = b_{P_0}^{m-1} - \left(\frac{\delta p^{m-1}}{\delta x_i} \right)_{P_0} \tag{9.34}$$

$$v_{i,P_0}^{m*} = \underbrace{\frac{b^{m-1} - \sum_j A_{P_j} v_{i,P_j}^{m*}}{A_{P_0}}}_{\tilde{v}_{i,P_0}^{m*}} - \frac{1}{A_{P_0}} \left(\frac{\delta p^{m-1}}{\delta x_i} \right)_{P_0}, \tag{9.35}$$

Velocity components v_i^{m*} do not satisfy the continuity equation, which is indicated with the symbol $*$. Therefore, the pressure must be updated to ensure mass conservation, and this is expressed as follows:

$$v_{i,P_0}^{m} = \tilde{v}_{i,P_0}^{m*} - \frac{1}{A_{P_0}} \left(\frac{\delta p^{m}}{\delta x_i} \right)_{P_0}. \tag{9.36}$$

Taking the divergence of Eq. (9.36) and using the continuity equation, we obtain the following expression (note that the density ρ is constant):

$$\underbrace{\left(\frac{\delta v_i^m}{\delta x_i} \right)_{P_0}}_{\text{vanish}} = \left(\frac{\delta (\tilde{v}_i^{m*})}{\delta x_i} \right)_{P_0} - \frac{\delta}{\delta x_i} \left(\frac{1}{A_P} \left(\frac{\delta p^m}{\delta x_i} \right)_{P_0} \right). \tag{9.37}$$

Consequently,

$$\frac{\delta}{\delta x_i} \left(\frac{1}{A_{P_0}} \left(\frac{\delta p^m}{\delta x_i} \right)_{P_0} \right) = \left(\frac{\delta (\tilde{v}_i^{m*})}{\delta x_i} \right)_{P_0}. \tag{9.38}$$

Solving Eq. (9.38) yields p^m. The corrected velocity field, v_i^m, can now be calculated from Eq. (9.36). Most of the velocity-pressure coupling methods use a pressure correction equation instead of the pressure equation. As stated above, v_i^{m*} and p^{m-1} do not satisfy the continuity equation and, therefore, need to be corrected:

$$v_i^m = v_i^{m*} + v' \qquad p^m = p^{m-1} + p'. \tag{9.39}$$

Subtracting the momentum equation:

$$A_{P_0} v_{i,P_0}^m + \sum_j A_{P_j} v_{i,P_j}^m = b_{P_0}^{m-1} - \left(\frac{\delta p^m}{\delta x_i}\right)_{P_0} \tag{9.40}$$

from Eq. (9.34) leads to the relation between the velocity and pressure corrections:

$$v'_{i,P_0} = \tilde{v}'_{i,P_0} - \frac{1}{A_{P_0}}\left(\frac{\delta p'}{\delta x_i}\right)_{P_0} \quad \text{with} \quad \tilde{v}'_{i,P_0} = -\frac{\sum_j A_{P_j} v'_{i,P_j}}{A_{P_0}}. \tag{9.41}$$

The continuity equation reads:

$$\frac{\delta(v_i^m)}{\delta x_i} = \frac{\delta(v_i^{m*})}{\delta x_i} + \frac{\delta(v_i')}{\delta x_i} = 0. \tag{9.42}$$

Taking the divergence of Eq. (9.41) and considering Eqs. (9.39) and (9.42) yields the pressure correction equation:

$$\frac{\delta}{\delta x_i}\left(\frac{1}{A_{P_0}}\left(\frac{\delta p'}{\delta x_i}\right)_{P_0}\right) = \left(\frac{\delta(v_i^{m*})}{\delta x_i}\right)_{P_0} + \underbrace{\left(\frac{\delta(\tilde{v}_i')}{\delta x_i}\right)_{P_0}}_{\text{unknown}}. \tag{9.43}$$

The Semi-Implicit Method for Pressure-Linked Equations (SIMPLE) neglects the unknown term on the right-hand side of Eq. (9.43) [10, 28]. Many outer iterations are often needed to obtain a converged solution.

The Pressure Implicit with Splitting of Operators (PISO) algorithm introduces a second correction of the velocities [10, 29]:

$$v''_{i,P_0} - \tilde{v}'_{i P_0} - \frac{1}{A_{P_0}}\left(\frac{\delta p''}{\delta x_i}\right)_{P_0}, \tag{9.44}$$

where $\tilde{v}'_{i P_0}$ is calculated based on Eq. (9.41).

Taking the divergence of Eq. (9.43) and considering that $\frac{\delta(\, v''_{i,P_0})}{\delta x_i} = 0$, leads to a second pressure correction equation:

$$\frac{\delta}{\delta x_i}\left(\frac{1}{A_P}\left(\frac{\delta p''}{\delta x_i}\right)_{P_0}\right) = \left(\frac{\delta(\tilde{v}_i')}{\delta x_i}\right)_{P_0}. \tag{9.45}$$

Solving the Eq. (9.45) obtains p''_{P_0}, and v''_{i,P_0} can then be calculated. The velocities, v_i^m, and the pressures, p^m, are obtained as follows:

$$v^m_{i,P_0} = v^{m*}_{i,P_0} + v'_{i,P_0} + v''_{i,P_0} \qquad p^m = p^{m-1} + p' + p''.$$

In PISO, the momentum equation is solved only once per time step. The velocity-pressure coupling algorithm combines the SIMPLE and PISO algorithms [4, 30, 31]. For maritime related applications, SIMPLE or PIMPLE algorithms are widely used.

9.9 Numerical Grids

Numerical grids may be classified as structured, block-structured, overlapping, or unstructured.

Three-dimensional structured grids consist (in case of finite-volume method) of control volumes having, except at the grid boundaries, exactly six neighbors. Grid points are indexed in lexicographical order to alleviate identifying neighboring points. The programming effort to solve the conservation equations is moderate, and the coefficient matrix of the resulting system of equations has a simple diagonal structure. Structured grids can be used only for simple geometries.

Block-structured grids split the computational domain into several blocks. Within each block, a structured grid is employed, and the whole domain is no longer structured. Block-structured grids offer higher flexibility if the individual grid blocks do not have to be matched at block interfaces. The programming effort for solving the conservation equations is higher than for structured grids.

Overlapping grids contain partially overlapping blocks of (typically) structured sub-grids. Computations on overlapping grids are performed alternately on each block, and the solutions are interpolated from one block to the others and imposed there as boundary conditions. Numerical methods are not conservative in overlapping grids. Their advantage is the much easier grid generation for flows around several bodies, especially if these are moving relative to each other. Examples are flows around ship hulls with propeller(s) and rudder(s). Interpolation errors may cause inaccuracies, slow convergence, or even divergence.

Unstructured grids offer in most cases, the highest flexibility and user-friendliness when generating grids around complex structures. Their control volumes may be arbitrary polyhedra, and different kinds of polyhedra may be combined. Local grid refinement and adaptive grid generation can be applied. The structure of the coefficient matrix of the associated system of equations is irregular. Thus, the computational effort (per control volume) for solving the conservation equations is higher compared to structured and block-structured grids. Unstructured grids consisting of hexahedral cells in the outer regions of the grid, and prism layers along the body, are recommended in case of solving Navier-Stokes or Euler equations.

In grids containing sliding interfaces, the solution domain consists of non-overlapping blocks that move relative to each other along a plane, cylindrical, or spherical interface. Unlike overlapping grids, implicit coupling of blocks with sliding interfaces is possible, allowing fully conservative computations. Along with the sliding interface, the face areas and their adjacent CVs must be updated at each time step. For details, see Muzaferija et al. [32].

The grid quality influences considerably the discretization errors and the convergence of the iteration. Therefore, grid quality is decisive for accuracy and required computational time. Important criteria for grid quality are: Control volumes should be as little warped as possible, grid lines should be as orthogonal as possible, and pronounced long side aspect ratios (e.g. higher than 1000) should be avoided.

For ships moving in waves, unstructured grids are widely used. Different techniques may be used to take account of ship motions: entire grid movements [33], morphed grids [1], and overlapping grids. A simple morphing grid technique was proposed by [4]. Recently, overlapping grids methods were implemented in CFD codes.

9.10 Free-Surface Flows

Free-surface flows may be modeled as one fluid bounded by a deformable surface, or as two or more immiscible fluids like air and water. In marine applications using field methods, usually, the latter model is applied. Computations of free-surface flows must determine also the interface between the fluids. Corresponding methods may be subdivided into front tracking and front capturing methods.

9.10.1 Front-Tracking Methods

The governing equations may be solved separately for each fluid. The interface between both fluid domains defines the free surface. The following boundary conditions apply at the free surface:

- *Kinematic condition*. This condition states that there is no convective mass transfer through the free surface. This means that the normal component of fluid velocity equals the normal component of the free surface velocity:

$$(\mathbf{v} - \mathbf{v}_b) \cdot \mathbf{n} = 0. \tag{9.46}$$

- *Dynamic condition*. This condition states that the forces acting on the free surface are in equilibrium. For the flow around ships, where surface tension is negligible, that means that the pressure and the tangential stress are equal on both sides of the free surface.

The advantage of this approach is that the interface remains sharp. However, if the free surface breaks up into droplets and/or bubbles, the method cannot be applied. Thus, in maritime applications, the method is hardly used.

9.10.2 Front-Capturing Methods

In front-capturing methods, the computation is performed on a grid that extends beyond the free surface. Grid surfaces must not be adjusted to the interface, which may cross control volumes arbitrarily.

9.10.2.1 Volume of Fluid (VoF) Methods

In these methods the free surface is determined indirectly by computing the fraction, with which each CV is filled by fluid 1, if a two-fluid flow (e.g., of water and air) is to be computed. In addition to the conservation equations for mass and momentum, a transport equation for the volume fraction C_1, or for C_i, $i = 1, 2, \ldots, N - 1$ in case of N fluids, must be solved [34, 35] because C_N can be determined from the condition

$$\sum_1^N C_i = 1. \tag{9.47}$$

The fluids are not considered separately, but are replaced by a mixture consisting of the different fluids in proportions C_i which vary in space and time. For the mixture, effective values of density and viscosity are assumed:

$$\rho = \sum_{i=1}^N C_i \rho_i, \qquad \mu = \sum_{i=1}^N C_i \mu_i. \tag{9.48}$$

The conservation equation for C_i is

$$\frac{\partial}{\partial t}\int_V C\,dV + \int_S C\mathbf{v}\cdot\mathbf{n}\,dS = 0. \tag{9.49}$$

In the region in which C_1 changes from 1 (water) to 0 (air), the discretization and the related numerical approximations of this equation are decisive for the accuracy of the results.

Examples of numerical schemes used to discretize the Eq. (9.49) are the Multi-dimensional Universal Limiter with Explicit Solution (MULES) algorithm [36–38], and the High-Resolution Interface-Capturing (HRIC) scheme [39].

The VoF method is widely used in maritime applications. It has proven to be suitable for handling complex free-surface phenomena.

9.10.2.2 Level-Set Method

To describe the free surface between two fluids, a scalar function Φ is defined within the entire fluid domain: the level-set function. It is negative in air, positive in water, and zero at the free surface. The free surface motion is described as a transport equation for Φ. Assuming the condition $D\Phi/Dt = 0$ everywhere in the fluid leads to the transport equation

$$\frac{\partial \Phi}{\partial t} + \frac{\partial (v_i \Phi)}{\partial x^i} = 0. \tag{9.50}$$

While we are interested only in the isocontour $\Phi = 0$, extending the definition of Φ to the entire fluid domain makes it easier to implement this approach. The level-set function is initialized at each control volume center as its distance (including sign) to the interface between the two fluids. In case of water and air, the density (ρ) and viscosity (ν) of the fluid should be either those of water (for positive Φ) or air (for negative Φ). However, to avoid numerical problems, the transition must be smoothed. Density and kinematic viscosity are thus determined as

$$\rho = (1 - C)\rho_{air} + C\rho_{water}; \quad \nu = (1 - C)\nu_{air} + C\nu_{water} \tag{9.51}$$

with

$$\begin{aligned} C &= 1 && \text{if } \Phi > \alpha \text{ (water);} \\ C &= 0 && \text{if } \Phi < -\alpha \text{ (air);} \\ C &= 0.5\left(1 + \sin\frac{\pi\Phi}{2\alpha}\right) && \text{otherwise.} \end{aligned} \tag{9.52}$$

α is the thickness of the transition zone.

For each time step, first, the equations for velocity (conservation of momentum), pressure correction (conservation of mass), and turbulence parameters are solved, using the previous Φ values. Then the transport equation (9.50) is used to update Φ. Points, for which $\Phi = 0$, define the new interface, but for other points, Φ is no longer the distance from the interface. However, this property is needed to keep the thickness of the transition zone constant and to prevent the loss of mass and momentum in the transitional zone. Thus, it is recommended to correct the Φ values to the distance from the surface $\Phi = 0$ at each time step. More details may be found in [40, 41].

9.11 Coupling Flow Equations and Rigid Body Motion Equations

To compute ship motions, the continuity and the Navier-Stokes equations for fluid motion must be coupled to the body motion Eqs. (2.40) and (2.41). Fluid forces and moments are obtained by integrating pressures and shear stresses over the wetted hull surface. Then, from (2.40) and (2.41) the body velocity at the center of gravity G and the angular velocity can be determined using explicit or implicit methods. Because explicit methods are unstable for large time steps, second-order implicit time integration schemes are recommended, e.g., the three-time-level scheme (9.25) or the Crank-Nicolson scheme (9.23), possibly with a suitable relaxation factor <1. They are applied once for each outer iteration of the procedure for fluid motion. From the angular velocity, the time derivative of the Euler angles $\dot{\alpha}$ is obtained using (2.35). The body velocity and $\dot{\alpha}$ are then used to determine the position of G and the Euler angles, again by applying any method for integrating ordinary differential equations. The outer iteration ends when convergence criteria for both fluid and body motion are satisfied. For more details see [1, 4, 42].

For some applications, it may be necessary to suppress individual degrees of freedom. This is the case, for example, when the speed or course of a ship is specified. To suppress selected degrees of freedom, special conditions must be fulfilled, see Sects. 12.6 and 11.8 as well as [4, 43].

9.12 Wave Generation and Damping in Field Methods

9.12.1 Preliminaries

An obvious method to introduce waves in nonlinear methods dealing with bodies in waves is described above in chapter 'Nonlinear Rankine Methods'. There the total flow is obtained by adding the wave flow and the disturbance caused by the body. If the wave flow is known, the numerical method is applied only to the disturbance. That can be done also in Finite-Volume Methods (FVM); see, e.g., [44].

This method may be impractical for typical CFD calculations because usually only the existing, generally applicable flow solvers are used. To apply these solvers to the disturbance instead of the total flow, not only the boundary conditions, but also the momentum conservation equations need to be changed. Thus, in the following, we presume that the numerical computation deals with the total flow.

We recommend starting the simulation not with the fluid at rest, as is typical in model tanks, but with the fully developed wave field to decrease the necessary simulation time. In case of parameter studies, an alternative option is to start with the total flow field obtained from a previous computation using the same waves as the initial solution. Closed formulas exist to determine the wave flow alone, e.g. Airy or Stokes wave theories. Other numerical methods are available for irregular seaways, e.g. the HOS method [45] or, for steep long-crested regular or irregular waves, e.g. a two-dimensional CFD computation.

9.12.2 An Intuitive Approach

CFD simulations for ships in head waves are often performed in analogy to model experiments in a wave tank. A wave maker is situated at one end of the rectangular computational grid; at the opposite end, there is a wave damping or absorption zone, and the side boundaries are situated parallel to the direction of wave propagation. However, in contrast to typical model experiments, a parallel flow is superimposed on the incident wave, directed opposite to the ship's forward speed. This allows confining the computational grid to a relatively small region (compared to a towing tank) surrounding the ship. For other wave directions, waves are generated at two sides of the rectangular grid and damped or absorbed at the two opposing sides.

In CFD simulations, regular and irregular waves are best generated not by a moving boundary, such as a flap or piston wave maker, but by specifying the oscillating normal velocities and the associated water volume fractions (or the level-set functions) at a fixed outer boundary. These quantities may be computed using a suitable linear or nonlinear wave theory. For long-crested waves, they may be obtained also from a two-dimensional, previously performed numerical computation, without the presence of a body. If the vertical distribution of the specified velocities deviate from the velocity distribution of a free-running wave (which would be so if a flap generates the waves), besides the progressive wave, a near-field wave is generated that fades out exponentially with increasing distance from the wave maker. The distance where the near-field wave may be neglected is decreased when the deviation of the specified velocity from that of the progressive wave is small. Additionally, the appropriate stationary parallel flow must be specified. Steep waves generate an additional mean mass flux, which is produced automatically if the periodic wave flow is specified correctly. In model tanks, this wave-generated mean mass flow, which occurs in the region between wave trough and wave crest, is compensated for by a backflow at greater depth. This flow seems not to substantially affect wave-induced ship motions and loads.

At the opposite outer boundaries, waves must be damped or absorbed. Active absorption by specifying a wave maker that cancels the generated waves is infeasible, because the disturbed waves, i.e., the radiated waves generated by the moving body and the incident waves diffracted by the body would be reflected. Furthermore, correctly determining the wave's phase is difficult. An alternative option of 'measuring' the waves ahead of the boundary, extrapolating these waves to the boundary, and generating an appropriate boundary condition reduces wave reflections. However, for three-dimensional flows, this procedure would be too inaccurate to eliminate these waves.

Therefore, in most instances, waves are damped. The simplest method to dampen waves is to use increasingly larger CVs (grid stretching). This causes numerical damping without at the same time greatly increasing the cell number as it lengthens the fluid region and thus delays waves reflected at the outer boundary before they reach the body. A disadvantage is that the reflection coefficient of the remaining waves is usually unknown. An increase of wave damping inevitably causes the damped

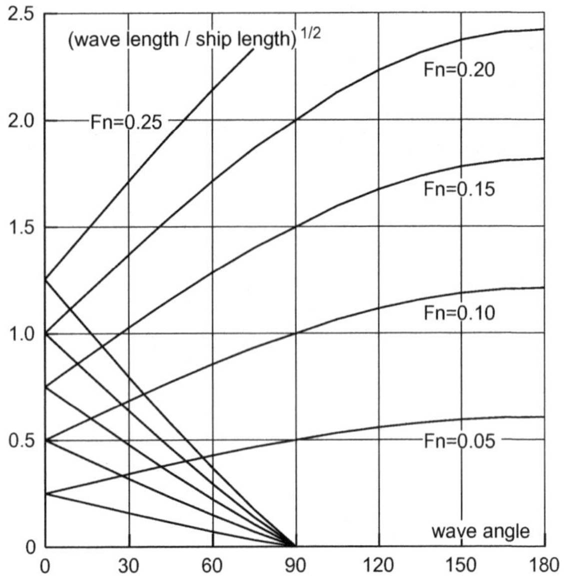

Fig. 9.2 Between the upper and lower curve segments for any Froude number F_n, the parameter τ exceeds 0.25

wave to split up into a reflected and a transmitted component. If waves are damped, in addition to the parallel inflow at the wave-generating boundaries, this means mass transport due to nonlinear waves must also flow out of the computational domain again.

One problem with this 'intuitive approach' is that the disturbed waves generated by the body are fully reflected at the wave-generating boundaries. Thus, the simulation should end before the reflected waves reach the body. For ships at zero speed or at small forward speeds, disturbed waves propagate away from its hull in all directions; at higher speeds, disturbed waves propagate only in transverse and aft directions. The limit between these cases is determined by the parameter

$$\tau = \frac{U\omega_e}{g}, \tag{9.53}$$

where U is ship speed, ω_e is the encounter angular frequency, and g is gravity acceleration. Disturbed waves ahead of the bow occur only if $\tau < \frac{1}{4}$. Figure 9.2 visualizes the combinations of Froude number, wavelength over ship length ratio, and wave heading angle μ that cause no progressive disturbed waves occurring ahead of the ship (i.e., $\tau \geq 0.25$).

Another problem of the 'intuitive approach' occurs for a ship advancing at certain wave headings and in natural seaways. For stationary flow, boundary conditions at inflow, outflow, and parallel side boundaries of the domain usually differ. Consequently, for regular waves, we must distinguish between luv boundaries (where waves enter the grid), lee boundaries (where waves leave the grid), and so-called leef boundaries (intermediate boundaries between lee and luv), where waves propagate parallel to the boundary. The distinction between these kinds of outer boundaries must be based not on the wave's phase velocity, but also on its group velocity, which, for deep water, is $\frac{1}{2}$ of the phase speed. For arbitrary wave directions relative to the ship's heading, a total of nine other kinds of boundaries exist, such as boundaries caused by waves from inflow luv, inflow lee, etc. For a ship in irregular and/or short-crested waves, this classification of boundaries completely breaks down. Thus, to analyze ships at forward speed in arbitrary wave fields, the condition at outer boundaries may not depend on wave direction. This can be accomplished using forcing zones.

The following three subsections are based on contributions by Robinson Perić [46–51].

9.12.3 Forcing Zones

In forcing zones, the source terms of the conservation equations for momentum are modified by adding

$$\int_V \gamma b(\mathbf{x})(v_{i,ref} - v_i)\, dV, \tag{9.54}$$

and the source term of volume fraction α by adding

$$\int_V \gamma b(\mathbf{x}) \left(\alpha_{ref} - \alpha\right) \mathrm{d}V. \tag{9.55}$$

Here, $v_{i,ref}$ and α_{ref} are the velocity components and the volume fraction, respectively, for the 'reference solution', which here is the flow caused by the incident wave, including the appropriate parallel flow. The additional source terms change the solution gradually to approximate the reference solution. Factor γ is used to scale the forcing, and function $b(\mathbf{x})$ is set equal to zero at the inner boundary of the forcing zone and unity at the outer boundary. In between, different (e.g., linear) functions of distance from the boundary may be applied. For arbitrary wave directions, forcing zones should be arranged along all outer boundaries of the computational grid. Suitable parameters of forcing zones can be determined analytically [49]. Too small γ causes reflections at the outer limit, too large γ causes reflections within the forcing zone (Fig. 9.3).

Test simulations without a body are recommended. Often, such test flows are two-dimensional; thus, they require no substantial computing time. They allow also to detect other sources of inaccuracies, e.g., a decrease of wave height outside of forcing zones caused by discretization errors. Figure 9.4 illustrates a case where, besides numerical wave damping, also unrealistic wiggles of the wave envelope occurred after a lengthy simulation. Here, the wiggles were produced by the interface sharpening method used in the interface capturing scheme. The problems may be

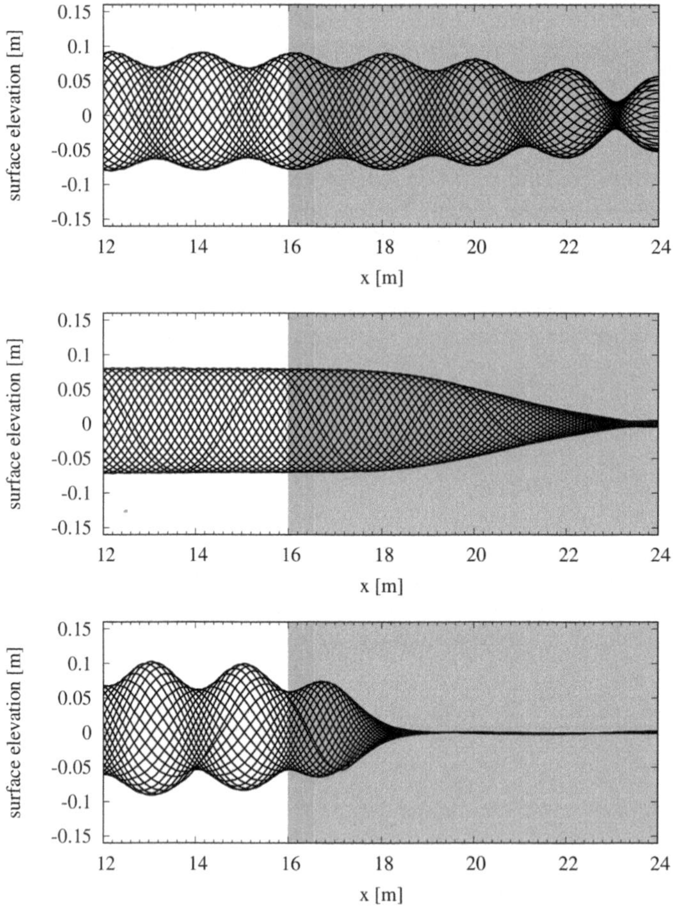

Fig. 9.3 Water surface elevation as a function of x, plotted for several time instants during the last simulated period. Regular waves travel in the $+x$ direction. Wave damping by a forcing zone (shaded gray area) using too small (top), optimum (middle), and too large parameter γ (bottom). From [49]

solved by specifying a smaller time step and using more outer iterations per time step. In other cases, it may be necessary to change the grid (especially the cell aspect ratio) and the parameters of the interface sharpening scheme.

Normally, the reference solution used in (9.54) and (9.55) should be as accurate as possible. For instance, if a linear wave approximation represents the reference solution for steep waves as the wave passes through the inner fluid domain outside of the forcing zones, a phase shift occurs between the computed and the reference solution. In the subsequent forcing zone, this mismatch may cause flow disturbances that may radiate waves back into the inner flow region. A phase shift occurring between the central flow region without forcing and forcing zones along the leef

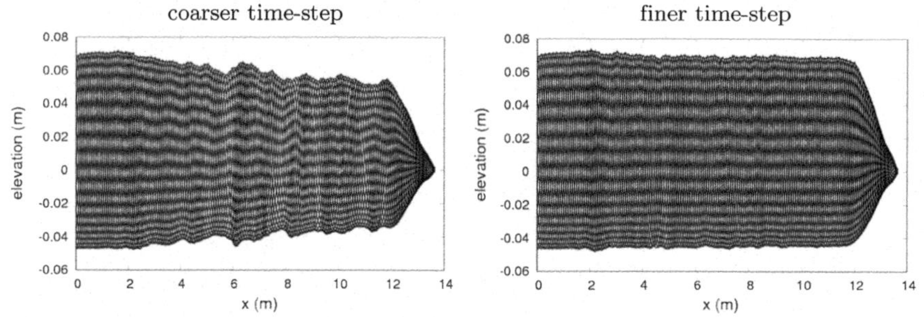

Fig. 9.4 Surface height after 40 wave periods T plotted at many time instants during the last simulated period. Regular waves propagate in the $+x$ direction to an absorption zone. Left: $\Delta t = T/500$, 5 outer iterations per time step. Right: $\Delta t = T/1000$, 8 outer iterations per time step

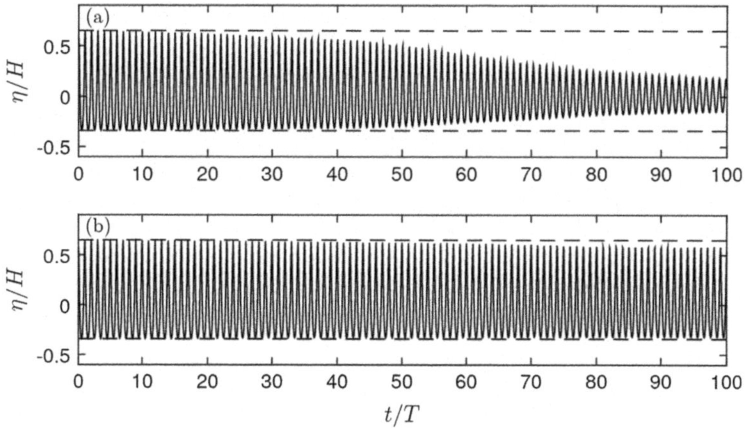

Fig. 9.5 Computed time series of surface elevations in the simulation of a periodic wave train using (top) the [53] k-ω-model and (bottom) a model optimized for long-duration flow simulations; from [52]

boundaries may be avoided by forcing only the velocity component normal to the wave propagation direction.

For some turbulence models and long-time simulations exceeding, say, 20 wave periods, the oscillatory motion of the water surface in waves may excite unrealistic turbulence near the water surface that dampens the wave (Fig. 9.5). Thus, it is recommended to check the turbulent viscosity ν_t at the water surface, where it should be zero, such as upstream of bodies if no wave breaking occurs. Values of ν_t/ν_{water} of the order of $10^3 - 10^4$ are certainly unrealistic and should be improved, e.g., by replacing the turbulence model [52].

If a suitable forcing zone is arranged along an outer boundary, no special outer boundary condition is required; if a conservation equation requires a function value outside of the grid, the value of the reference solution is used.

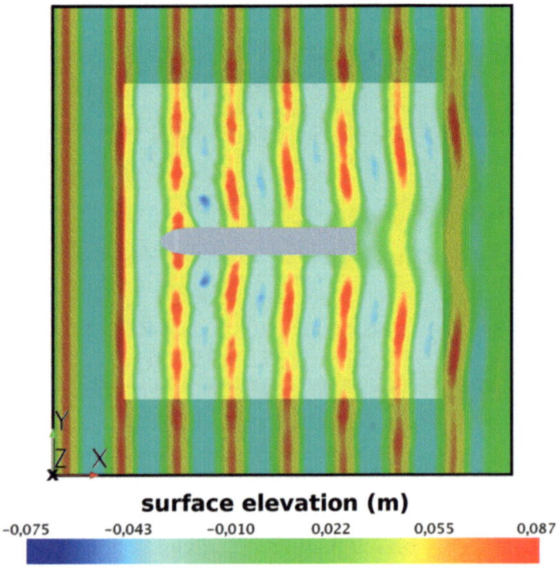

Fig. 9.6 Top view of a box-shaped simulation domain with a ship in regular waves. Forcing zones are shaded gray

Based on experience with the nonlinear Rankine source method described above, another use of forcing zones may prove advantageous also for field methods: to specify the total fluid region as a forcing zone, using the incoming waves plus the stationary flow as the reference solution. γ is then chosen appropriately smaller, and $b(\mathbf{x}) = 1$ everywhere.

9.12.4 Simulation Setup

In the following, the focus is on wave generation and wave damping via forcing zones for finite volume-based flow simulations of floating bodies subject to long-crested incoming waves, as illustrated in Figs. 9.6 and 9.7.

The computational domain is typically a box or a cylinder filled with water and air, where both substances are modeled as incompressible fluids. (For special applications, such as slamming with entrapped air, the compressibility of air may be relevant.) To account for the free surface, the volume of fluid method [54] combined with the HRIC [39] scheme may be used. The flow equations may be solved via commercial codes (e.g. ANSYS fluent or SIEMENS STAR-CCM+) or via research codes, such as OpenFOAM.

If viscous effects can be neglected, turbulence modeling and fine meshing along the body boundary may be omitted to reduce the computational effort. Gravity must be modeled, whereas surface tension can be neglected.

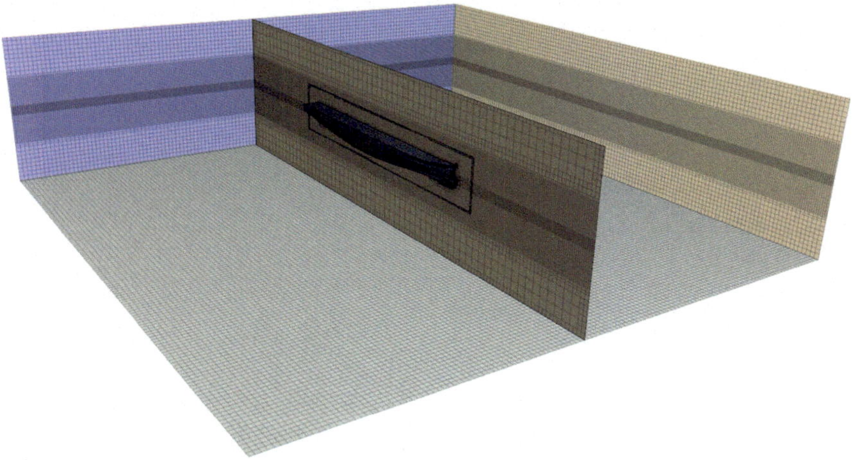

Fig. 9.7 Computational grid with local mesh refinement near the ship and the water surface. The box surrounding the ship indicates the boundary of the overset mesh

The overset grid in Fig. 9.7 is convenient to analyze ship motions. To restrain deviations from the planned ship motion, a spring and damper system or a steering and propulsion control mechanism may have to be modeled.

9.12.5 Space and Time Discretization

As an example, Fig. 9.8 shows a computational grid highly refined near the water surface. When doubling the cell size, there should be at least about 5 cells before the cell size is changed again; the mesh in Fig. 9.8 could be improved in this respect.

The coarsest discretization used in engineering practice consists of roughly 25–30 cells per wavelength λ and 5–6 cells per wave height H. It is recommended not to change the cell size in the direction of wave propagation up to a water depth of 0.25λ. In the air region and in the vertical direction greater than 0.5λ below the water surface, the grid may be coarser.

The time step size Δt should be chosen such that the local Courant number

$$Co = \frac{v_i \Delta t}{\Delta x_i} < 1 \tag{9.56}$$

for all CVs, where v_i is the local velocity, and Δx_i is the cell size in direction i. Even $Co < 0.25$ and less may be required to avoid instability and to reduce numerical dissipation at the water surface [55].

Typical are 500–1000 time steps per wave period. The number of iterations per time step with the SIMPLE algorithm is typically 5–8, depending on the under-

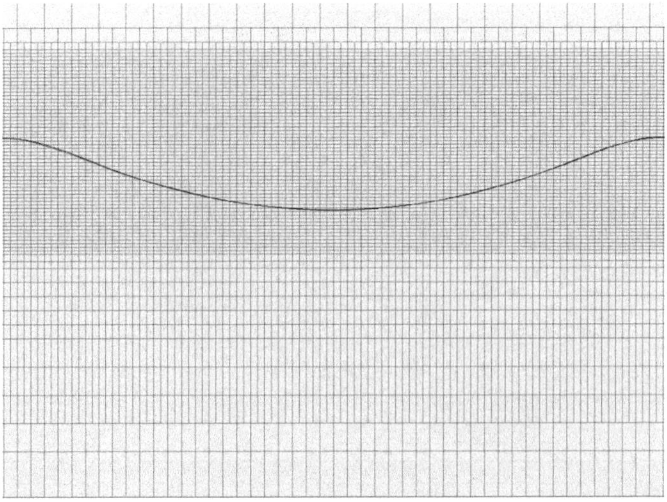

Fig. 9.8 Mesh section in the vicinity of the water surface (black line) for comparatively fine mesh, with 95 cells per wavelength λ and 20 cells per wave height H

relaxation used and on the required accuracy. All-time and space discretization schemes should be of order two or higher because first-order schemes introduce excessive numerical diffusion.

A two-dimensional flow simulation with domain length of 10 wavelengths can be simulated using, say, $10^4 - 10^5$ cells. Simulating 20 wave periods requires a computing time of an hour or less when computed in parallel on four cores of a present (2020 model) desktop computer. For a three-dimensional computation using, on average, n cells in the transverse direction, the computing times would increase roughly n-fold.

9.13 Numerical Errors

The accuracy of numerical methods for fluid flow is sensitive to the discretization of the conservation equations. The relevance of inaccuracies depends on the involved phenomena. The flow around ships and offshore structures in waves is dominated by the pressure. Hydrostatic effects play an important role, and wave-induced ship motions and sectional loads are less affected by discretization errors than many other applications [1]. Nevertheless, resonant roll motions do depend on the discretization because the accurate prediction of the damping of roll motions requires an appropriate resolution of viscous effects [56]. Furthermore, hydrodynamic damping of hull vibrations is particularly sensitive to discretization errors because it is mainly caused by short radiated waves [1]. Wave-induced and sloshing-induced impact loads are especially influenced not only by the spatial, but also by the temporal discretization [57].

Uncertainty quantification comprises verification of the codes used. Verification involves demonstrating that a computer code correctly implements the model. The code should be verified before obtaining the solution. Usually, as developers verify the code, users then need to focus on verifying the solution. To verify the solution, the numerical error for which the exact solution is not known should be estimated, and this must be done before validation. Validation is concerned with solving the appropriate equations. The purpose of validation is to estimate modeling errors, for which well-founded experimental data are essential.

Ferziger et al. [10] and Perić [50] classified numerical errors in different categories. Numerical solutions are always approximations and naturally include systematic and avoidable errors. Systematic errors comprise discretization errors, modeling errors, iteration errors, and round-off errors. Discretization errors identify the difference between the solution of the discretized equations and the exact solution of the differential equations. Modeling errors denote the difference between the real flow and the exact solution of the mathematical model, e.g., turbulence, cavitation, combustion, etc. Iteration errors specify the difference between the exact solution and an iterative solution of the discretized equations, including errors caused by the finite arithmetic accuracy of computers. Round-off errors designate the difference between a calculated approximation of a certain quantity and its exact mathematical value. Avoidable errors include programming errors and user errors. Programming errors are the so-called bugs. Every code has them, and they are difficult to spot. User errors, like with every tool, easily occur when setting up the flow simulation. Such errors occur when mixing up geometric units when specifying incorrect material data, such as wrong fluid properties, when defining boundary conditions, or when interpreting the solution, etc. Finally, it is important to always carefully examine the results. Not just the numbers themselves but also terms and definitions, such as forces, velocities, and pressures need to be critically monitored, Perić [50].

A numerical error, δ, is here defined as the difference between a given solution, ϕ_i, and the exact value,ϕ_0.

$$\delta = \phi_i - \phi_0 \; . \tag{9.57}$$

The associated uncertainty, U_{ϕ_i}, specifies the interval that contains the exact solution with a certain degree of confidence. It is related to the error as follows:

$$\phi_i - U_{\phi_i} \leq \phi_0 \leq \phi_i + U_{\phi_i} \; . \tag{9.58}$$

9.13.1 Discretization Errors

Discretization errors depend on not only the grid properties but also on the discretization process itself. The shape of control volumes used to generate the numerical grid (tetrahedral, hexahedral, polyhedral, etc.), the grid's quality (non-orthogonality, skewness, aspect ratio, etc.), its fineness (size of control volumes), its distribution of control volumes and, for non-uniform grids, its local grid refinements constitute grid properties. The discretization process as such always relies on approximations of surface and volume integrals, on gradients of cell centers at cell faces and at applied boundaries, and on interpolations necessary to compute changing values at locations other than cell centers. The order of the approximation is important as well.

9.13.2 Modeling Errors

Modeling errors are caused by a number of sources. For example, they arise when the equations being solved do not exactly describe the flow, because the model used depends on the assumptions made. Even if the equations can be solved analytically, the solution may not necessarily match reality. If the flow must account for turbulence, major sources of modeling errors arise because all turbulence models involve assumptions and empirical data that are not universally valid. Often, flow simulations at high Reynolds numbers are limited by excessive requirements for grid spacing and time step size, which means that Navier-Stokes equations cannot be solved with sufficient accuracy to obtain statistical measures of quantities of interest.

Models of physical phenomena involving multi-phase flows, such as cavitating flows, are also a source of modeling errors as are approximations of boundary conditions needed to specify inlet, outlet, far-field, and wall effects. They also affect the solution and belong to modeling errors. Another frequent source of errors is when the geometry is simplified, where even small features may lead to tolerance uncertainties. Applying symmetry conditions for a symmetric geometry may result in significant errors when the flow is not symmetrical. Often this occurs, for example, for inflows behind cylinders. Uncertainty in fluid properties that depend on temperature, pressure, concentration, etc., can also be significant.

9.13.3 Iteration Errors

The iterative solution method for linear equations relies on inner iterations; the iterative solution method for nonlinear and coupled equations, on outer iterations. Convergence criteria need to be separately specified for inner and outer iterations to precisely keep track of arithmetic operations (single or double) and round-off errors.

Discretization, modeling, and iteration errors are by far the largest numerical errors. However, for unsteady simulations, also statistical errors may be important.

9.13.4 Reference Procedures to Determine Discretization Errors and Uncertainties

Numerical errors are hardly avoidable. However, it is important to quantify them. Various procedures are available in the literature. These include the Guidelines of the International Towing Tank Conference [58]. The following procedures of Eça and Hoekstra [59], Xing and Stern [60], and Oberhagemann [4] also document procedures to estimate such errors, and several other authors have applied these methods successfully for flow problems in the maritime context, e.g., Burmester et al. [56]. Many guidelines and procedures define the convergence ratio, R, as follows:

$$R = \frac{\phi_2 - \phi_1}{\phi_3 - \phi_2} \,. \tag{9.59}$$

Here ϕ_1, ϕ_2, and ϕ_3 denotes numerical solutions obtained with fine, medium, and coarse discretizations, respectively.

Monotonic convergence is achieved if $0 < R < 1$; oscillatory convergence, if $-1 < R < 0$; monotonic divergence if $R > 1$; and oscillatory divergence if $R < -1$.

The basic idea of error estimation from discretization refinement studies is to approximate the numerical solution by a power series expansion of the spatial and temporal grid. With decreasing grid spacing and time step, the discretization error is reduced, and the solution values asymptotically approach the discretization-independent solution. The following approaches may be applied (e.g., [56]).

9.13.4.1 Procedure by Eça and Hoekstra [59]

The discretization error, δ, is simply the difference between the solution ϕ_i obtained on grid i and the grid-independent solution, ϕ_0. The discretization error (ϵ) may be estimated as follows:

$$\epsilon \approx \delta = \phi_i - \phi_0 = \alpha h_i^p \tag{9.60}$$

where α is a constant, h_i is the relative grid spacing, which depends on the grid spacing, Δx_i. p is the order of approximation. To compute the discretization error based on Eq. (9.60), at least three discretization levels are required. Δx_i and h_i are estimated as follows:

$$\Delta x_i = \left(\frac{1}{N_{CV,i}}\right)^{\frac{1}{n_D}} \quad , \quad h_i = \frac{\Delta x_i}{\Delta x_1} \tag{9.61}$$

where $N_{CV,i}$ is the number of control volumes of the grid identified by subscript i, and n_D is the grid's spatial dimension. For three-dimensional computations, $n_D = 3$. Subscript 1 marks the solution obtained on the finest grid and the corresponding time step.

For unsteady flows (9.60) is extended to read as follows:

$$\delta = \phi_i - \phi_0 = \alpha_x h_i^{p_x} + \alpha_t t_i^{p_t} \tag{9.62}$$

where subscripts x and t identify the associated space and time-dependent values, respectively. t_i is the relative time step calculated as follows:

$$t_i = \frac{\Delta t_i}{\Delta t_1} . \tag{9.63}$$

Equation (9.62) contains five unknowns and, therefore, to solve this equation, at least five solutions obtained on five discretization levels are required.

Equations (9.64) or (9.65) or (9.66) may be used to estimate discretization errors if (9.60) yields unreliable or unrealistic solutions. If this occurs, the order of grid convergence, p, is then either too small or too large. The three estimators associated with these solutions are expressed as follows:

$$\delta_1 = \phi_i - \phi_0 = \alpha h_i , \tag{9.64}$$

$$\delta_2 = \phi_i - \phi_0 = \alpha h_i^2 , \tag{9.65}$$

$$\delta_{12} = \phi_i - \phi_0 = \alpha_1 h_i + \alpha_2 h_i^2 . \tag{9.66}$$

Equation (9.64) or (9.65) are applicable for monotonic converging solutions. Equation (9.66) may be applied for non-monotonic converging solutions.

The authors recommend using one discretization level more than the number of unknowns, thereby allowing determining grid-independent solution from least-squares fitting of the available solution values. Least-squares method may be used for Eqs. (9.60), (9.64)–(9.66) to minimize.

$$S^w(\phi_0, \alpha, p) = \sqrt{\sum_{i=1}^{n_g} w_i(\phi_i - (\phi_0 + \alpha h_i^p))^2} . \tag{9.67}$$

Finally, over the number of grids, n_g, the least-squares solutions must be summed. The weighting factor, w_i, in this expression allows weighting a particular solution according to its relative importance. If a solution obtained on the finer discretization level is considered to be more reliable, its weighting factor should be higher than the weighting factors for the solutions obtained on the coarser discretization levels. If all solutions are equally reliable, the weighting factors are unity for all grids.

For a non-weighted approach, the product $n_g \,.\, w_i$ is replaced by unity, and the uncertainty can be obtained after estimating the error.

9.13.4.2 Procedure by Oberhagemann and el Moctar [4, 61]

This approach follows a similar principle as the method by Eça and Hoekstra. However, both methods differ in some details. A non-dimensional scalar refinement ratio, Υ, is defined as follows:

$$\begin{aligned}\Upsilon_i &= \sqrt{\frac{1}{4}\left[\left(\frac{\Delta x_i}{\Delta x_1}\right)^2 + \left(\frac{\Delta y_i}{\Delta y_1}\right)^2 + \left(\frac{\Delta z_i}{\Delta z_1}\right)^2 + \left(\frac{\Delta t_i}{\Delta t_1}\right)^2\right]} \\ &= \sqrt{\frac{1}{4}\left[\left(\frac{1}{r_x}\right)^{2(i-1)} + \left(\frac{1}{r_y}\right)^{2(i-1)} + \left(\frac{1}{r_z}\right)^{2(i-1)} + \left(\frac{1}{r_t}\right)^{2(i-1)}\right]}.\end{aligned} \tag{9.68}$$

Subscript 1 denotes a reference discretization (in space and time) which is coarsest (in contrast to Eça and Hoekstra). Index i is the level of refinement. r_x, r_y, r_z, and r_t are the refinement factors in x, y, z directions and in time, respectively.

Term $\Upsilon_{i\approx\infty} = 0$ corresponds to a grid representation with infinitely small grid spacing in all directions leading to the grid-independent solution.

The discretization error, δ, is defined as follows:

$$\epsilon \approx \delta = \phi_i - \phi_0 = a_1\Upsilon_i + a_2\Upsilon_i^2 + \cdots . \tag{9.69}$$

As at least one discretization solution is required for each unknown, at least three discretization levels are needed for a second-order approximation.

The least-squares minimization of Eça and Hoekstra [59] is applied to first (Eq. (9.70))- and second (Eq. (9.71))-order polynomial formulation of Eq. (9.69):

$$S_1(\phi_0, a) = \sqrt{\sum_{i=1}^{n_g} (\phi_i - (\phi_0 + a\Upsilon_i))^2} \tag{9.70}$$

$$S_2(\phi_0, a_1, a_2) = \sqrt{\sum_{i=1}^{n_g} (\phi_i - (\phi_0 + a_1\Upsilon_i + a_2\Upsilon_i^2))^2}, \tag{9.71}$$

where n_g denotes the number of refinement levels.

The authors recommend that simulations be carried out on spatial and temporal grid uniformly refined in all directions. In this case, the refinement factor is kept the same for all three spatial directions and for time. Hence, the Courant number (9.19) is supposed to be constant for each discretization level.

9.13.4.3 Procedure by Xing and Stern [60]

The authors developed a procedure known as the factors of safety (FS) method.

The observed order of approximation is obtained as follows:

$$p = \frac{\ln\left(\frac{\phi_3-\phi_2}{\phi_2-\phi_1}\right)}{\ln(r)} . \tag{9.72}$$

The fine grid is defined as the reference grid. Consequently, ϕ_1, ϕ_2, and ϕ_3 denote the fine, medium, and coarse grid solutions, respectively. Denoting grid spacing by Δx_i, the refinement factor, r, is written as follows:

$$r = \frac{\Delta x_2}{\Delta x_1} = \frac{\Delta x_3}{\Delta x_2} . \tag{9.73}$$

The discretization error of the solution on the fine grid is estimated as follows:

$$\epsilon \approx \delta = \frac{\phi_2 - \phi_1}{r^p - 1} . \tag{9.74}$$

However, the authors suggest multiplying δ with the ratio of convergence, P, here defined as follows:

$$P = \frac{p}{p_{th}}, \tag{9.75}$$

where p_{th} is the theoretical order of approximation, which is limited by the discretization schemes.

This method requires simulations performed on a set of three grids to obtain the sought-after errors and uncertainties.

The above-described procedures were applied and compared by Burmester et al. [56].

9.14 Application

In most applications related to motions of and loads on ship and offshore structures, methods based on potential theory are used. These methods are not only efficient in terms of computing time but also robust and, therefore, they are still the preferred choice in practice. However, under certain conditions, they have their limitations in terms of accuracy because highly nonlinear physical phenomena, such as green water, slamming in mixed two-phase flows, violent sloshing, and breaking waves, must be considered. These phenomena occur often in extreme seas. In such cases, field methods are then the preferred choice, which solve the Reynolds-averaged Navier-Stokes equations coupled with a Volume of Fluid (VOF) method, and the nonlinear ship motion equations. The literature includes procedures that combine boundary element methods, and field methods to calculate long-term values of ship responses, see el Moctar et al. [62]. The documented procedure was based on the combined use of a Boundary Element Method (BEM), a statistical theory to analyze random processes, and a field method to solve the Reynolds-averaged Navier-Stokes (RANS) equations. The BEM computed the ships Response (sectional loads, motions) Amplitude Operators (RAOs) needed for a probabilistic long-term analysis to identify relevant seaways. The extended RANS equations were coupled with nonlinear equations of motion to obtain ship motions and the corresponding pressure distribution acting on ships in the identified relevant seaways. Reliable predictions of short-term statistics of responses of ship responses in severe irregular seaways were estimated to show the best capability to assess the applied methods. We present results of computational methods to assess motions and sectional loads on ships in severe irregular waves. The numerical methods coupled the RANS equations in an implicit way with the nonlinear rigid body motion equations. The free surface was computed based on the VOF method. A Finite-Volume-Method discretized the solution domain, using hexahedral control volumes. The velocity and pressure fields were coupled based on the SIMPLE algorithm. Boundary conditions provided the appropriate time-dependent velocity field and free surface elevation at the inlet and a hydrostatic pressure at the outlet. At the inlet boundary, non-uniform velocity profiles were specified to define irregular waves based on linear wave theory (see Chap. 4). Details may be found in Ley [63]

9.14.1 Test Case Description

Table 9.2 lists principal particulars of the investigated cruise ship tested at a full-scale speed of 6 knots. Model tests at a scale of 1:50 were performed at the model basin CEHIPAR, Klein et al. [64]. Computations and model tests treated the ship hull as rigid. The ship model was self-propelled and tested in deep water in long-crested seaways having a JONSWAP spectrum, (see Chap. 4, Eq. (4.75)). Here, for the sake of simplicity, only long-crested head seaways ($\mu = 180$ deg) were investigated. Based on the JONSWAP spectrum formulation, these random sea states were generated by varying the peak enhancement factor, γ. Table 9.3 lists the parameters identifying the irregular sea states investigated for comparative purposes. In addition to ship responses, the incident wave elevation, ζ, at the ships' midship position was measured to assess the numerical prediction of wave propagation. This was of interest in sea conditions where significant nonlinearities of the wave process were expected. Furthermore, surface elevation monitoring enabled assessing the numerical damping of waves likely to occur in field method simulations.

H_s is the significant wave height, γ is the peak enhancement factor, T_p is the peak period, and T_d is the test duration. The wave steepness, s, is defined as $s = \frac{2\pi H_s}{gT_z^2}$. T_z is the zero-up-crossing period.

Table 9.2 Main particulars of the cruise ship

Length bet. perpendiculars [m]	216.80
Moulded breadth [m]	32.20
Design draft [m]	7.20
Block coefficient C_B [-]	0.65
Displacement [t]	34087
Mass moment of inertia (I_{xx}) [tm^2]	5.619E+06
Mass moment of inertia (I_{yy}) [tm^2]	1.003E+08
Metacentric height (GM) [m]	2.50
Longitudinal Centre of Gravity [m]	99.60
Vertical Centre of Gravity [m]	15.30

Table 9.3 Parameters of investigated irregular wave (full-scale)

H_s [m]	T_p [s]	γ	s [–]	T_d [s]
10.5	12.22	3.3	0.075	1600

9.14.2 Results

The computed free-surface elevation and the ship motions and sectional loads in irregular waves (see Table 9.3) were compared with experimental data [64]. Figure 9.9 (left) shows that free surface elevation compares closely to the measured signal, except for large values (e.g., at times of 575 and 990 s) with wave heights of about 15 and 20 m, respectively. Numerical dissipation of wave energy may have prevented missing these peaks. However, the midship vertical bending moment, Fig. 9.9 (right), is less affected. General agreement of the wave process appeared to be more important than exactly replicating the wave peak. Generally, hogging moment peaks (negative values) are underpredicted. Computed pitch, Fig. 9.10 (left), favorably matches experiments, but heave deviates randomly, Fig. 9.10 (right).

Figure 9.11 (left) shows exceedance rates (unit [s^{-1}]) of incident wave elevation, based on a time interval of 26.7 min in irregular waves (see Table 9.3). Computed free-surface elevation compare well with experiments. Figure 9.11 (right) shows exceedance rates of wave-induced vertical bending moments at midship. Except for large absolute values of bending moments, the numerical method yields a fair agreement with experiments and was able to predict properly the sagging-hogging asymmetry of bending moment amplitudes. The numerical method replicates the trend for moderate response levels. Figure 9.12 (left) and (right) show computed and measured exceedance rates of heave and pitch motions, respectively. The method obtains fair overall agreement with experiments.

To check the ability of the numerical method to predict the influence of the peak enhancement factor, γ, on midship bending moments, simulations were performed in sea states of significant wave height and peak period listed in Table 9.3, relevant for the long-term extreme values. The peak enhancement factor was varied from 1 to 6. Figure 9.13 shows exceedance rates of measured (left) and computed (right)

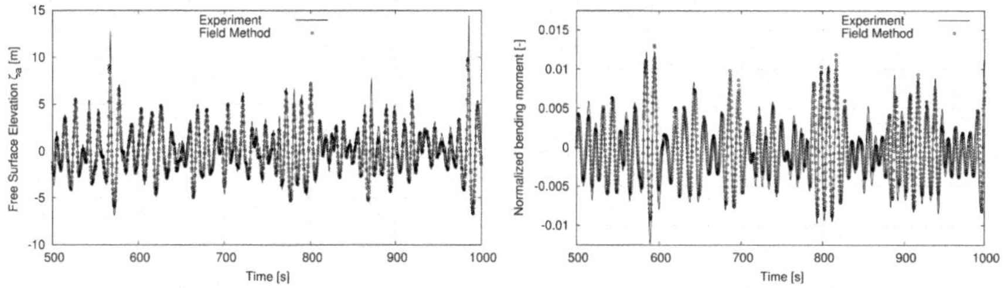

Fig. 9.9 Computed and measured time histories of wave elevation (left) and non-dimensional midship vertical bending moment M_y (right). M_y was made non-dimensional with $\rho B L_{PP}^2 H_s/2$

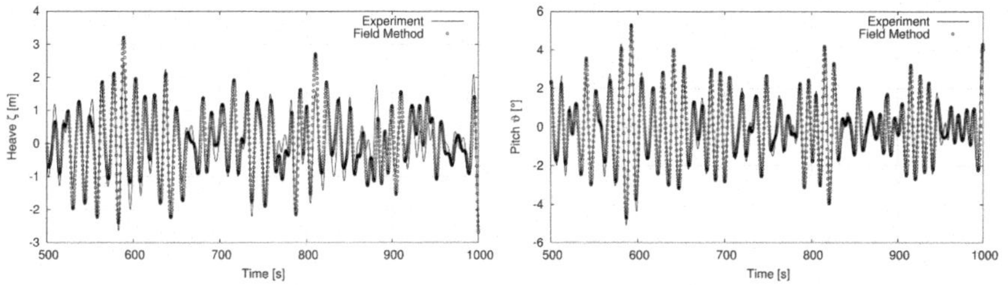

Fig. 9.10 Computed and measured time histories of pitch (left) and heave motions (right)

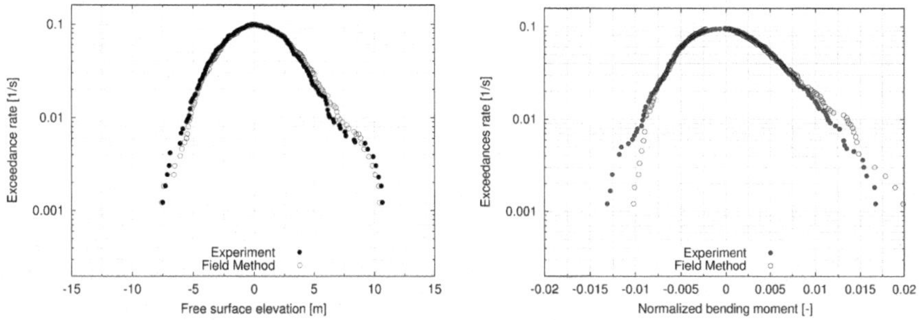

Fig. 9.11 Exceedance rate of wave elevation (left) and non-dimensional midship vertical bending moment (right)

midship vertical bending moment peaks for peak enhancement factors γ of 1.0, 3.3, and 6.0. The sampling time duration was here 5000 s. These results indicate a general trend of increasing bending moment peaks for the same exceedance rate as values of γ increase. This is more pronounced for extreme values. Experiments and computations show similar trends.

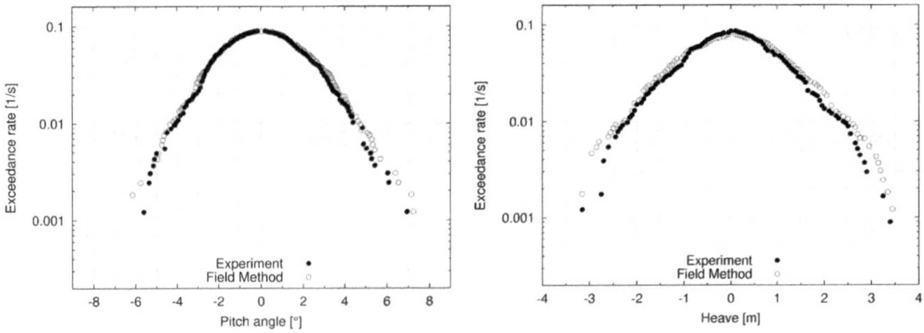

Fig. 9.12 Exceedance rate of pitch (left) and heave (right) motions

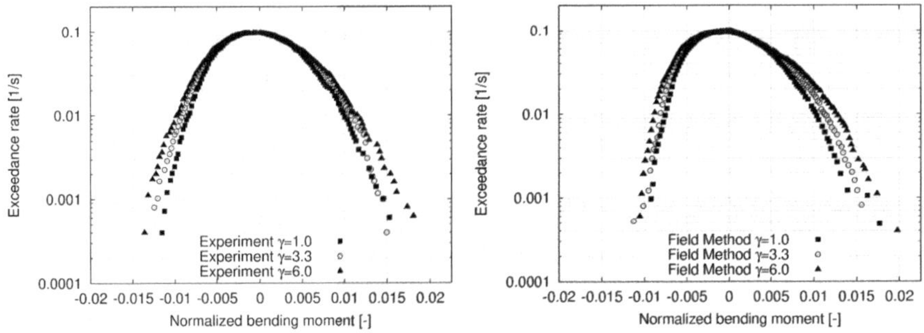

Fig. 9.13 Exceedance rate of measured (left) and computed (right) non-dimensional midship vertical bending moment for different peak enhancement factors γ

References

1. O. el Moctar, J. Ley, J. Oberhagemann, T.E. Schellin, Nonlinear computational methods for hydroelastic effects of ships in extreme seas. Ocean Eng. **130**, 659–673 (2017)
2. O. el Moctar, J. Oberhagemann, T.E. Schellin, Free surface RANS method for hull girder springing and whipping. SNAME Trans. **119**, 48–66 (2011)
3. S. Seng, J.J. Jensen, Slamming simulations in a conditional wave, in *Proceedings of the 31st ASME International Conference on Ocean, Offshore and Arctic Engineering, Rio de Janeiro* (2012)
4. J. Oberhagemann, On prediction of wave-induced loads and vibration of ship structures with finite volume fluid dynamic methods. Ph.D. thesis, University of Duisburg-Essen (2016)
5. J. Oberhagemann, J. Ley, O. el Moctar, Prediction of ship response statistics in severe sea conditions using RANSE, in *Proceedings of ASME 31st International Conference on Ocean, Offshore and Arctic Engineering, Rio de Janeiro* (2012), pp. 461–468
6. J. Ley, O. el Moctar, An enhanced 1-way coupling method to predict elastic global hull girder loads, in *Proceedings of ASME 2014 33rd International Conference on Ocean, Offshore and Arctic Engineering, OMAE, San Francisco* (2014)
7. J. Ley, O. el Moctar, J. Oberhagemann, T.E. Schellin, Assessment of loads and structural integrity of ships in extreme seas, in *Proceedings of the 30th Symposium on Naval Hydrodynamics* (2014)
8. C. Jiang, O. el Moctar, G.M. Paredes, T.E. Schellin, Validation of a dynamic mooring model coupled with a RANS solver. Mar. Struct. **72**, 102783 (2020)

9. J.H. Ferziger, M. Perić, *Computational Methods for Fluid Dynamics* (Springer, Berlin, 2008)
10. J.H. Ferziger, M. Perić, R.L. Street, *Computational Methods for Fluid Dynamics*, vol. 3 (Springer, Berlin, 2002)
11. J.F. Wendt, *Computational Fluid Dynamics: an Introduction* (Springer Science & Business Media, New York, 2008)
12. H. Lomax ANS, T.H. Pulliam, D.W. Zingg, *Fundamentals of Computational Fluid Dynamics* (Springer Science & Business Media, New York, 2013)
13. H.K. Versteeg, W. Malalasekera, *An Introduction to Computational Fluid Dynamics: the Finite Volume Method* (Pearson Education, Harlow, 2007)
14. M. Perić, Numerische Strömungsmechanik. Technical report, Hamburg University of Technology (1997). Lecture Notes
15. O. el Moctar, Numerical computations of flow forces on maneuvering ships (in German). Ph.D. thesis, Hamburg University of Technology (2001)
16. B.E. Launder, D.B. Spalding, The numerical computation of turbulent flows. Comput. Methods Appl. Mech. Eng. **3** (1974)
17. G.J. Yoo, B.C. Hwang, Calculation of developing turbulent flows in a rotating pipe. J. Turbomach. **113** (1991)
18. F.R. Menter, Two-equation eddy-viscosity turbulence models for engineering applications. AIAA J. **32**(8), 1598–1605 (1994)
19. C. Fureby, S.L. Toxopeus, M. Tormalm, K. Petterson, A computational study of the flow around the KVLCC2 model hull at straight ahead conditions and at drift. Ocean Eng. **118**, 1–16 (2016)
20. I. Shevchuk, Study of unsteady hydrodynamic effects in the ship stern area under shallow water conditions. Ph.D. thesis, University of Rostock (2016)
21. Z. Lilek, M. Perić, A fourth-order finite volume method with collocated variable arrangement. Comput. Fluids **24** (1995)
22. H. Jasak, Error analysis and estimation for the finite volume method with applications to fluid flows. Ph.D. thesis, University of London, Imperial College (2013)
23. I. Demirdžić, S. Muzaferija, Numerical method for coupled fluid flow, heat transfer and stress analysis using unstructured moving meshes with cells of arbitrary topology. Comput. Methods Appl. Mech. Eng. **125** (1995)
24. J.H. Ferziger, M. Perić, *Computational Methods for Fluid Dynamics* (Springer Science and Business Media, New York, 2012)
25. P.K. Kohsla, S.G. Rubin, A diagonally dominant second-order accurate implicit scheme. Comput. Fluids **2** (1974)
26. S. Muzaferija, Adaptive finite volume method for flow predictions using unstructured meshes and multigrid approach. Ph.D. thesis, University of London (1994)
27. P. Sames, Ein Beitrag zur Berechnung turbulenter Schiffsumströmungen. Ph.D. thesis, University of Hamburg, Institute for Naval Architecture (1996)
28. L.S. Caretto, A.D. Gosman, S.V. Patankar, D.B. Spalding, Two calculation procedures for steady, three-dimensional flows with recirculation, in *Proceedings of the 3rd International Conference on Numerical Methods in Fluid Mechanics* (Springer, 1973), pp. 60–68
29. R. Issa, Solution of the implicitly discretized fluid flow equations by operator splitting. J. Comput. Phys. **113** (1991)
30. H. Weller, Pressure-velocity solution algorithms for transient flows. Technical report, OpenCFD Ltd. (2005)
31. A. Peters, O. el Moctar, Numerical assessment of cavitation-induced erosion using a multi-scale Euler-Lagrange method. J. Fluid Mech. **894**, A19 (2020)
32. S. Muzaferija, M. Perić, I. Demirdžić, Numerical method for simulation of flow problems involving moving and sliding grids, in *Proceedings of the 7th International Symposium on Computational Fluid Dynamics* (1997)
33. O. el Moctar, A. Brehm, T.E. Schellin, Prediction of slamming loads for ship structural design using potential flow and RANSE codes, in *Proceedings of the 25th Symposium on Naval Hydrodynamics* (2004)

34. S. Muzaferija, M. Perić, Computation of free-surface using finite volume method and moving grids. Numer. Heat Transf. Part B **32**, 39–384 (1997)
35. H.G. Weller, A new approach to VOF-based interface capturing methods for incompressible and compressible flow. Technical report TR/HGW, OpenCFD Ltd. (2008)
36. H.G. Weller, Bounded explicit and implicit second-order schemes for scalar transport. Technical report, OpenCFD Ltd. (2006)
37. S.T. Zalesak, Fully multidimensional flux-corrected transport algorithms for fluids. J. Comput. Phys. **31**(3), 335–362 (1979)
38. S.M. Damian, An extended mixture model for the simultaneous treatment of short and long scale interfaces. Ph.D. thesis, Universidad Nacional Del Littoral (2013)
39. S. Muzaferija, M. Perić, Computation of free surface flows using interface-tracking and interface-capturing methods, in *Nonlinear Water Wave Interaction*, ed. by O. Mahrenboltz, M. Markiewicz (WIT Press, Southampton, 1999), pp. 59–100
40. H. Zhang, L.L. Zheng, V. Prasad, T.Y. Hou, A curvilinear level set formulation for highly deformable free surface problems with application to solidification. Numer. Heat Transf. Part B **34**(1), 1–30
41. M. Vogt, A. Cura-Hochbaum, Numerical simulation of ship motions in head waves with a RANSE method. Technical report 1649, Hamburg Ship Model Basin (HSVA)
42. J. Brunswig, O. el Moctar, Prediction of ship motions in waves using RANSE, in *Proceedings of the 5th Numerical Towing Tank Symposium, Hamburg* (2004)
43. M. Klemt, Bewegungssimulation schwimmender Körper in viskoser Strömung. Ph.D. thesis, Technical University Hamburg-Harburg (2004)
44. P. Ferrant, L. Gentaz, B. Alessandrini, D. Le Touze, A potential/RANSE approach for regular water wave diffraction about 2-D structures. Ship Technol. Res. **56**(4), 165–171 (2003)
45. G. Ducrozet, F. Bonnefoy, P. Ferrant, *HOS-Ocean: Open-Source Solver for Nonlinear Waves in Open Ocean Based on High-Order Spectral Method* (Elsevier, Amsterdam, 2016)
46. R. Perić, M. Abdel-Maksoud, Assessment of uncertainty due to wave reflections in experiments via numerical flow simulations, in *The Twenty-Fifth International Ocean and Polar Engineering Conference (ISOPE)*. International Society of Offshore and Polar Engineers (2015)
47. R. Perić, M. Abdel-Maksoud, Generation of free-surface waves by localized source terms in the continuity equation. Ocean Eng. **109**, 567–579 (2015)
48. R. Perić, M. Abdel-Maksoud, Reliable damping of free-surface waves in numerical simulations. Ship Technol. Res. **63**(1), 1–13 (2016)
49. R. Perić, M. Abdel-Maksoud, Analytical prediction of reflection coefficients for wave absorbing layers in flow simulations of regular free-surface waves. Ocean Eng. **147**, 132–147 (2018)
50. R. Perić, Lecture notes on practical computational fluid dynamics, University of Duisburg-Essen (2019)
51. R. Perić, M. Abdel-Maksoud, Analytical estimate of the reflection coefficient for forcing zones in 3D-flow simulations with nonlinear free-surface waves, in *Submitted Proceedings of the 33rd Symposium on Naval Hydrodynamics, Osaka, Japan* (2020)
52. B.E. Larsen, D.R. Fuhrman, On the over-production of turbulence beneath surface waves in Reynolds-averaged Navier-Stokes models. J. Fluid Mech. **853**, 419–460 (2018)
53. D.C. Wilcox, *Turbulence Modeling for CFD*, vol. 2 (DCW Industries Inc., La Canada, 1993)
54. C.W. Hirt, B.D. Nichols, Volume of fluid (VOF) method for the dynamics of free boundaries. J. Comput. Phys. **39**(1), 201–225 (1981)
55. O. Ubbink, R.I. Issa, Method for capturing sharp fluid interfaces on arbitrary meshes. J. Comput. Phys. **153**, 26–50 (1999)
56. S. Burmester, G. Vaz, O. el Moctar, Towards credible CFD simulations for floating offshore wind turbines. Ocean Eng. Accepted (2020)
57. W. Lyu, M. Riesner, A. Peters, O. el Moctar, A hybrid method for ship response coupled with sloshing in partially filled tanks. Mar. Struct. **67**, 102643 (2019)
58. ITTC, Uncertainty analysis in CFD-verification and validation methodology and procedures, *ITTC Recommended Procedures and Guidelines 7.5-03-01-01* (2017)

59. L. Eça, M. Hoekstra, A procedure for the estimation of the numerical uncertainty of CFD calculations based on grid refinement studies. J. Comput. Phys. **262**, 104–130 (2014)
60. T. Xing, F. Stern, Factors of safety for Richardson extrapolation. J. Fluids Eng. **132**(6) (2010)
61. J. Oberhagemann, O. el Moctar, An extrapolation method for discretization independent solution of uncertainty quantification procedure for time domain CFD computations. Ship Technol. Res. Submitted for Publication (2020)
62. O. el Moctar, T.E. Schellin, T. Priebe, CFD and FE methods to predict wave loads and ship structure response, in *Proceedings of the 26th Symposium on Naval Hydrodynamics* (2006)
63. J. Ley, Assessment of fluid-structure-interaction for ship and offshore structures in extreme waves. Ph.D. thesis, To be submitted to the University of Duisburg-Essen (2021)
64. M. Klein, A. Maron, G. Clauss, F. Miguel, Specifications of models tests. Technical report. EU Project Extreme Seas Del. 5.4, Technische Universität Berlin, Berlin (in German) (2012)

Chapter 10
Wave-Induced Hull Vibrations

Abstract In this chapter, we describe modern numerical methods used to compute global hull girder vibrations. We also discuss the causes and effects of such vibrations and the importance of the associated damping. In addition, we present sample applications.

10.1 Overview

Reliable predictions of loads acting on ships in extreme seas are required to assure the integrity of the ship structure. At least for larger ships, accurate predictions must include hull girder vibrations, as these lead to higher extreme loads and more load cycles. Hull vibrations increase also accelerations, which may lead to damage of cargo or even loss of containers stacked on deck. Especially on passenger ships and yachts, wave-induced (and other) vibrations should be avoided as much as possible as they impair the comfort of people on board.

In rigid bodies, the largest (linear) part of loads and stresses varies with the encounter frequency between ship and waves. Here, of main importance are wavelengths of the order of ship length; much shorter and longer waves cause only small rigid-body wave loads and stresses. For larger ships of the typical, moderate Froude numbers, the encounter periods exceed, usually, 8 s. In the corresponding frequency range, the influence of hull flexibility is small. The excitation by bending moments is slightly reduced because the vertical hull deformation is nearly in phase with the wave contour; the effect of this is partly canceled by the deformation accelerations. Altogether, the effect of hull flexibility on loads and stresses is usually negligible in the frequency range below 1/8 Hz in large ships.

In a higher frequency range, however, the flexibility of ship hulls is important for wave loads and stresses, and it dominates wave-induced vertical accelerations. This is caused by the low damping of hull vibrations, which results in strong resonance peaks around the eigenfrequencies of the elastic ship hull. For large ships, the lowest vibration eigenperiod is, typically, not far from 2 s. Thus, there is a large gap in frequency between rigid body loads and elastic hull vibrations. That allows to separate both contributions by filtering measured time tracks.

B. O. el Moctar et al., *Numerical Methods for Seakeeping Problems*,
https://doi.org/10.1007/978-3-030-62561-0_10

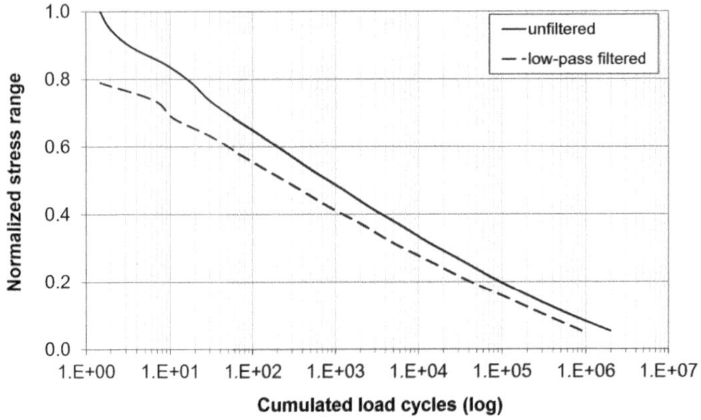

Fig. 10.1 Cumulative distribution of total (unfiltered) and rigid body (low-pass filtered) non-dimensional stress ranges measured on board a Panamax containership (Kahl and Menzel, [1])

The time function of wave loads or stresses in an elastic hull is, thus, a stochastic process spanning a wide frequency range. In such a case, the effect on fatigue damage or fatigue crack propagation cannot be described correctly by the distribution of stress amplitudes, or of ranges between a maximum and the following minimum of the stress. Instead, distribution of stress ranges resulting from a range-pair count must be used. That means that the difference between the largest maximum and the lowest minimum of the total time track constitutes the largest range. Within the time interval between two already paired extremes, again the largest maximum and the lowest minimum are paired, and so forth until each maximum is paired to a minimum.

Figure 10.1 shows the cumulative frequency distribution of range-pair counted stress ranges measured on a Panamax containership [1] over about a year. The broken curve was derived from the low-pass filtered stress record using a limit frequency well below the lowest eigenfrequency of the hull; thus it contains mostly rigid body stresses. For the continuous curve, the limit frequency was much higher, so that stresses caused by hull vibrations are included. Obviously, the vibration contributes substantially to both the fatigue and to the total extreme loads.

The amplitude of wave-induced hull vibrations, and of the stresses caused by them, increases with ship length and with speed, especially in head waves. The lowest eigenfrequency belongs to the vertical two-node bending mode, in which hull cross-sections move in opposite direction between the midship section and at the ship's ends (top in Fig. 10.3). Head waves, the encounter frequency of which is near $\omega_1 = 2\pi f_1$, or $\frac{1}{2}\omega_1$, $\frac{1}{3}\omega_1$, ... for second- and third-order ... excitation, often excite this so-called springing vibration continuously. The accompanying loads add on to the loads which would occur on a rigid hull. In this case, the waves which excite the vibration are short compared to ship length: for a large ship at 20 knots in head waves, first-order excitation at the lowest hull eigenfrequency is caused by waves of, typically, length $L_{pp}/10$, whereas second-order and third-order excitation requires wavelengths of, roughly, $L_{pp}/4$ and $L_{pp}/2$, respectively. Higher order excitation per

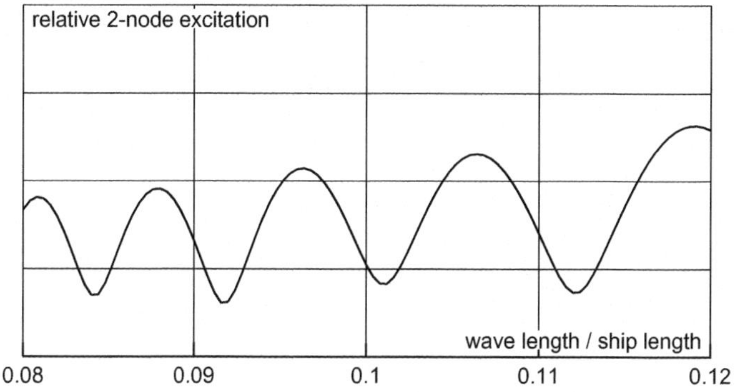

Fig. 10.2 Two-node vertical vibration excitation in regular head waves depending on wavelength. The indicated encounter angular frequencies ω_e correspond to a ship of 300 m length at 20 knots speed

wave height is weaker than the first-order excitation per wave height; however, the second-order or third-order excitation may be substantial or even dominant because of the larger wavelength and correspondingly larger amplitude of the exciting waves.

Another feature complicates especially predictions of first-order excited hull vibrations: the variation of vibration excitation with wavelength. Especially for the two-node mode, the excitation by regular waves results mainly from pressure oscillations at bow and stern, where the eigenform has larger absolute values than elsewhere. Depending on wavelength, bow and stern excitation may be in phase or of opposite phase, which results in oscillations of total excitation over wavelength (Fig. 10.2). In a natural, irregular seaway, substantial amplitudes of the two-node hull vibration are caused only by those wave contributions the encounter frequency of which is close to the corresponding eigenfrequency. If, in the case plotted in Fig. 10.2, the eigenfrequency ω_1 is 3.46 rad/s, for the waves causing resonance the excitation and thus the vibration amplitude are small; for frequencies 3.35 and 3.60 rad/s, on the other hand, excitation and amplitude are larger. Because the eigenfrequency varies with the loading case, both maximum and minimum excitation will occur in practice, even for a single ship in a given seaway. Thus, if predictions of springing amplitudes use only a single eigenfrequency for a given mode, the result may be far from reality. If that simplification cannot be avoided to reduce the number of investigated cases, it is recommended to adjust the eigenfrequency to a nearby value of maximum excitation and to regard the smaller amplitudes for slightly different frequencies as a safety allowance.

But also much longer (and consequently higher) waves may generate hull vibrations following slamming events. Slam-induced hull vibrations are called whipping.

Most often, slamming occurs when the raised bow impacts the water surface. The slam is followed by vertical hull girder vibrations, which gradually fade out. In sharp ships, bow flare slamming is typical, which occurs when the upper part of the flaring bow sections enters the water, generating a spectacular 'spray curtain'. In full ships, on the other hand, bottom slamming occurs when a flat forward region of the ship's bottom, which was raised out of the water, hits the water surface. Also, stern slamming can occur, especially in containerships and yachts which have a broad, nearly flat stern overhang.

In both springing and whipping, the two-node vertical vibration mode is most important; other modes appear less important for maximum and fatigue loads. Springing occurs in low to moderate seas, whereas whipping requires moderate to high waves. Correspondingly, the more frequent springing contributes significantly to the fatigue loads, whereas whipping increases substantially extreme loads.

Predicting springing and whipping and the combined rigid body and vibration-induced loads requires applying numerical methods for fluid flow, for rigid body motions, and for hull deformation. Wave-induced hull deformations are calculated by a linear finite-element (FE) method, using either a three-dimensional model or a one-dimensional (beam) model of the ship structure. To predict linearly excited springing, a linear method for fluid flow can be used to determine transfer functions of springing motion, loads, and stresses. For nonlinearly excited springing and for whipping, simulations are required, in which the fluid flow computation includes at least nonlinear corrections of the otherwise linear flow and ship motion calculations. An alternative is to use fully nonlinear methods for flow and motion analysis, e.g., those described in previous chapters. For finite-volume methods, the required computer time is higher for elastic than for rigid body calculations because the fluid grid must be finer to predict springing. And/or the time step must be smaller to predict whipping.

When using a Rankine source potential flow method for springing, for short waves the free-surface panel mesh can be omitted if the exact free-surface condition is substituted by its limit for high frequency $\phi = 0$. This condition is satisfied simply by using a mirror sink above the water surface for each source below the water surface. The lower limit of non-dimensional frequency, above which this limit is sufficiently accurate, is not yet known. If this simplification is used, an additional vibration damping by waves radiated by the vibrating hull must be added, because the limit condition $\phi = 0$ eliminates radiated waves.

For usual merchant ship shapes, it is important to take account of the steady stern wave, because vibration damping and added mass (in case of springing also excitation) are sensitive to the submergence of the overhanging hull region above the rudder. If the steady deformation of the water surface is taken into account, the above-described use of mirror sinks to satisfy the condition $\phi = 0$ is not exact, but appears sufficiently accurate.

If the fluid flow is modeled by Euler or Navier-Stokes equations, it proved best to alternate in the outer iteration within each time step between improving the fluid flow and the body motion. The body motion part comprises the rigid-body motion, the elastic deformation, and the adaptation of the fluid grid.

10.2 Modeling Stiffness and Mass

For some ships, detailed three-dimensional FE models of the steel structure are generated for strength and vibration analyses. When such a model shall be used for wave-excited vibrations, it appears advantageous to reduce the numerous degrees of freedom by applying a modal method, i.e., by superimposing the deformation from that of a small set of eigenvectors. Both dry modes (computed neglecting fluid effects) and wet modes may be used. The latter requires additional effort to compute the added mass terms in the mass matrix, but they allow to proceed then with fewer modes; often it will suffice to use, besides rigid body motions, only one deformation mode: the two-node vertical bending.

More often the hull structure is modeled as a straight beam. Here, the difficulty is to determine correct approximations of shear area and bending moment of inertia. Neglecting shear deformations, i.e., modeling the hull as a Bernoulli beam, is too inaccurate at least for three-node and higher modes of vibrations. Thus, a Timoshenko beam, which models bending and shear deformations, is usually applied for vertical bending. Horizontal bending vibrations are of less practical interest because a ship hull is, usually, much stiffer for horizontal than for vertical vibrations and because horizontal wave excitation is weaker. Torsional vibrations are of practical interest only in ships with large deck openings, especially containerships. In these ships, there is strong coupling with horizontal vibrations. In our view, coupled horizontal and torsional vibrations of open-deck ships cannot be adequately modeled using the Wlassov theory for torsion of thin-walled prismatic beams, but require a three-dimensional FE model. Longitudinal vibrations are usually omitted because their eigenfrequencies are high and their coupling with transverse and vertical bending is small.

In the following, a Timoshenko beam model of the ship structure for vertical vibrations is described based on [2].

10.2.1 Finite-Element Discretization

The ship structure is subdivided in x (longitudinal) direction into finite elements. In each element, the hull girder is approximated as a prismatic structure. The continuous hull deformation $\mathbf{u}(x, t)$ is represented numerically by (a) the vertical shift and (b) the rotation of the cross section about the transverse (y) axis, at the FE nodes, i.e., the ends of the finite elements. The complex amplitudes of (a) and (b) for all nodes are combined to the deformation vector $\hat{\mathbf{u}}$. Then, the deformation equation is

$$(\underline{\mathbf{K}} + i\omega_e\underline{\mathbf{D}} - \omega_e^2\underline{\mathbf{M}})\hat{\mathbf{u}} = \hat{\underline{\mathbf{F}}}, \tag{10.1}$$

where $\underline{\mathbf{K}}$, $\underline{\mathbf{D}}$ and $\underline{\mathbf{M}}$ are the stiffness, damping, and mass matrices, respectively, and $\hat{\underline{\mathbf{F}}}$ is the excitation vector.

10.2.2 Use of Approximate Modes

The matrices $\underline{\mathbf{K}}$, $\underline{\mathbf{D}}$, and $\underline{\mathbf{M}}$ depend on the mode shape. For instance, in $\underline{\mathbf{K}}$ the shear lag is responsible for a reduction of the bending stiffness depending on the mode. (Shear lag is the decrease of longitudinal stresses in the flanges of a bent girder with distance from the web.) And in $\underline{\mathbf{M}}$, the added masses decrease with decreasing distance between vibration nodes. Thus, in the following the quantities $\underline{\mathbf{K}}$, $\underline{\mathbf{D}}$, and $\underline{\mathbf{M}}$ are written with an index j to identify a certain mode. To take account of this dependence on mode shape, a modal approach is required when modeling the structure as a beam. That means that the deformation $\hat{\mathbf{u}}$ is superimposed from real mode vectors $\mathbf{w}_j$:

$$\hat{\mathbf{u}} = \sum_{j=1}^{J} \mathbf{w}_j \hat{v}_j. \tag{10.2}$$

where J is the number of modes considered. $\mathbf{w}_j$ are real vectors, the components of which are the translations and rotations at all nodes for mode j, and $\hat{v}_j$ are complex scalars, i.e., superposition factors of the different modes.

Using (10.2), the motion Eq. (10.1) becomes

$$\sum_{j=1}^{J} \underbrace{(\underline{\mathbf{K}}_j + i\omega_e \underline{\mathbf{D}}_j - \omega_e^2 \underline{\mathbf{M}}_j)}_{-\underline{\mathbf{V}}_j} \mathbf{w}_j \hat{v}_j = \hat{\underline{\mathbf{F}}}. \tag{10.3}$$

If J is smaller than the number of degrees-of-freedom of the FE system, the linear system for the unknown amplitudes $\hat{v}_j$ is over-determined. To obtain a system of equations containing as many equations as unknowns, and to approximately decouple these equations, the Galerkin method is used; i.e., (10.3) is left-multiplied by $\mathbf{w}_k^T$, where T indicates the transpose, and k is another mode index:

$$\sum_{j=1}^{J} (\mathbf{w}_k^T \underline{\mathbf{K}}_j \mathbf{w}_j + i\omega_e \mathbf{w}_k^T \underline{\mathbf{D}}_j \mathbf{w}_j - \omega_e^2 \mathbf{w}_k^T \underline{\mathbf{M}}_j \mathbf{w}_j) \hat{v}_j = \mathbf{w}_k^T \hat{\underline{\mathbf{F}}}, \quad k = 1 \ldots J. \tag{10.4}$$

For fixed k and j, the expression $\mathbf{w}_k^T \underline{\mathbf{K}}_j \mathbf{w}_j$ is a real scalar: the virtual work produced by the stiffness force $\underline{\mathbf{K}}_j \mathbf{w}_j$ acting along the virtual deformation $\mathbf{w}_k$. For all possible values of k and j, these scalars are combined to the modal stiffness matrix $\mathbf{K}$ of size $J \times J$:

$$\mathbf{K} = (\mathbf{w}_k^T \underline{\mathbf{K}}_j \mathbf{w}_j, \quad k = 1 \ldots J, \quad j = 1 \ldots J). \tag{10.5}$$

Correspondingly, modal stiffness matrices for damping ($\mathbf{D}$) and mass ($\mathbf{M}$) are defined as well as the modal excitation vector

$$\hat{\mathbf{F}} = (\mathbf{w}_k^T \hat{\underline{\mathbf{F}}}, \quad k = 1 \ldots J)^T. \tag{10.6}$$

Table 10.1 Main particulars of the containership used as example

L_{pp}	304 m	Draft at AP	13.69 m
Breadth	42.8 m	Draft at FP	12.56 m
Total mass	115 500t	Investigated speed	20 knots

This transforms (10.4) into the matrix equation

$$\underbrace{(\mathbf{K} + i\omega_e \mathbf{D} - \omega_e^2 \mathbf{M})}_{-\mathbf{V}} \hat{\mathbf{v}} = \hat{\mathbf{F}}, \qquad (10.7)$$

where the matrices are square $J \times J$ matrices, and the column vectors $\hat{\mathbf{v}}$ and $\hat{\mathbf{F}}$ have J components.

If $\mathbf{w}_j$ were the exact mode shapes, and if $\underline{\mathbf{K}}_j$ and $\underline{\mathbf{M}}_j$ were the same for all modes, then $\mathbf{K}$ and $\mathbf{M}$ would be diagonal matrices. Because of using estimated mode vectors $\mathbf{w}_j$, this condition is not satisfied. To determine exact mode vectors would require to iteratively change the mode vectors $\mathbf{w}_j$ and the matrix V resulting from them. It appears easier to avoid this iteration by keeping the approximate mode vectors unchanged.

The $\mathbf{w}_j$ may be determined as 'dry modes' neglecting added mass and possibly shear lag, or one may determine them as eigenvectors of the 'vibration matrix' $\underline{\mathbf{V}}_j$ given in Eq. (10.2), where j corresponds to the most important two-node eigenform, neglecting the damping. For the ship described in Table 10.1, the first six approximate mode shapes are shown in Fig. 10.3, together with the shapes of wave-excited

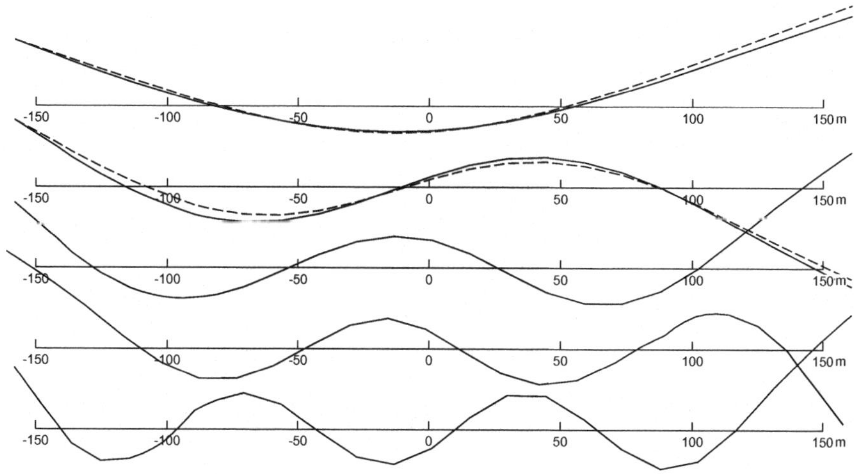

Fig. 10.3 Mode approximations (continuous) and deformations at resonance peaks for $\mu = 180°$ (broken lines)

vibrations at resonance. Small differences between approximate and correct mode shapes allow reducing the number J of approximate modes required to accurately model the actually interesting modes, especially the two-node mode.

10.2.3 Mass and Stiffness Matrix

At any cross section, the bending stiffness of the hull is characterized by EI, where E is Young's modulus, and I is the moment of inertia of the cross section of the longitudinal members around their cross-sectional center of gravity. At least for higher modes, I must be reduced for the shear lag. The shear stiffness is GA_s, where G is the shear modulus, and A_s is the effective web area for vertical shear. The non-dimensional ratio κ between bending and shear rigidity is defined as

$$\kappa = \frac{12EI}{GA_s l_E^2}, \tag{10.8}$$

where l_E is the length of the finite element.

According to [3], the stiffness matrix $\underline{\mathbf{K}}_E$ of a finite element of a Timoshenko beam follows from its bending and shear rigidity as

$$\underline{\mathbf{K}}_E = \frac{EI}{(1+\kappa)l_E^3}\left[\begin{pmatrix} 12 & -6l_E & -12 & -6l_E \\ & 4l_E^2 & 6l_E & 2l_E^2 \\ & & 12 & 6l_E \\ & & & 4l_E^2 \end{pmatrix} + \kappa\begin{pmatrix} 0 & 0 & 0 & 0 \\ & 1l_E^2 & 0 & -1l_E^2 \\ & & 0 & 0 \\ & & & 1l_E^2 \end{pmatrix}\right] \tag{10.9}$$

Here and in the following, the lower left part of the symmetrical matrices is omitted. The four rows of the element matrices correspond to displacement and then rotation at the left end (smaller x), and then displacement and rotation at the right end of the element.

The mass matrix of a Timoshenko beam finite element depends on the mass per length μ and on the mass moment of inertia about the transverse axis y of the cross section per length $\bar{\mu}$:

$$\underline{\mathbf{M}}_E = \frac{1}{(1+\kappa)^2}\left[\frac{\mu l_E}{420}\begin{pmatrix} 156 & -22l_E & 54 & 13l_E \\ & 4l_E^2 & -13l_E & -3l_E^2 \\ & & 156 & 22l_E \\ & & & 4l_E^2 \end{pmatrix} + \frac{\kappa\mu l_E}{120}\begin{pmatrix} 84 & -11l_E & 36 & 9l_E \\ & 2l_E^2 & -9l_E & -2l_E^2 \\ & & 84 & 11l_E \\ & & & 2l_E^2 \end{pmatrix}\right.$$

$$\left. + \frac{\kappa^2\mu l_E}{120}\begin{pmatrix} 40 & -5l_E & 20 & 5l_E \\ & 1l_E^2 & -5l_E & -1l_E^2 \\ & & 40 & 5l_E \\ & & & 1l_E^2 \end{pmatrix} + \frac{\bar{\mu}}{30l_E}\begin{pmatrix} 36 & -3l_E & -36 & -3l_E \\ & 4l_E^2 & 3l_E & -1l_E^2 \\ & & 36 & 3l_E \\ & & & 4l_E^2 \end{pmatrix}\right.$$

$$
\left. + \frac{\kappa\bar{\mu}}{6l_E} \begin{pmatrix} 0 & 3l_E & 0 & 3l_E \\ & 1l_E^2 & -3l_E & -1l_E^2 \\ & & 0 & -3l_E \\ & & & 1l_E^2 \end{pmatrix} + \frac{\kappa^2\bar{\mu}}{6l_E} \begin{pmatrix} 0 & 0 & 0 & 0 \\ & 2l_E^2 & 0 & 1l_E^2 \\ & & 0 & 0 \\ & & & 2l_E^2 \end{pmatrix} \right] . \tag{10.10}
$$

The formula results from assuming
(a) that μ and $\bar{\mu}$ are constant over x within the finite element and
(b) that the x dependence of shift and rotation of cross sections are coupled to each other as in a static deformation.
Assumption (b) is uncritical, but assumption (a) is violated that the above formula appears overly complicated.

The total structural stiffness matrix $\underline{\mathbf{K}}$ and the total mass matrix $\underline{\mathbf{M}}$ are constructed from the above element matrices as usual in FE methods. From these 'non-modal' matrices, the modal matrices $\mathbf{K}$ and $\mathbf{M}$ are determined using (10.5), and its mass matrix analogously. However, further contributions to the modal vibration matrix $\mathbf{V}$ are required.

10.2.4 Other Contributions

Several other effects contribute to the 'vibration matrices' $\underline{\mathbf{V}}_j$ and $\mathbf{V}$. In the Following, they are only described qualitatively; details to their computation are given, among others, in [2].

The static pressure increase with depth generates a positive contribution to the stiffness matrix $\mathbf{K}$. Another contribution is caused by weight: If the mass center of gravity of a ship section is located above or below the neutral axis of the ship's longitudinal members, the weight force together with the rotation of the cross section causes a (small) moment which decreases or increases, respectively, the stiffness.

Added masses must be included in μ. They are determined using potential flow methods. Accurate results may be computed by a linear three-dimensional Rankine source method using the approximate modal deformation $\mathbf{w}_j$ and the condition $\phi = 0$ at the free surface. If the two-dimensional flow around heaving ship sections is used to compute the added mass per length, a reduction factor must be applied that depends on the mode shape. Rather than to apply reduction factors from literature, we recommend computing the added mass per length for a horizontal, partly immersed cylinder performing vertical vibrations of small amplitude with a sinusoidal mode shape. The cross section of the cylinder should correspond to that of the actual ship section. The computation corresponds closely to a two-dimensional computation with one exception. Instead of constant source strength over longitudinal coordinate x, the source strength should vary as $\cos(kx)$, where k is the wave number of the vibration. The potential of such a source line is given by the Hankel function of the second kind of order zero.

10.3 Vibration Damping

Vibrations $z(t)$ of a mechanical system having one degree of freedom may be characterized by the equation of free motions

$$kz + d\dot{z} + m\ddot{z} = 0, \tag{10.11}$$

where k is stiffness (negative force per deviation from equilibrium), d is damping (negative force per deviation velocity), and m is (physical and added) mass. Inserting the ansatz

$$z(t) = \text{Re}(\hat{z}e^{i\omega t}) \tag{10.12}$$

gives (for $\hat{z} \neq 0$)

$$k + i\omega d - \omega^2 m = 0, \tag{10.13}$$

the solution of which is

$$\omega = \frac{id}{2m} \pm \underbrace{\sqrt{-\frac{d^2}{4m^2} + \frac{k}{m}}}_{\omega_r}. \tag{10.14}$$

Critical damping, d_c, is defined as the damping resulting in $\omega_r = \text{Re}(\omega) = 0$:

$$d_c = 2\sqrt{km}. \tag{10.15}$$

In case of smaller actual damping, oscillations of declining amplitude occur, as may be shown by inserting the complex ω of (10.14) into (10.12):

$$z(t) = \text{Re}(\hat{z}e^{i\omega_r t})e^{-\frac{d}{2m}t}. \tag{10.16}$$

Thus, successive maxima z_1 and z_2 of $z(t)$ are related as

$$\frac{z_2}{z_1} = e^{-\frac{d}{2m}\frac{2\pi}{\omega_r}} \approx e^{-2\pi d/d_c}, \tag{10.17}$$

where the approximation is valid for small d. Equation (10.17) is used to determine the nondimensional damping ratio

$$D = \frac{d}{d_c} = \frac{1}{2\pi}\ln\left(\frac{z_1}{z_2}\right) \tag{10.18}$$

from a time track of free vibrations.

Because resonant ship vibrations are nearly uncoupled to vibrations at other eigenfrequencies or to the rigid-body motions, they behave similarly to a one-degree-of-freedom system. Thus, their damping can be quantified by the damping ratio D, which can be determined from measured vibration decay.

Damping ratios of gross hull vibrations are, typically, in the range of 1–5%. For instance, Price [4] documents vibratory whipping-induced response measurements of a frigate (p. 296) and of the vessel 'City of Plymouth' (p. 381). The decay of these vibrations indicates damping ratios of about 0.9 and 1.2%, respectively. For the vessel of Table 10.1, damping ratios between 1.76 and 5.08% were found [5] using the same method. (On ships without forward speed in sheltered waters, much smaller damping ratios are found, but these are irrelevant here.)

Vibration damping of ship hull girders is the topic of ship dynamics, which is, apparently, least understood, and erroneous conclusions are published frequently. Often only 'structural damping' is discussed. However, listening to the sound of a tuning fork or a piano string demonstrates that the material damping of steel is extremely small if the stress amplitudes are far below the yield stress; thus, the ship's steel structure does not contribute noticeably to hull vibration damping. The following is an attempt to quantify roughly other physical processes that contribute to the observed damping. It is based on [2]. The topic requires more attention and theoretical, numerical, and experimental investigations before a sound prediction method for hull vibration damping can evolve.

10.3.1 Transom Damping

If an immersed transom moves vertically, it generates a vertical momentum in the surrounding fluid. According to slender-body theory, the momentum remains in the water when the water separates smoothly from the hull at the transom contour. For a vibrating ship, the change in periodic vertical momentum between the inflow (ahead of the ship, excluding waves) and the outflow corresponds to a vertical oscillatory force acting on the hull that induces damping. The vibration energy of the hull is transferred to kinetic energy of the oscillatory fluid motion behind the transom. A slender-body approximation of the virtual work produced by this force is

$$\mathrm{Im}(V_{jjT}) = U\omega_j w_{jT}^2 \mu_{zT}, \tag{10.19}$$

where U = ship speed, w_{jT} = vertical component of the eigenform at the transom, ω_j = eigenfrequency of mode j, and μ_{zT} = two-dimensional added mass of the transom cross section in the vertical motion. For springing frequencies, μ_{zT} is well approximated by the mass of water in a semicircle of diameter B_T, the waterline breadth at the transom:

$$\mu_{zT} = \rho\pi B_T^2/8. \tag{10.20}$$

Equations (10.19) and (10.20) show that transom vibration damping is proportional to ship speed, thus it cannot be measured in a ship at quay. And it is proportional to B_T^2; thus, it is sensitive to the stern wave produced by ship speed, and it is expected to vary, even in moderate waves, between zero (if the transom is emerged) and substantial values.

From (10.19) follows the transom contribution to the damping ratio of mode j as

$$D_{jT} = \frac{\mathrm{Im}(V_{jjT})}{2\omega_j^2 \int_L w_{jz}^2 \mu \, dx} \tag{10.21}$$

where μ is the sum per unit length of ship mass and added mass for vertical motion as functions of x. A corresponding equation holds for the other damping contributions.

10.3.2 Wave Radiation Damping

In case of linearization with respect to vibration amplitude, the potential ϕ of the flow around a vibrating ship satisfies the boundary condition at the free surface

$$g\phi_z = \left(-U\frac{\partial}{\partial x} + i\omega\right)^2 \phi. \tag{10.22}$$

The condition cannot be satisfied in three-dimensional flow, even for the lowest hull eigenfrequency, because it requires a fine panel grid to resolve the short waves generated by the vibration. Thus, it is replaced by its limit for $\omega \to \infty$, i.e. $\phi = 0$. Because this condition does not allow the generation of waves, the damping caused by wave radiation cannot be determined directly.

For the two-dimensional flow around oscillating ship sections, however, condition (10.22) (with ship speed $U = 0$) can be satisfied easily and accurately for the frequencies of interest here, for instance, by the method described in Chap. 3, using a very fine discretization along the waterline. An improvement would be to assume a periodical variation along the x-axis using a Hankel function instead of the logarithm as a source potential. The computation gives d, the vertical damping force amplitude per length, divided by the vertical velocity amplitude. It was found that d becomes large if the flare angle α between the vertical and the section contour tangent at the waterline approaches 90°. d can be approximated as

$$d = \frac{\rho g \sqrt{A} \tan\alpha}{0.75\omega}, \tag{10.23}$$

where A = section area below the waterline. The formula shows that vibration damping by wave radiation is generated nearly only at the stern, where the hull part above propeller and rudder may have sections of extreme flare at the waterline. Also here, the damping is very sensitive to stern submergence; thus, it depends on the steady stern wave (and therefore on speed U) and changes during the passage of longer waves along the hull.

The contribution of the total hull to the vibration matrix is

$$\mathrm{Im}(V_{jjW}) = \omega_j \int_L w_{jz}^2 d\, dx. \tag{10.24}$$

In the relevant literature, the extreme dependence of wave radiation damping on flare angle α is, at least usually, not noticed. This explains that this kind of damping is erroneously held to be negligible.

10.3.3 Other Causes of Damping

In [2], other causes of vibration damping were roughly estimated:

- Bilge keels. In a natural seaway, bilge keels may cause flow separation in case of upward motion of a hull cross section relative to the surrounding water. Additional to the larger, low-frequency, wave-induced, rigid body motions, there are small-amplitude, high-frequency vibratory motions. The part of the vertical force per length, which oscillates with the vibration frequency, may be estimated as

$$f = \rho C_d B |v_{rel1}| v_2. \tag{10.25}$$

  Here, B is ship breadth, v_{rel1} is the relative vertical velocity between water (orbital velocity) and cross section (heave and pitch) excited by longer waves acting on the rigid hull, and v_2 is the vibration velocity of the cross section. The resistance coefficient C_d was, tentatively, estimated as 0.8 for sections with bilge keels.
- Propeller. Its vibratory vertical velocity produces, together with the inflow speed, an oblique inflow. (Compared to this, the vibration-induced tilt of the propeller shaft is small.) The vertical force produced by this is opposite to the velocity; it is estimated from several sources as approximately

$$f = 0.25 \rho n_P D P^2 \frac{A_E}{A_0} v_2, \tag{10.26}$$

  where n_P = rotation rate, D = diameter, P = pitch, A_E/A_0 = expanded blade area ratio.
- Hatch cover friction produced by combined vibratory and low-frequency hull deformations.

10.3.4 Results

For the ship the main particulars of which are given in Table 10.1, vibration damping was estimated using the above relations. The ship was assumed sailing at service

Table 10.2 Contribution to the damping ratio D in percent for the ship of Table 10.1 due to different damping processes

Number of vibration nodes	2	3	4	5	6	7
Eigenfrequency [Hz]	0.505	1.040	1.578	2.067	2.636	3.061
Transom	1.850	0.910	0.580	0.580	0.300	0.210
Wave radiation	0.400	0.075	0.028	0.016	0.007	0.004
Bilge keels	0.080	0.034	0.027	0.021	0.022	0.016
Propeller	0.030	0.022	0.006	0.004	0.001	0.001
Hatch cover friction	0.006	0.003	0.002	0.001	0.001	0.001
Sum	2.366	1.044	0.643	0.622	0.331	0.232

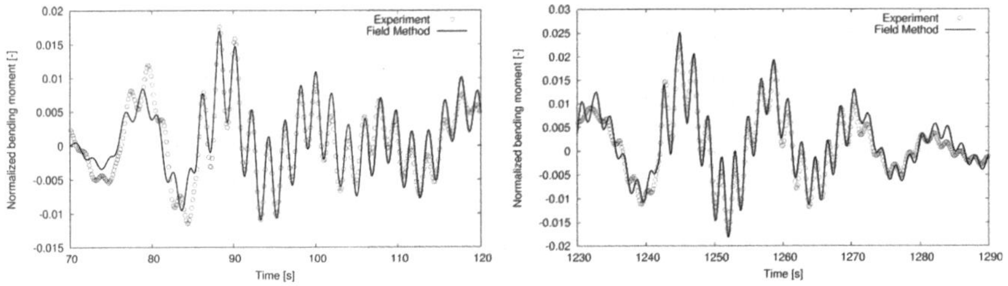

Fig. 10.4 Computed and measured time histories of non-dimensional midship vertical bending moment for different time intervals

speed in a moderate seaway. Table 10.2 shows the results. For the two-node vibration mode, the estimated total damping ratio (2.4%) is well within the range (1.75–5.08%) found in measurements on that ship during service [5]. In this example, substantial contributions to vibration damping are only due to the transom effect and wave radiation; however, in cases of ships without an immersed transom in a heavy seaway, bilge keel and even propeller damping were found to be substantial.

The table demonstrates also a decreasing total damping ratio for the higher modes, which is roughly proportional to the inverse eigenfrequency. That means that in a given time interval, vibration amplitudes of all modes decrease, approximately, by the same factor. This contradicts, however, vibration measurements of ships in service, where whipping vibrations of higher modes seem to fade out after a slam much faster than the two-node vibration (see Fig. 10.4). Thus, it seems that there is still another, unknown damping mechanism the effect of which increases with frequency. It was speculated repeatedly that in bulkers the cargo may cause substantial vibration damping, as well as the shock absorbers of cars, trucks, and wagons in ferries, ropax vessels, etc. However, no quantitative studies of these effects are known to the authors.

10.4 Comparison Between Computed and Measured Loads

Computed sectional loads on a large containership in high irregular long-crested waves were compared with the results of model tests. The spectral energy distribution $S(\omega)$ was chosen according to the JONSWAP formulation (see Chap. 4). Table 10.3 lists principal particulars of the ship and the seaway parameters. The investigations were carried out at a full-scale speed of 15 knots. Model tests, using a scale of 1:80, were performed at the model basin CEHIPAR, Klein et al. [6]. The self-propelled ship model consisted of four segments, which were fixed to an aluminum backbone that resulted in the correct vibration modes and natural frequencies.

The computations were performed using a Reynolds-averaged Navier-Stokes (RANS) solver coupled (in an implicit way: two-way coupling) with equations of rigid body motions and equations for structural deformations based on Timoshenko beam theory. The basic concept of the method is to split the ship motions into rigid body motions and elastic deformations. The rigid body motions can be of large amplitude and are, consequently, formulated nonlinearly, while the elastic deformations are of small amplitude. The latter are linearized and computed using a Timoshenko beam to model the ship structure. The advantages of this approach are its simplicity and the few required input data to establish a sufficient structural representation. As a drawback compared to a three-dimensional structure model, only the global structural vibration responses can be obtained, and of these only planar (typically vertical) bending modes. The numerical grid for RANS simulations was morphed to account for rigid body motions and structural deformations. A detailed description of the numerical method can be found in el Moctar et al. [7], Oberhagemann [8], and Ley [9].

Computed and measured normalized midship vertical bending moment s reveal significant effects of whipping following bow flare slamming events, Fig. 10.4. Experiments and computations agree fairly well. Computed and measured vibration amplitudes match favorably immediately after the impacts, and the vibration frequency is well reproduced numerically.

Figure 10.5 shows the number of exceedances of range-pair counted ranges of the non-dimensional midship vertical bending moment both for the original time tracks and after low-pass filtering with a cut-off frequency of 0.25 Hz to remove the vibratory loads. The difference between the filtered and unfiltered results shows the hull elasticity effects on loads. There is a significant difference between filtered and unfiltered signals. First, the number of response ranges increases dramatically due to hull girder vibrations. Second, the maximum response range obtained in the unfiltered signal is about 50% larger than the maximum response in the filtered signal.

Table 10.3 Principal particulars of the containership [9] and the seaway parameters

Ship data:	L_{pp} 333.44 m	Breadth 42.80 m	Draft 13.10 m	Block coeff. 0.655
Seaway data:	sign. height 12.5 m	Peak period 11.8 s	Peak enhancem. 5.0	Duration 1400 s

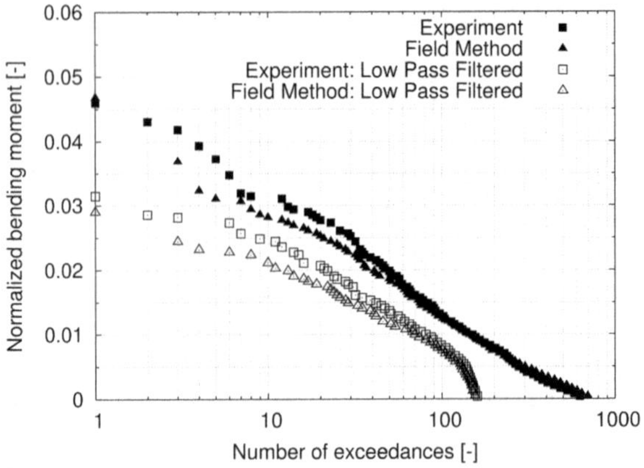

Fig. 10.5 Number of exceedances of range-pair counted ranges of the non-dimensional vertical midship bending moment exceeding the limit value given on the abscissa from model experiments and computations, with and without low-pass filtering (time duration 1400 s)

Numerically and experimentally determined number of exceedances show a similar increase in vibration numbers. Numerical results underpredict the responses at a low number of exceedances. In general, numerical predictions of bending moments in the elastic body compare favorably to experimental results.

References

1. A. Kahl, W. Menzel, Full-scale measurements on a PanMax containership, in *Ship Repair Technology Symposium Proceedings, Newcastle upon Tyne* (2008), pp. 59–66
2. H. Söding, Computation of springing transfer functions. J. Eng. Marit. Environ. JEME 158 (2010)
3. G. Jensen, H. Söding, Ship vibrations (in German). Technical report, Technical university Hamburg-Harburg (2003)
4. W.G. Price, The dynamics of ships (1991)
5. J. Kaufmann, V. Shigunov, Know-how springing. Technical report no. 2009.035, Germanischer Lloyd (2009)
6. M. Klein, A. Maron, G. Clauss, F. Miguel, Specifications of models tests. Technical report EU Project Extreme Seas Del. 5.4, Technische Universität Berlin, Berlin (in German) (2012)
7. O. el Moctar, J. Ley, J. Oberhagemann, T.E. Schellin, Nonlinear computational methods for hydroelastic effects of ships in extreme seas. Ocean Eng. **130**, 659–673 (2017)
8. J. Oberhagemann, On prediction of wave-induced loads and vibration of ship structures with finite volume fluid dynamic methods. Ph.D. thesis, University of Duisburg-Essen (2016)
9. J. Ley, Assessment of fluid-structure-interaction for ship and offshore structures in extreme waves. Ph.D. thesis, To be submitted to the University of Duisburg-Essen (2021)

Chapter 11
Additional Forces and Moments

Abstract Whereas previous chapters dealt with the water flow and the corresponding pressure force and moment on the hull, this chapter describes approximations of other influences. Especially the small moments in excitation, restoring, and damping of roll motions require taking account of additional influences, e.g., fins, of the lift effect of hull and rudder in oblique flow, of the roll damping by bilge keels, and of the control forces required to keep the ship's intended speed and course both in reality and during simulations. Also, recommendations are derived for the required control forces to attain stable motions not overly influencing the free motions. Often neglected effects are included, e.g., the influence of the stern wave on metacentric height, which may amount to about 1 m at 15 knots speed.

Previous chapters about computing motion transfer functions considered in detail only instationary forces caused by fluid pressure. Here, other time-harmonic forces, which may influence body motions in waves, are considered as well as force corrections applicable to different computational methods. Often, the given formulae are linearized with respect to motion and wave amplitudes. In special cases improved nonlinear formulas may be necessary to attain the required accuracy.

In this chapter, the inertial coordinate system x, y, z follows the mean forward motion of the ship. Its origin is in height of the undisturbed water surface. Coordinate axes x, y, z point forward, to starboard, and downward, respectively. As in previous chapters, time-harmonic quantities will be characterized by complex amplitudes designated by the hat symbol. The real part of complex amplitudes is the value occurring at time $t = 0$ (when a wave trough is at the coordinate origin) and the imaginary part a quarter encounter period earlier (normally) or later (in case of negative encounter frequency ω_e, i.e., if the ship overtakes the waves from behind).

Forces and moments will be given in this chapter as they result from wave and body motions. To include them in the motion equation, they are separated into
(a) contributions of the 6 unit body motions (these contribute to the complex added-mass matrix) and
(b) contributions of the wave motion (they influence the exciting force and moment on the right-hand side of the motion equation system).

B. O. el Moctar et al., *Numerical Methods for Seakeeping Problems*,
https://doi.org/10.1007/978-3-030-62561-0_11

11.1 Weight

The ship's weight gm causes no oscillatory force, but an oscillatory moment

$$\mathbf{M}_w = \mathbf{x}_G \times (0, 0, gm) = gm(y_G, -x_G, 0), \tag{11.1}$$

where $\mathbf{x}_G$ is the mass center of gravity. $\mathbf{x}_G$ results from its ship-fixed position $\underline{\mathbf{x}}_G$:

$$\mathbf{x}_G = \mathbf{T}\underline{\mathbf{x}}_G + \mathbf{u}, \tag{11.2}$$

where $\mathbf{u}$ is the translation and $\mathbf{T}$ the matrix representing the ship rotations given in (2.33) and, linearized, in (5.33). (In Chap. 8, the weight moment was eliminated by computing all moments as turning around the point $\mathbf{x}_G$.)

11.2 Roll Restoring Moment

The restoring moment M_1 about the longitudinal axis is the sum of contributions from hull pressure and weight. Usually, it is characterized by the metacentric height $\overline{GM}$ defined by the relation

$$\frac{dM_1}{d\varphi} = gm\overline{GM}, \tag{11.3}$$

where φ is the static heel angle (around the x axis). This equation will be used here as a definition of $\overline{GM}$.

$\overline{GM}$ determines the roll eigenfrequency. Because of the small damping of roll motions for single-hull ships, the roll transfer function has a significant peak at the roll eigenfrequency. Thus, an accurate determination of the metacentric height is important for obtaining correct roll motions. On the other hand, $\overline{GM}$ is small compared to the ship dimensions breadth and draft. Thus, inaccuracies in computing GM of, say, 0.1 m may influence roll amplitudes substantially.

Additionally to typical textbooks of naval architecture, we will consider here also the squat (change of draft and trim due to forward speed), the steady ship waves, and the steady pressure distribution around the hull. This refers especially to Chaps. 7 and 8.

In the steady flow determination described, for instance, in Chap. 7, the squat is taken into account by modifying the positions $\mathbf{x}$ of the nodes of the hull patches. These patches are used to determine the buoyancy force and moment. To remain consistent, also the vertical coordinate of the mass center of gravity must be changed for squat; without this, the metacentric height would be underestimated.

Compared to typical discretizations in computer-aided design, hydrodynamic hull panel meshes are coarser to avoid excessively large systems of equations for determining the source strengths. Nonetheless, the hull volume within the panel mesh must approximate the real hull volume within the curved hull surface accurately to minimize errors of $\overline{GM}$ computed from the panel mesh. Therefore, panel corners should be located a little outside of the hull in convex hull regions, such that the panel includes approximately the same volume as the real, curved hull surface. This is easily attained by shifting the panels from an initial, approximate location to their final, improved position. For triangular panels, the midpoints of the improved panel sides should be on the hull surface, because these midpoints are supports for the three-point Gauß integration rule over a triangular region.

Both the squat and the steady ship waves result in a waterline that differs from that of the ship at rest. For the containership of Fig. 7.1, the transom is not immersed for the ship at rest, but is immersed to nearly full breadth at 15 knots speed ahead. This causes an increase of metacentric height of about 1 m ($L_{pp}=$ 286 m).

In Chap. 7, the static disturbance potential ϕ^0 and the accompanied pressure distribution p^0 is assumed to follow the ship motions. Thus, the roll moment vector generated by the pressure p^0 acting on the hull patches remains constant in body-fixed coordinates, and its longitudinal component remains also constant in the inertial system in case of heel (rotation about the $\underline{x}$ axis). However, the lower (due to heel) side of the ship, an additional area of the hull will be immersed, and on the other side, an area will emerge from the water. This will produce a moment proportional to the heel angle φ. Also, the static pressure $\rho g z$ will produce a heel moment due to immersing and emerging areas along the waterline. Because the sum $p^0 + \rho g z$ is zero at the waterline, both contributions cancel in the limit of small φ. Thus, for determining $\overline{GM}$ by (11.3), one must consider the moment due to the static pressure change by heel, $\rho g \varphi y$, only by adding the contributions on the (body-fixed) patches; there is no influence of the change of the immersed hull area.

The moment around the x-axis due to the static pressure change is, therefore (in the limit $\varphi \to 0$):

$$M_1 = \rho g \varphi \sum \int y(\mathbf{x} \times \mathbf{n})_1 \, da, \tag{11.4}$$

where the sum is taken over all body patches, and the integration covers the area a of a single patch. $\mathbf{n}$ is the unit normal vector on the patch pointing into the body. The index 1 designates the x-component. Combining (11.3) and (11.4) gives:

$$\overline{GM} = \frac{\rho}{m} \sum \int y(\mathbf{x} \times \mathbf{n})_1 \, da. \tag{11.5}$$

For integrations over the panel area, usually the integrand at the midpoint of the patch is multiplied by the panel area. Here, however, we deal with zeroth-order quantities, which require a more accurate treatment than first-order quantities. Because the integrand depends quadratically on y, the midpoint rule is only an approximation. The following integration formula gives the exact result for a plane triangular patch:

$$\int y(\mathbf{x} \times \mathbf{n}) \, da = \sum_{i=1}^{3} y_i \left(\frac{1}{4}\mathbf{x}_m + \frac{1}{12}\mathbf{x}_i \right) \times \mathbf{a}, \tag{11.6}$$

where **a** is the patch area vector: normal on the patch, directed into the body, magnitude = patch area. $\mathbf{x}_m$ is the midpoint of the patch, $\mathbf{x}_i$ and y_i refer to corner point i.

11.3 Additional Roll Damping

To consider roll damping independent of sway motion, using instead a 'roll axis', is not expedient, because the height of the roll axis results from the motion calculation (from coupled sway, roll, and possibly yaw motion) and is initially not known accurately. Nonetheless, most publications on roll damping consider the roll motion independent of the other degrees of freedom. Therefore, this section describes how the axis of free roll motions is determined and how a known additional scalar roll damping (determined, e.g., by Ikeda's method [1]) is introduced into the motion equation. It would not be correct to modify only the element $B_{4,4}$ of matrix **B**, because that refers to a rotation about the x-axis, not the roll axis. Further, the additional roll damping may not contain contributions contained already in the matrix **B** of the motion Eq. (7.84), like wave radiation damping or the foil effect of the hull.

Except for a special coordinate system, the free roll motion is a combination of translations of the body-fixed coordinate origin and rotations about the coordinate axes. Most important is the coupling of roll and sway motion; less important than with the yaw motion; and in case of linearization and a symmetrical body including its mass distribution, there is no coupling with the surge, heave, and pitch motion. Nonetheless, in the following coupled motions of all six degrees of motions are considered because the result can easily be simplified if less degrees of freedom are coupled with roll.

Free motions of a rigid floating body (i.e., in the absence of excitation, starting from a non-equilibrium position) obey, in case of linearization, to the equation

$$[-\omega_e^2(\mathbf{M}+\mathbf{A})+\mathbf{S}]\hat{\mathbf{u}}=0, \tag{11.7}$$

where **M** is the mass matrix, **A** is the complex added mass matrix, **S** is restoring matrix (all 6 by 6), and $\hat{\mathbf{u}}$ is the complex motion (translation and rotation) amplitude vector interpreted as a column vector. The equation corresponds to (5.39) if the excitation is set to zero and $\omega_e^2\mathbf{A}=\mathbf{B}$. The negative imaginary part of **A** constitutes damping.

Equation (11.7) constitutes a generalized complex, possibly non-symmetrical eigenvalue problem. Its solution consists of six complex eigenvalues ω_j^2 and six eigenvectors $\hat{\mathbf{u}}_j$, $j=1$ to 6 of arbitrary magnitude. Of the six eigenvalues, usually, three will be zero; they correspond to the free surge, sway, and yaw motion, which lack a restoring force or moment. Of the j values with non-zero eigenvalues ω_j, that for which $|\hat{u}_{4,j}|/|\hat{\mathbf{u}}_j|$ (4 designates the fourth component) is maximum, designates the roll motion. In the following, that j is designated as r (for roll).

The vector equation (11.7) is changed to scalar equations by left multiplication by any eigenvector $\hat{\mathbf{u}}_k$ transposed to a row vector:

$$-\omega_j^2 \hat{\mathbf{u}}_k^T (\mathbf{M}+\mathbf{A})\hat{\mathbf{u}}_j + \hat{\mathbf{u}}_k^T \mathbf{S}\hat{\mathbf{u}}_j = 0. \tag{11.8}$$

The eigenvectors are $(\mathbf{M}+\mathbf{A})$-orthogonal; that means: For all $j \neq k$, there is

$$\hat{\mathbf{u}}_k^T (\mathbf{M}+\mathbf{A})\hat{\mathbf{u}}_j = 0 \tag{11.9}$$

and

$$\hat{\mathbf{u}}_k^T \mathbf{S}\hat{\mathbf{u}}_j = 0. \tag{11.10}$$

After these preliminaries, (11.7) is now complemented by adding a damping term (additionally to the damping contained in **A**) to model the additional roll damping:

$$[-\omega_e^2(\mathbf{M}+\mathbf{A}) + i\omega_e\mathbf{D} + \mathbf{S}]\hat{\mathbf{u}}_j = 0. \tag{11.11}$$

The term (11.8) changes to

$$-\omega_j^2 \hat{\mathbf{u}}_k^T (\mathbf{M}+\mathbf{A})\hat{\mathbf{u}}_j + i\omega_j \hat{\mathbf{u}}_k^T \mathbf{D}\hat{\mathbf{u}}_j + \hat{\mathbf{u}}_k^T \mathbf{S}\hat{\mathbf{u}}_j = 0. \tag{11.12}$$

D should change only the roll motion $j = r$; free motions in the other five degrees of freedom should be unchanged. This is attained if

$$\hat{\mathbf{u}}_k^T \mathbf{D}\hat{\mathbf{u}}_j = 0 \tag{11.13}$$

for $k = 1$ to 6 and $j = 1$ to 6 except for the case $k = j = r$. The case $k = j = r$ describes the free roll motion:

$$-\omega_r^2 \underbrace{\hat{\mathbf{u}}_r^T (\mathbf{M}+\mathbf{A})\hat{\mathbf{u}}_r}_{m_r} + i\omega_r \underbrace{\hat{\mathbf{u}}_r^T \mathbf{D}\hat{\mathbf{u}}_r}_{d_r} + \underbrace{\hat{\mathbf{u}}_r^T \mathbf{S}\hat{\mathbf{u}}_r}_{s_r} = 0. \tag{11.14}$$

This and the following is correct, independent of the scaling of the eigenvectors by arbitrary complex factors. However, the terms m_r, d_r, and s_r can be interpreted as roll inertia, additional damping, and restoring constants expressed in the usual measuring units only if the eigenvector $\hat{\mathbf{u}}_r$ is scaled such that its fourth component is 1.

If the additional roll damping is given by its damping ratio D_0 (= actual/critical damping of free roll motions), d_r in (11.14) follows from the usual relation for systems with one degree of freedom:

$$d_r = \hat{\mathbf{u}}_r^T \mathbf{D}\hat{\mathbf{u}}_r = 2\sqrt{m_r s_r} D_0. \tag{11.15}$$

The additional roll damping matrix **D** shall be determined such as to satisfy the 35 scalar equations (11.13) (for $j, k = 1$ to 6 except $k = j = r$) and the 1 scalar equation (11.15). To this end, the modal 6 by 6 matrix

$$\mathbf{U} = (\hat{\mathbf{u}}_1, \hat{\mathbf{u}}_2, \hat{\mathbf{u}}_3, \hat{\mathbf{u}}_4, \hat{\mathbf{u}}_5, \hat{\mathbf{u}}_6) \tag{11.16}$$

is introduced. Using **U**, the scalar equations (11.13) and (11.15) can be combined to a single matrix equation:

$$\mathbf{U}^T\mathbf{D}\mathbf{U} = \mathbf{V}, \tag{11.17}$$

where **V** is a 6 by 6 matrix having elements 0 everywhere except for element $V_{r,r}$, which is d_r. Thus, **D** can be determined as

$$\mathbf{D} = (\mathbf{U}^{-1})^T\mathbf{V}\mathbf{U}^{-1}, \tag{11.18}$$

where the relation $(\mathbf{U}^T)^{-1} = (\mathbf{U}^{-1})^T$ is used. Because only the element r, r of **V** is $\neq 0$, this can be simplified to

$$D_{k,j} = (\mathbf{U}^{-1})_{r,k}(\mathbf{U}^{-1})_{r,j}d_r \tag{11.19}$$

for k and j from 1 to 6. This matrix **D** must be used in (11.11) to represent the additional roll damping.

For a symmetric body, the eigenvector $\hat{\mathbf{u}}_r$ defines the roll axis in case of small roll amplitude. The body is assumed to perform free roll motions according to the roll eigenvector

$$\mathbf{u}_r = (\underbrace{u_1, u_2, u_3}_{\mathbf{t}}, \underbrace{u_4, u_5, u_6}_{\boldsymbol{\alpha}}). \tag{11.20}$$

The shift **v** (in the inertial frame) of a body-fixed point at location $\mathbf{x} = (x, y, z)$ due to this motion is

$$\mathbf{v} = \mathbf{t} + \boldsymbol{\alpha} \times \mathbf{x}, \tag{11.21}$$

where translation **t** and rotation $\boldsymbol{\alpha}$ are defined in (11.20). The y component of **v** is

$$v_2 = t_2 + \alpha_3 x - \alpha_1 z. \tag{11.22}$$

For points on the roll axis, $v_2 = 0$. This gives the z coordinate of the roll axis as a function of x:

$$z(x) = \frac{u_2 + u_6 x}{u_4}. \tag{11.23}$$

$\mathbf{u}_r$ is a complex eigenvector; thus $z(x)$ is complex. Because roll damping is small, the imaginary part of $z(x)$ will be extremely small and should be omitted; it indicates that the roll axis will vary a little during a roll period.

11.4 Additional Surge Damping

Contrary to roll damping, an additional surge damping cannot be specified as a percentage of the critical surge damping, because there is no restoring force for surge; thus, the quantity s_r in the equivalent of (11.15) for surge motion would be zero. Therefore, an additional surge damping is specified as a negative longitudinal force per increment of speed u:

$$d_{surge} = \frac{d}{du}[R(u) - (1-t)T(u)], \tag{11.24}$$

where R is resistance, T propeller thrust, and t is thrust deduction fraction. $T(u)$ must be determined for a realistic change in the rate of rotation n of the propeller with changing torque due to changing u; e.g., in model experiments, n remains approximately constant. In any case, T will decrease with increasing speed u.

A further difference between surge and roll is that surge is only slightly coupled to other motions. Thus, it appears sufficient to let the damping matrix **D** unchanged except for element 1, 1, which is increased by d_{surge}.

11.5 Fins

11.5.1 Preliminaries

Here, approximate formulae for the force produced by fixed (to the body) or movable foils like rudders, roll damping fins, etc., are presented. Their shape is characterized by a mean chord length c (measured in mean flow direction; Fig. 11.1) and a span s measured in the y, z-plane. For foils of non-rectangular planform, c is defined as foil area divided by span; then foil forces based on c and s are nearly independent of the shape of the planform. Only the fin influence on motion transfer functions is discussed; thus, the relatively small drag force and nonlinear lift forces (especially stall) will be neglected.

A fin generates waves and vortices. Its influence on other fins and on the hull forces is also neglected. That may be important if one fin operates behind another one, e.g., a rudder behind the keel fin of a sailing yacht.

The lift force of a fin is assumed to act at $\mathbf{x}_F$. Normally, a good guess of the pressure center is at half span, $\frac{1}{4}c$ behind the profile nose. To estimate the lift of a fin, an angle of attack is determined at the reference point $\mathbf{x}_{F1}$ located $\frac{1}{2}c$ behind $\mathbf{x}_F$. Added mass forces generated by a fin, on the other hand, act more or less at the fin area center $\mathbf{x}_{F2}$, and they depend on the acceleration of that point.

The direction of the axis of a movable fin is characterized by the unit vector $\mathbf{a}_F$. Its x-component a_1 is neglected; thus, $\mathbf{a}$ is in the yz plane. Fin rotation angles δ_F

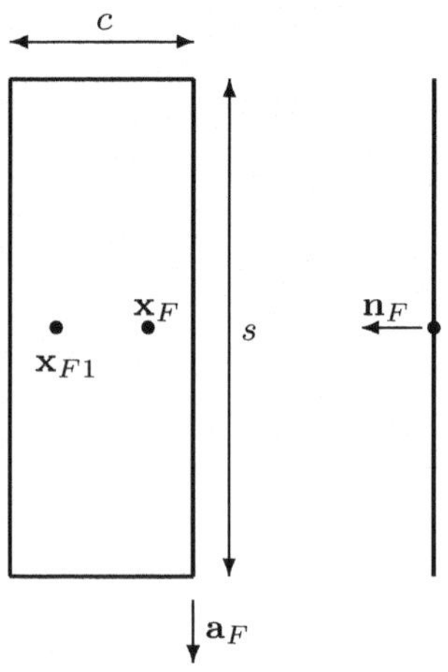

Fig. 11.1 Definition of fin quantities: fin seen from starboard (left) and from behind (right)

are regarded as positive if they are turning right-hand about **a**. They are assumed as proportional to body motions:

$$\hat{\delta}_F = c_{F1}\hat{u}_1 + \cdots + c_{F6}\hat{u}_6 = \mathbf{c}_F \cdot \hat{\mathbf{u}}. \tag{11.25}$$

Here, $\hat{\mathbf{u}}$ is the 6-component vector of ship translation and rotation amplitudes, and $\mathbf{c}_F$ is a complex vector modeling the steering of the fin by a PD controller: The real parts of $\mathbf{c}_F$ cause (positive or negative) restoring forces and the imaginary parts (positive or negative) damping forces. Usually, most components of $\mathbf{c}_F$ are zero; for roll damping fins, e.g., only the imaginary part of c_{F4} is non-zero.

Let $\mathbf{n}_F$ be the unit vector in the y, z-plane and normal to the fin center plane (Fig. 11.1):

$$\mathbf{n}_F = (1, 0, 0) \times \mathbf{a}_F. \tag{11.26}$$

Then the periodic fin force has direction $\pm\mathbf{n}_F$, and it is proportional to the relative (between fluid and fin) flow speed or acceleration in direction $\mathbf{n}_F$.

In deep water, an undisturbed incident wave of complex amplitude 1 generates a flow speed in direction $\mathbf{n}_F$ of amplitude

$$\hat{v}_{FW} = \frac{\omega}{ik}\, \boldsymbol{\nu} \cdot \mathbf{n}_F\, e^{\boldsymbol{\nu} \cdot \mathbf{x}_{F1}}, \tag{11.27}$$

where

$$\boldsymbol{\nu} = ik(-\cos\mu, \sin\mu, i), \tag{11.28}$$

and μ is wave direction (0 for following wave, 90° for wave from starboard). The corresponding formula for shallow water (bottom at $z = z_B$) is

$$\hat{v}_{FW} = \frac{\omega}{\sinh(kz_B)} e^{i(-kx_{F1}\cos\mu + ky_F \sin\mu)} \mathbf{n}_F \begin{pmatrix} 0 \\ \sin\mu \cosh(k(z_F - z_B)) \\ -i \sinh(k(z_F - z_B)). \end{pmatrix} \quad (11.29)$$

The complex amplitude of the fin velocity in direction $\mathbf{n}_F$ due to ship motions relative to the inertial coordinate system is

$$\hat{v}_{FS1} = i\omega_e(\hat{\mathbf{t}} + \hat{\boldsymbol{\alpha}} \times \mathbf{x}_{F1}) \cdot \mathbf{n}_F. \quad (11.30)$$

Here, $\hat{\mathbf{t}}$ is the complex amplitude of the translational motion of the ship-fixed coordinate origin, and $\hat{\boldsymbol{\alpha}}$ that of the (linearized) ship rotation.

An additional motion at the reference point $\mathbf{x}_{F1}$ is caused by the active fin rotation $\hat{\delta}_F$. If l is the longitudinal distance between $\mathbf{x}_{F1}$ and the fin axis (positive if the axis is in front of $\mathbf{x}_{F1}$), the additional velocity component in direction $\mathbf{n}_F$ is

$$\hat{v}_{FS2} = i\omega_e l \hat{\delta}_F. \quad (11.31)$$

In many cases this contribution will be small; however, for so-called zero-speed fins, this contribution is essential if there is no substantial forward speed.

The presence of the hull is assumed to change the orbital velocity by a factor C_r to

$$C_r \hat{v}_{FW}. \quad (11.32)$$

C_r must be estimated or computed separately for the case in question. It is about 1 for fins at a sufficient distance from the hull; it is >1 for fins attached to the hull and pointing to the sides or downward, and it is nearly zero for fins attached at the aft end of the hull if they do not reach out of the hull's wake. The same factor C_r is applied to the velocity $\hat{v}_{FS1}$ due to the ship motions to take account of the hull influence on $\hat{v}_{FS1}$.

11.5.2 Force Due to Accelerations

The added mass of a fin for accelerations in direction $\mathbf{n}_F$ is determined as

$$m_F = C_M \frac{\pi}{4} \rho c^2 s. \quad (11.33)$$

C_M is ≈ 1 for $s/c \gg 1$, and $C_M \approx \Lambda$ for $s/c \ll 1$. The aspect ratio is defined here as:

$$\Lambda = \left\{ \begin{pmatrix} 2s/c & \text{if the fin is attached to the hull without gap,} \\ s/c & \text{if the flow around both ends of the span is not restricted.} \end{pmatrix} \right. \quad (11.34)$$

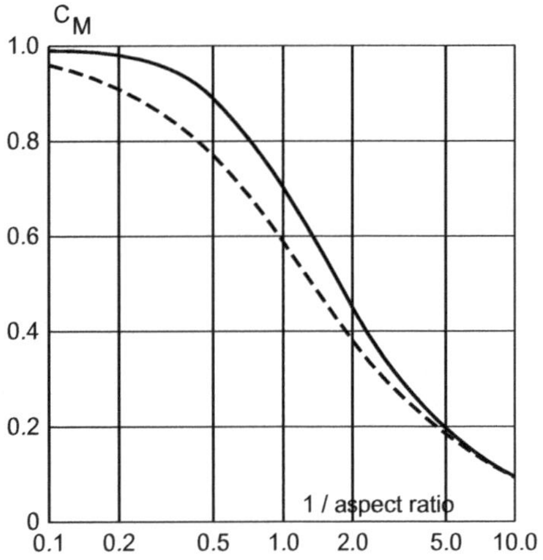

Fig. 11.2 Added mass coefficient C_M depending on the *inverse* aspect ratio Λ. Full line for elliptical, broken line for rectangular planform. Data from Yu [2]

If the planform is not rectangular, use A/s instead of c. Figure 11.2 may be used to estimate C_M.

The complex amplitude of the acceleration due to orbital (wave) motion and ship motion is $i\omega C_r \hat{v}_{FW}$ and $i\omega_e C_r \hat{v}_{FS}$, respectively. Thus, the added-mass force at the fin due to the wave is

$$\hat{F}_{FW} = m_F i\omega C_r \hat{v}_{FW}, \tag{11.35}$$

and due to the ship motion

$$\hat{F}_{FS} = -m_F i\omega_e (C_r \hat{v}_{FS1} + \hat{v}_{FS2}). \tag{11.36}$$

11.5.3 Force Due to Velocity and Angle of Attack

The following estimation of the fin force due to (flow and fin) velocity is largely based on [3]. The paper deals with fin forces in a steady flow; corrections for time-harmonic motions are discussed below.

The complex amplitude of the lift force in direction $\mathbf{n}_F$ due to the time-varying angle of attack α_F of the fin is

$$\hat{L} = \frac{\rho}{2} v^2 s c \hat{C}_L = v^2 b_F \hat{\alpha}_F. \tag{11.37}$$

Here, $\hat{\alpha}_F$ is the complex amplitude of the angle of attack, and

$$b_F = \frac{\rho}{2} s c \frac{\mathrm{d}C_L}{\mathrm{d}\alpha_F} \tag{11.38}$$

is the lift coefficient assumed proportional to the angle of attack.

To improve the accuracy for larger angles of attack, the lift coefficient may be approximated as

$$C_L = \frac{\mathrm{d}C_L}{\mathrm{d}\alpha_F}\alpha_F + C_d\alpha_F|\alpha_F|. \tag{11.39}$$

C_d is a coefficient of fin resistance in transverse flow. For fins with sharp ends of the span, $C_d = 1$ is a good choice. As the second term depends nonlinearly on the angle of attack, an equivalent linearization based on the estimated real amplitude $|\alpha_{FE}|$ of the angle of attack (in the real wave, not for wave amplitude 1) may be used. Thus, instead of (11.38), the following formula may be used alternatively:

$$b_F = \frac{\rho}{2}sc\left(\frac{\mathrm{d}C_L}{\mathrm{d}\alpha_F} + \frac{8}{3\pi}C_d|\alpha_{FE}|\right). \tag{11.40}$$

For arbitrary area ratio Λ, a good approximation for $\mathrm{d}C_L/\mathrm{d}\alpha_F$ in free flow around a foil is

$$\frac{dC_L}{d\alpha_F} = 2\pi\,\frac{\Lambda(\Lambda+0.7)}{(\Lambda+1.7)^2}\,. \tag{11.41}$$

However, $\mathrm{d}C_L/\mathrm{d}\alpha_F$ may be influenced substantially by various effects:

- Exceeding the stall angle will decrease the lift.
- Gaps between fixed and movable parts of the fin, or between hull and fin, decrease the lift.
- In model experiments, i.e., for small Reynolds number, the lift of fins is smaller.
- The effect of the boundary layer along the hull may be taken into account by decreasing $dC_L/d\alpha_F$. In model experiments, the effect may be extremely large.

The influence of these effects is discussed extensively by Bohlmann [4].

11.5.4 Effect of Inflow Conditions

In many cases, the time-averaged fluid speed v at a fin may be approximated as the negative ship speed U. For a rudder behind a ship, but excluding the effect of the propeller slipstream, the wake fraction w gives

$$v = -U(1-w). \tag{11.42}$$

However, one should be aware that the wake fraction at the rudder is different from that at the propeller in front of the rudder. Whereas the viscous component of the wake fraction may be similar at propeller and rudder, the potential component of the wake fraction is, typically, in the range of 0.1 for single-screw ships, but it is nearly zero for the usual arrangement of a rudder.

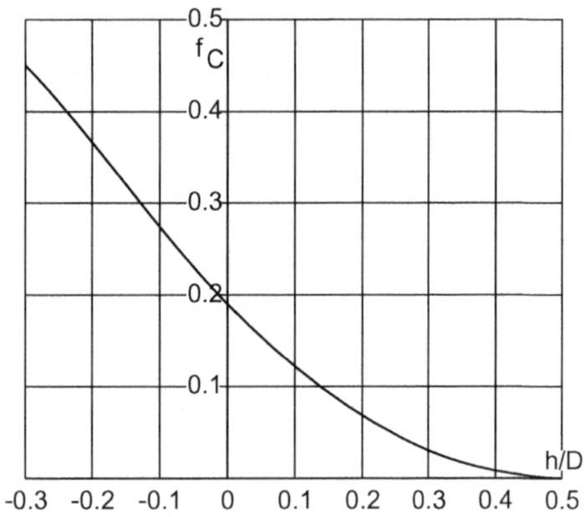

Fig. 11.3 Dependence of f_C on the vertical exceedance of the rudder beyond the propeller circle

If a fin is located behind a propeller (as is typical for rudders), the propeller slipstream increases the flow speed at the fin and thus the lift generated by the fin. Using the speed of the slipstream as v would neglect effects of the finite horizontal and vertical extent of the slipstream, which reduces its effect. Instead, it may be recommended to determine the *additional* lift due to the propeller in front of a fin by the formula [3]

$$T\left(1+\frac{1}{\sqrt{1+C_{th}}}\right)C\sin\delta. \tag{11.43}$$

Here, T is the propeller thrust, δ is the rudder angle relative to the propeller axis, and C_{th} is the thrust loading coefficient of the propeller:

$$C_{th}=T/\left(\frac{1}{2}\rho v_A^2\frac{\pi}{4}D^2\right), \tag{11.44}$$

where v_A is the inflow speed to the propeller. The factor

$$C=1-f_C(h_{top})-f_C(h_{bottom}) \tag{11.45}$$

depends on the vertical extent of the rudder beyond the upper (h_{top}) and lower (h_{bottom}) limit of the propeller circle. The function f_C is given in Fig. 11.3.

To determine the angle of attack of a fin, in the following body and wave orbital motions are taken into account, but radiation and diffraction waves will be neglected. Then the complex amplitude of the angle of attack $\hat{\alpha}_F$ is the sum of the following contributions (positive if they produce a force in direction $\mathbf{n}_F$):

- Due to the wave, for $v\neq 0$:

$$\hat{\alpha}_{FW}=\frac{C_r\hat{v}_{FW}}{v}. \tag{11.46}$$

For $v=0$, the expression is not evaluated.

- Due to the fin moving with the ship in direction $\mathbf{n}_F$:

$$\hat{\alpha}_{FS1} = -\frac{C_r \hat{v}_{FS1} + \hat{v}_{FS2}}{v}. \tag{11.47}$$

- Due to the rotation of the fin with the ship:

$$\hat{\alpha}_{FS2} = -C_r \mathbf{a}_F \cdot \hat{\boldsymbol{\alpha}}. \tag{11.48}$$

- Due to fin control:

$$\hat{\alpha}_{FS3} = -C_\delta \mathbf{c}_F \cdot \hat{\mathbf{u}}. \tag{11.49}$$

C_δ is a factor:

$C_\delta = 1$ for fins turning as a single body (e.g., spade rudder);
$C_\delta < 1$ if the front part of the fin is fixed and only the aft part rotates (semi-balanced rudder);
$C_\delta > 1$ if the front part moves by the angle $\mathbf{c}_F \hat{\mathbf{u}}$ and an aft flap moves by larger angle (flap rudder).

The total complex amplitude of the fin force becomes

$$\hat{\mathbf{F}}_F = \left[\hat{F}_{FW} + \hat{F}_{FS} + v^2 b_F (\hat{\alpha}_{FW} + \hat{\alpha}_{FS1} + \hat{\alpha}_{FS2} + \hat{\alpha}_{FS3}) + T \left(1 + \frac{1}{\sqrt{1 + C_{th}}} \right) C \hat{\delta} \right] \mathbf{n}_f. \tag{11.50}$$

The first two terms on the right-hand side are mass forces acting at $\mathbf{x}_{F2}$; the remaining lift forces act at $\mathbf{x}_F$. Thus the complex amplitude of the moment of the fin force is

$$\hat{\mathbf{M}}_F = (\hat{F}_{FW} + \hat{F}_{FS}) \mathbf{x}_{F2} \times \mathbf{n}_F + \left[v^2 b_F (\hat{\alpha}_{FW} + \hat{\alpha}_{FS1} + \hat{\alpha}_{FS2} + \hat{\alpha}_{FS3}) + T \left(1 + \frac{1}{\sqrt{1 + C_{th}}} \right) C \hat{\delta} \right] \mathbf{x}_F \times \mathbf{n}_F. \tag{11.51}$$

11.5.5 Influence of Oscillation Frequency

The dependence of fin force on oscillation frequency ω_e is discussed using classical theoretical results of potential theory [5, 6] for the aspect ratio $\Lambda = \infty$. Corresponding results for finite aspect ratio are not known.

For a time-harmonic translation v of a fin (relative to the surrounding, undisturbed fluid) in direction of the fin normal vector $\mathbf{n}_F$ (Fig. 11.1), the complex amplitude of the force in direction $\mathbf{n}_F$ is

$$\hat{L}_v = -\pi \rho c s U i \omega_e \hat{v} \left(i \frac{K}{2} + C(K) \right), \tag{11.52}$$

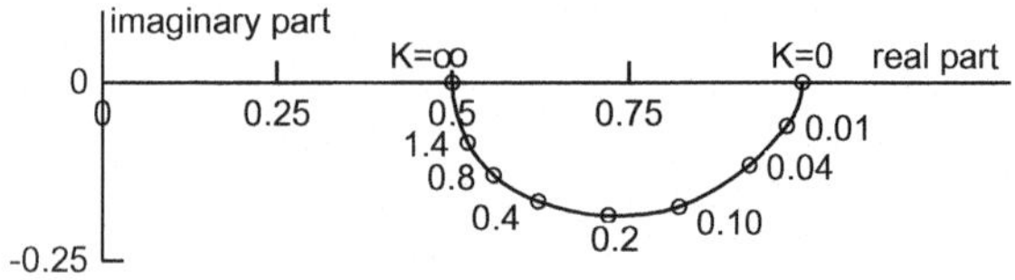

Fig. 11.4 Complex function C(K) depending on reduced frequency K

where c and s designate the chord length and span, respectively, and K is the 'reduced' (non-dimensional) frequency:

$$K = \frac{\omega_e c}{2U}. \tag{11.53}$$

The first term in (11.52) is the added mass term; it is independent of U. Here, the second term is important. The complex function $C(K)$ is given in Fig. 11.4; it is 1 for stationary flow ($K = 0$) and approaches $\frac{1}{2}$ for large K.

Here, we need the foil lift for a combined transverse shift and a rotation about an axis parallel to $\mathbf{a}_F$. In the following, the transverse shift v is defined to hold at point $\mathbf{x}_{F1}$; thus the rotation α_F is interpreted as turning around the axis through x_{F1}. In that case, the rotation generates an additional lift of complex amplitude

$$\hat{L}_\alpha = -\pi \rho c s U^2 \hat{\alpha}_F \left(C(K) + \frac{iK}{2} + \frac{K^2}{4} \right). \tag{11.54}$$

$C(K)$ and K are defined as above. The first term is due to the angle of attack, the second due to its time rate of change, and the term is last due to the foil's added mass.

Equations (11.52) and (11.54) show that only the lift due to transverse velocity and foil rotation angle must be reduced by factor $C(K)$; mass terms (first term in (11.52) and last term in (11.54)) and the term due to angular velocity (middle term in (11.54)) need no reduction for non-zero frequency. For roll damping fins and roll motion, in many cases, K is so small that no correction for oscillation frequency is required. However, in other cases, especially for other ship motions, a correction of the in-phase force according to Re(C) and the out-of-phase force according to Im(C) appears appropriate.

In case of wave-induced fin forces, there is also a spacial variation of the inflow. It results in smaller C factors for K values exceeding about 0.1 compared to Fig. 11.4.

11.5.6 Special Cases

Fins that represent an extension of the hull, especially skegs and rudders, may be treated as parts of the hull. If a movable fin (e.g., a rudder) is treated as part of the hull (by neglecting the gap between stern and rudder), then only the forces due to the fin motion relative to the hull (the rudder angle) must be accounted for as a fin force, because the force on the fin in its average position is already contained in the hull force.

Fins having a span s which is in the same size range or larger than the distance between fin center and the roll axis (Sect. 11.3) will be treated more accurately if they are subdivided into two or more 'part fins' so that the sum of their spans s is equal to that of the total fin. In that case, the lift gradient $dC_L/d\alpha_F$ of the part fins has to be determined using the aspect ratio Λ of the total fin, and not that of the part fins.

11.6 Foil Effect of the Hull

If hull force and moment are determined by potential methods as described in Chaps. 5–8, they must be corrected for the influence of flow separation at the aft end of the body (rudder, skeg, transom). It causes the so-called foil effect of the hull in case of oblique motion. Without that correction, for a body with steady forward speed U and transverse speed v relative to the water far from the body, potential theory gives a longitudinal and transverse force of zero and a 'Munk moment' about the vertical axis of size

$$N_M = -(m_y - m_x)Uv, \tag{11.55}$$

where m_x and m_y are added mass in longitudinal and transverse directions, respectively. Because, for slender bodies, m_x is much smaller than m_y, m_x will be neglected.

The effect of flow separation at the aft end can be taken into account by assuming, besides sources, also vortices to model the periodic flow. However, because the foil effect is relatively small compared to other transverse forces in seakeeping, it appears sufficient to use a simpler method: the slender-body theory (Ref. [7], chapter Body Forces). It is applicable only to slender bodies like ships. Here, slender-body theory is further simplified by neglecting periodic changes of the free-surface height, and by assuming a constant transverse added mass μ_y per length for all body sections from stem to stern. (Practically, for μ_y the maximum value near to the stern should be taken.) Then, slender-body theory gives the following expressions for the transverse force Y and the yaw moment N around the coordinate origin:

$$Y = -\mu_y U v; \quad N = -\mu_y U v x_f, \tag{11.56}$$

where x_f is the longitudinal coordinate of the front end of the body.

The difference between slender-body theory and potential theory without flow separation will be added to correct the transverse force and the yaw moment for the foil effect:

$$Y = -\mu_y U v \tag{11.57}$$

$$N - N_M = -\mu_y U v x_v + m_y U v. \tag{11.58}$$

For constant μ_y over x, we have

$$m_y = \mu_y L, \tag{11.59}$$

where L is body length. This results in the correction of the yaw moment

$$N - N_M = (-x_v + L)\mu_y U v = -x_a \mu_y U v, \tag{11.60}$$

where x_a is the x coordinate of the aft end of the body including its rudder(s). The transverse added mass per length μ_y can be approximated as

$$\mu_y = \rho \frac{\pi T^2}{2}, \tag{11.61}$$

where T is section draft. The formula is correct for elliptical section shapes of arbitrary ratios of breadth to draft.

Corresponding considerations for a body performing a yaw motion show that, in this case, the above corrections of transverse force and yaw moment should be made using the transverse speed v of the body at its aft end.

The above formulas refer to stationary flow. For the time-harmonic flow of interest here, a correction appears necessary [5, 6]. For foils, this correction was discussed already in Sect. 11.5.5. Here, however, the reduced frequency K defined in (11.53) is usually (except for cases near encounter wave frequency zero) so large that $C = \frac{1}{2}$ is appropriate. For diffraction (wave-induced) forces, C is even smaller; therefore, in our programs, the foil effect is not applied to diffraction forces and moments.

Whereas the foil effect corrections have a relatively small influence on sway and yaw motions, for ships with speed ahead the influence on roll is substantial or even dominant. Therefore, the variation of transverse speed and force at x_a over the vertical coordinate z (measured downward from the waterline) is considered. If v_{sy} is the transverse speed of the stern at longitudinal position x_a due to sway and yaw motion only, the total transverse speed is

$$v(z) = v_{sy} - z\dot{\varphi}, \tag{11.62}$$

where φ is roll angle. The distribution of Y over z is assumed elliptic because this is the actual lift distribution over span in low-aspect-ratio foils (however, in case of no foil rotation). Assuming this distribution also here, the following expression for Y is derived from (11.57):

$$Y = -\mu_y U C \frac{4}{\pi T} \int_0^T (v_{sy} - z\dot{\varphi})\sqrt{1 - z^2/T^2}\, dz. \tag{11.63}$$

The factor $4/\pi T$ is required to cancel the factor $\int \sqrt{1 - z^2/T^2}\, dz$ modeling the elliptic z-dependence of transverse force. The roll moment of Y follows then as

$$M_1 = \mu_y U C \frac{4}{\pi T} \int_0^T z(v_{sy} - z\dot{\varphi})\sqrt{1 - z^2/T^2}\, dz. \tag{11.64}$$

Evaluating the integral gives the additional roll moment due to the foil effect:

$$M_1 = \mu_y UCT \left(\frac{4}{3\pi} v_{sy} - \frac{1}{4} T\dot{\varphi} \right). \tag{11.65}$$

Both terms contribute to the roll damping: the first by coupling between sway and roll, the second directly.

11.7 Bilge Keels

Bilge keels are necessary for single-hull ships to increase roll damping, especially in case of no forward speed, where fins and the foil effect of the hull do not cause substantial roll damping. Many papers, most prominent those of Ikeda [1], derived regression formulas from model experiments for the roll moment generated by bilge keels as a function of roll angular velocity. This section is based on a paper by Schumann [8], which itself is a further development of the method [1].

The moment about the roll axis (Sect. 11.3) due to bilge keels on *both* ship sides is approximated by the formula

$$M_{BK} = 2\, \frac{4}{3\pi} \rho l J^2 r^2 \omega_0 \varphi_a (bC_D r \cos\alpha + I_N). \tag{11.66}$$

The meaning of the symbols is

ρ density of water
l length of one bilge keel
J factor for velocity increase at the bilge keel due to the ship's section shape
r distance from roll axis to midpoint of bilge keel
ω_0 roll eigenfrequency for small roll
φ_a roll amplitude
b bilge keel breadth (normal to the hull)
C_D drag coefficient of bilge keel
α inclination of bilge keel against radius to the roll axis
I_N influence of hull pressure modified by the bilge keel

If these data vary substantially over the length of a bilge keel, it should be subdivided into several parts.

The drag coefficient C_d depends on the non-dimensional amplitude of the relative motion a between bilge keel and fluid:

$$a = J\varphi_a r/b; \tag{11.67}$$

$$C_D = \frac{78.7}{2\pi a + 4.43} + 1.6. \tag{11.68}$$

The factor J should be determined by a two-dimensional potential flow computation around a ship section without bilge keels, which rotates with an arbitrary angular velocity $\dot{\varphi}$ about the roll axis. The tangential flow velocity v is evaluated at a distance $b/2$ from the section contour to determine

$$J = \frac{|v - \dot{\varphi} r|}{|\dot{\varphi} r|}. \tag{11.69}$$

Regression formulas for J found in relevant papers appear inaccurate.

For roll amplitudes beyond 10–15°, the effect of the hull pressure change due to a bilge keel (I_N in Eq. (11.66)) often exceeds that of the fluid pressure on the bilge keel itself. To determine the non-dimensional pressure change on the hull, the drag coefficient C_D is separated into a positive part $C_D^+ = 1.2$ for the forward (in roll motion direction) side of the bilge keel, and a negative part for the backward side (reduced by the empirical factor 0.9):

$$C_D^- = 0.9(1.2 - C_D). \tag{11.70}$$

The distribution of the pressure change has been investigated in model experiments and by RANS calculations. On the forward side, the non-dimensional pressure change is approximated as decreasing linearly with distance (measured along the section contour) from the bilge keel, where it is 1.2, to zero at

$$s^+ = 2.5 a_m^{0.3} b, \tag{11.71}$$

where

$$a_m = \text{Minimum}(r\varphi_a/b,\ 65). \tag{11.72}$$

On the backward side, the non-dimensional pressure change is assumed constant $= C_D^-$ for a contour length of $\frac{1}{2}s^-$, and decreasing linearly from there to zero at the contour length distance s^-, where

$$s^- = (2a_m - 0.72 a_m^{1.2}) b. \tag{11.73}$$

The moment about the roll axis caused by the non-dimensional pressure change is integrated numerically along the section contour. Because forward and backward side changes within one roll cycle, the moment occurring for positive and for negative rotation direction are averaged to obtain the quantity I_N.

The formulas are given above hold for ships without forward speed. For $U > 0$, the roll damping of bilge keels can increase (in some cases by >100%) or decrease (at least up to 30%) depending on Froude number and roll amplitude. Further changes are possible if the profile of the bilge keel is not rectangular.

11.8 Control Forces

In dynamic positioning systems, thrusters are used to suppress horizontal translations and possibly yaw motions. Similarly, thrusters (e.g., Voith-Schneider propellers) are used sometimes to suppress or dampen roll motions of ships or by swinging a hanging load on a crane barge. To determine motion transfer functions in such cases, it is assumed that a thruster force

- acts at the location $\mathbf{x}_P$ of the thruster,
- oscillates harmonically over time with complex amplitude $\hat{P}$, and
- acts in direction $\mathbf{a}_P$ (for positive $P(t)$), where $|\mathbf{a}_P| = 1$.

$\hat{P}$ is set proportional to the six-component motion (three translations, three rotations) amplitude $\hat{\mathbf{u}}$:

$$\hat{\mathbf{P}} = \sum (\mathbf{f} \cdot \hat{\mathbf{u}})\mathbf{a}_p, \tag{11.74}$$

where the sum is taken over all thrusters. The complex six-component vector $\mathbf{f}$ specifies the PD (proportional and differential) control of the thruster: the real parts of its components are proportionality factors between the respective body motion component and the thrust, and the imaginary parts are proportionality factors between thrust and body velocity divided by ω_e. These factors should be chosen such that neither the maximum thruster force P_{max} nor the maximum rate of its change $\dot{P}_{max}$ are exceeded under realistic wave conditions (not for wave amplitude 1). The moment produced by the thrusters is

$$\hat{\mathbf{M}}_P = \sum \mathbf{x}_P \times \hat{\mathbf{P}}. \tag{11.75}$$

In simulations of nonlinear ship motions, one needs control forces (meant to include also moments) to keep, approximately, the intended mean forward speed and course angle. Possibly, one wants to keep also, averaged over time, the intended longitudinal and transverse position and/or the ship's path. In the latter case, a PID control is required. The control forces should be small to decrease their influence on ship motions, but large enough to avoid extreme deviations from the intended ship motion. Normally, only the degrees of freedom that lack a restoring force, i.e., surge, sway, and yaw, need a control force.

To alleviate the selection of suitable control parameters for each of the different motion degrees-of-freedom to be controlled, in our simulation program the following non-dimensional parameters are defined and specified by the user:

- The P (restoring) parameter γ is defined as the square of the ratio ω/ω_e, where ω_e is the wave encounter frequency (significant value in case of a natural seaway), and ω is the eigenfrequency that results for free oscillations of the hull in air (without added mass and other restoring and damping forces) under the action of the control force.
- The D (damping) parameter δ is defined as the ratio between actual and critical damping, that results under the action of the control force alone for the idealized free oscillation described above.
- The I (integral) parameter ϵ is defined as an integral part of the control force divided by (a) the time integral of the deviation from the intended position (from the beginning of the simulation up to the current time) and (b) by $\gamma^2 m\omega_e^3$. (In case of rotations, ship mass m must be substituted by the moment of inertia.)

In these definitions, the motion degrees of freedom are considered as un-coupled and are described in the intermediate coordinate system $(\tilde{x}, \tilde{y}, \tilde{z})$ of Chap. 8 and then converted into the inertial coordinate system.

The integral part of the control force must be limited to avoid instability of the control. To determine the stability limit for ϵ, and to better understand the effect of γ and δ, the one degree-of-freedom motion equation is investigated for a ship in a regular wave. Three angular frequencies will be distinguished:

- ω_e is the frequency of encounter between wave and ship.
- Ω is the eigenfrequency of the ship without control; for surge, sway, and yaw it is usually zero.
- ω is any frequency.

Using these symbols, the one degree of freedom motion equation for the ship with control force, but without excitation, is

$$\left[-\omega^2 + \gamma^2\omega_e^2 + \Omega^2 + i\omega(d/m + 2\delta\gamma\omega_e - \epsilon\gamma^2\omega_e^3/\omega^2)\right]\hat{u} = 0. \tag{11.76}$$

As above, quantities with the hat symbol designate complex amplitudes. For free motions of non-zero amplitude $\hat{u}$, the complex expression between brackets is zero. Its real part is zero for the eigenfrequency

$$\omega = \sqrt{\Omega^2 + \gamma^2\omega_e^2}. \tag{11.77}$$

Thus, for heave, roll, and pitch motions, for which Ω is >0, the eigenfrequency will change due to the control force by 1%, 3%, or 10% if γ is chosen 0.14, 0.24, or 0.45 times ω_e/Ω, respectively. For surge, sway, yaw ($\Omega = 0$), $\gamma = 1$ would cause resonance of the exciting waves and thus severe influence of the control force on the motions. Therefore, γ should be ≤ 0.1 for these motions.

A sway control force, which does not act in height of the roll axis, has a substantial influence on roll motion even for $\gamma = 0.1$ in case of non-zero δ because other means of roll damping are small. Thus, it is recommended to use $\delta = 0$ or very small for sway. The alternative to use $\gamma = 0$ would lead to strong transverse drift motion in case of high oblique waves. In other cases, $\delta = 0.4$ may be recommended.

In (11.76), the expression in parentheses is the damping of the ship motion, including the control force. For stable motion, it must be positive:

$$d/m + 2\delta\gamma\omega_e - \epsilon\gamma^2\omega_e^3/\omega^2 > 0. \tag{11.78}$$

Using (11.77) and the relation between damping constant d and the non-dimensional damping ratio D,

$$d = 2m\Omega D, \tag{11.79}$$

the stability condition is transformed to

$$\epsilon < 2(D\Omega/\omega_e + \delta\gamma)[1 + (\Omega/\gamma\omega_e)^2]. \tag{11.80}$$

For $\Omega = 0$, this results in the simple condition,

$$\epsilon < 2\delta\gamma. \tag{11.81}$$

Using half of the critical value, $\epsilon = \delta\gamma$, was found to work well also for the cases with $\Omega > 0$.

References

1. Y. Ikeda, Y. Himeni, N. Takana, Components of roll damping of ship at forward speed. Trans. JSNA **143** (1978)
2. Y.T. Yu, Virtual masses of rectangular plates and parallelepipeds in water. J. Appl. Phys. **16**, 724–729 (1945)
3. H. Söding, Limits of potential theory in rudder flow predictions. Ship Technol. Res. 141–155 (1998). Weinblum lecture
4. H.J. Bohlmann, Computation of hydrodynamic coefficients of submarines to predict their motions (in German). Ph.D. thesis, Hamburg University, Institute for Naval Architecture (1990)
5. T. von Karman, W.H. Sears, Airfoil theory for non-uniform motion. J. Aeronaut. Sci. **5/10**, 379ff (1936)
6. O. Grim, Hydrodynamic inertial and damping forces and excitation forces (in German). Technical report (1975)
7. J.E. Brix (ed.), *Manoeuvring Technical Manual* (Seehafen-Verlag, Hamburg, 1993)
8. C. Schumann, Computing roll damping of bilge keels (in German). Technical report, Schiffsrat, Unpublished Report (2019)

Chapter 12
Special Topics

Abstract This chapter deals with the influence of sails and suspended loads on roll motions and how to take account of forces and moments generated by roll damping tanks. Also, the dangerous wave-riding condition is discussed, in which fast ships may be accelerated by steep following waves up to the wave celerity. The chapter explains also how motion restraints, for instance, in case of contact of a ship with a fixed obstacle, can be handled in computations.

12.1 Sails

Here, we deal only with the influence of sails on seakeeping. In high-performance sailing yachts, the influence of sail forces may be more complex than handled here, but in conventional sailing boats and cutters, their influence on transfer functions may be roughly approximated as described here.

Sails may cause substantial roll damping; the influence on other motions, however, is small. Wind on the superstructure and on deck load (e.g., containers) has an influence on roll damping also in ships without sails. However, even in strong wind, the influence appears small.

Sails cannot be taken into account like fins in the water simply by using air density ρ_{air} instead of water density ρ; the important difference is that the wind may blow from any direction, not only lengthwise as the steady water flow along with a moving ship.

At first, damping forces are considered. The velocity-dependent force on a sail, acting normal to the 'sail plane', is approximated as

$$\mathbf{F}_S = \frac{1}{2}\rho_{air}\mathbf{s}(\mathbf{u}_W - \mathbf{u}_S)^2 C_N(\alpha). \tag{12.1}$$

α (to be distinguished from ship rotations $\boldsymbol{\alpha}$) is the angle of attack of the wind relative to the sail approximated as a plane. C_N is the normal force coefficient of the wind force on the sail. Only forces normal to the sail plane are considered. $\mathbf{s}$ is a vector normal to the sail plane; magnitude = sail area. $\mathbf{u}_W$ is the 'apparent' wind velocity, i.e., the wind velocity relative to the ship sailing with its average speed ahead. $\mathbf{u}_S$ is

B. O. el Moctar et al., *Numerical Methods for Seakeeping Problems*,
https://doi.org/10.1007/978-3-030-62561-0_12

the oscillating velocity of the sail speed reference point x_{S1} due to ship motions. Its complex amplitude is

$$\hat{\mathbf{u}}_S = i\omega_e(\hat{\mathbf{t}} + \hat{\boldsymbol{\alpha}} \times \mathbf{x}_{S1}), \tag{12.2}$$

where $\hat{\mathbf{t}}$ and $\hat{\boldsymbol{\alpha}}$ are the complex amplitudes of the translation (at the ship's coordinate origin) and rotation of the ship, respectively. $\mathbf{x}_{S1}$ is located 0.5 times average chord length behind the sail force reference point $\mathbf{x}_S$:

$$\mathbf{x}_{S1} = \mathbf{x}_S + 0.5\mathbf{c} \tag{12.3}$$

$\mathbf{x}_S$ may be assumed to lie between the sail area center of gravity and half the height of the sail with respect to height, and about $\frac{1}{4}c$ behind the front end of the sail at that height. $\mathbf{c}$ is the 'chord vector'. Its magnitude is the average chord length of the sail, and the direction is from the forward to the backward edge of the sail. (For aft wind, the point of attack of the wind force $\mathbf{x}_S$ and the velocity reference point $\mathbf{x}_{S1}$ exchange their positions.)

$\mathbf{c}$ and the 'mast vector' $\mathbf{m}_S$ (points downward; magnitude = height of the sail) determine the sail area vector $\mathbf{s}$:

$$\mathbf{s} = \mathbf{c} \times \mathbf{m}_S. \tag{12.4}$$

For a small absolute value of the angle of attack, the normal force coefficient C_N is approximated as [1]

$$C_N = 2\pi \frac{\Lambda(\Lambda + 0.7)}{(\Lambda + 1.7)^2}\,\alpha. \tag{12.5}$$

Λ is the aspect ratio, i.e., height over average chord length. If wind flow under the lower edge of the sail is restricted, Λ must be increased; for full suppression of the flow below the lower edge of the sail, by a factor of 2. The angle α, where the maximum C_N of about 1.5 is attained, is called stall angle. Beyond that angle, a value of $C_N = 1.2$ may be assumed correspondingly for negative α.

In (12.1), due to ship motions the quantities $\mathbf{u}_S$ and α change. Linearizing this change and omitting stationary quantities, the complex amplitude of $(\mathbf{u}_W - \mathbf{u}_S)^2$ is

$$-2\mathbf{u}_W \cdot \hat{\mathbf{u}}_S, \tag{12.6}$$

and the complex amplitude of α is approximately

$$\hat{\alpha} = \frac{-\hat{\mathbf{u}}_S \cdot \mathbf{s}}{|\mathbf{u}_W||\mathbf{s}|}. \tag{12.7}$$

For a large absolute value of $\hat{\alpha}$, the expression for $\hat{\alpha}$ is inaccurate, but in that case the term containing $\hat{\alpha}$ is small anyway.

This gives the complex amplitude of the sail force as

$$\hat{\mathbf{F}}_S = \frac{1}{2}\rho_{air}\mathbf{s}\left(-2\mathbf{u}_W \cdot \hat{\mathbf{u}}_S C_N(\bar{\alpha}) + |\mathbf{u}_W|^2 \frac{dC_N}{d\alpha}\hat{\alpha}\right). \tag{12.8}$$

The second term is zero beyond the stall angle, i.e. if $C_N = \pm 1.2$. $\bar{\alpha}$ is the average angle of attack (positive, if u_W has a positive component in direction **s**):

$$\bar{\alpha} = \arcsin \frac{\mathbf{u}_W \cdot \mathbf{s}}{|\mathbf{u}_W||\mathbf{s}|}. \tag{12.9}$$

The moment due to the sail force is

$$\hat{\mathbf{M}}_S = \mathbf{x}_S \times \hat{\mathbf{F}}_S. \tag{12.10}$$

For light ships with large sails, the added air mass of the sails may contribute considerably to the moment of inertia around the longitudinal axis. Therefore, also added-mass forces due to the sails should be taken into account. Their complex amplitude is

$$\hat{\mathbf{F}}_{Sa} = -c_M \frac{\pi}{4}\rho_{air}|\mathbf{s} \times \mathbf{c}|\ i\omega_e \hat{\mathbf{u}}_S \tag{12.11}$$

c_M can be estimated as described for fins. The moment due to this force corresponds to (12.10).

Like for underwater fins, to assess the roll-damping and added-mass effect of a sail accurately, it may be necessary to subdivide it into horizontal strips each of which is handled as described above, using Λ for the total sail.

12.2 Suspended Load

In crane ships or crane barges, resonant rolling, or pitching influenced by swinging motions of a hanging load may be a problem. Thus, the following case is considered.

A mass m is assumed to hang on a rope of length l. The upper end of the rope is attached to a ship-fixed point $\underline{\mathbf{x}}_l$. (Quantities expressed in ship-fixed coordinates will be indicated by an under-bar.) Here, only the motions of the load in y-direction are considered because these are often more important than those in x-direction; however, extending the case to load motions in x and y is straightforward. Ship and load motions will be linearized.

Damping of the swing motion of the load, e.g., due to air resistance, is neglected. Thus, the only damping mechanism is the coupling between load and ship motions.

In the fundamental linear motion equation

$$[\mathbf{S} - \mathbf{B} - \omega_e^2 \mathbf{M}]\hat{\mathbf{u}} = \hat{\mathbf{F}}_e, \tag{12.12}$$

S is assumed to describe the ship with load fixed at point $\underline{\mathbf{x}}_l + (0, 0, l)$; correspondingly for **M**. $\hat{\mathbf{u}}$ is the ship motion in six degrees of freedom.

The motion of the load is described in the ship-fixed coordinate system by the complex motion amplitude $\hat{\underline{y}}_m$. We will linearize also with respect to $\hat{\underline{y}}_m$. The motion of the load induces an excitation in addition to the wave excitation $\hat{\mathbf{F}}_e$:

$$\hat{l} = \begin{pmatrix} 0 \\ \omega_e^2 \\ 0 \\ g - \omega_e^2(\underline{z}_l + l) \\ 0 \\ \omega_e^2 \underline{x}_l \end{pmatrix} m\hat{\underline{y}}_m. \tag{12.13}$$

The motion of the load is described in the inertial system by the linearized motion equation

$$m\ddot{y}_m = \frac{y_l - y_m}{l} mg. \tag{12.14}$$

Expressed in complex amplitudes, the motion equation is

$$-\omega_e^2 m\hat{y}_m = \frac{\hat{y}_l - \hat{y}_m}{l} mg. \tag{12.15}$$

From the relation between body-fixed and inertial coordinates,

$$\mathbf{x} = \underline{\mathbf{x}} + \mathbf{t} + \boldsymbol{\alpha} \times \underline{\mathbf{x}}, \tag{12.16}$$

follow expressions for $\hat{y}_l$ and $\hat{y}_m$:

$$\hat{y}_l = \hat{u}_6\underline{x}_l - \hat{u}_4\underline{z}_l + \hat{u}_2; \tag{12.17}$$

$$\hat{y}_m = \hat{u}_6\underline{x}_l - \hat{u}_4(\underline{z}_l + l) + \hat{u}_2 + \hat{\underline{y}}_m. \tag{12.18}$$

Inserting these terms into (12.15) yields:

$$\hat{\underline{y}}_m[1 - g/(l\omega_e^2)] = \hat{u}_4(\underline{z}_l + l - g/\omega_e^2) - \hat{u}_6\underline{x}_l - \hat{u}_2. \tag{12.19}$$

The system of equations to calculate the ship and load motions must then be extended for the additional degree of freedom $\hat{\underline{y}}_m$:

$$\begin{pmatrix} & 0 \\ & -\omega_e^2 m \\ \mathbf{S}-\mathbf{B}-\omega_e^2\mathbf{M} & 0 \\ (6\times 6) & -[g-\omega_e^2(\underline{z}_l+l)]m \\ & 0 \\ & -\omega_e^2\underline{x}_l m \\ 0\ 1\ 0\ -(\underline{z}_l+l-g/\omega_e^2)\ 0\ \underline{x}_l & 1-g/(l\omega_e^2) \end{pmatrix} \begin{pmatrix} \\ \hat{u} \\ (6\times 1) \\ \\ \hat{\underline{y}}_m \end{pmatrix} = \begin{pmatrix} \\ \hat{\mathbf{F}}_e \\ (6\times 1) \\ \\ 0 \end{pmatrix}. \tag{12.20}$$

The additional (seventh) column corresponds to (12.13), the seventh row to (12.19).

12.3 Roll Damping Tanks

Tanks partly filled with fluid (normally water) are often used to dampen roll motions. The lowest eigenfrequency of the sloshing motion in the tank should be similar to the roll eigenfrequency of the floating body. Like for suspended loads, it will be assumed that the mass and restoring matrices describe the body, including the mass of the tank fluid fixed in its rest position.

Further, it is assumed that the relation between tank motion (relative to the inertial system) and the force and moment exerted by the sloshing motion of the fluid on the tank can be determined. For instance, one may move the tank in six degrees-of-freedom, determine (by computation or experiment) the force and moment (together, the six components) of the sloshing fluid, subtract force and moment which would occur for the frozen fluid, linearize the relation between force and motion if it is nonlinear, and add the resulting 6 by 6 complex matrix of force amplitude per motion amplitude to the matrix **B** in the motion Eq. (12.12).

In case of symmetry of ship and tank with respect to $\underline{y} = 0$, only the asymmetrical motions sway, roll, and yaw, and the asymmetrical components of force and moment must be considered. It would not be correct (and no reasonable approximation) to neglect the horizontal force exerted by the fluid motion on the tank. Also, a restriction of tank motions to rotations about the roll axis of the ship would not be correct, because that axis refers to free roll motions, not to motions under the influence of waves, and because transverse motions of ship and tank cause sloshing.

12.4 Negative Encounter Frequency

When using a coordinate system that follows the ship with its mean forward velocity U (relative to the fluid far from the ship), the water surface coordinate $\zeta(t)$ of a regular wave with wave number k, direction μ, and amplitude $\hat{\zeta}$ is

$$\zeta(t) = \mathrm{Re}(\hat{\zeta} e^{i(\omega t - k[x+Ut]\cos\mu + ky\sin\mu)}) = \mathrm{Re}(\hat{\zeta} e^{i(\omega_e t - kx\cos\mu + ky\sin\mu)}), \tag{12.21}$$

where

$$\omega_e = \omega - kU \cos \mu. \tag{12.22}$$

If the ship overtakes following (i.e., $|\mu| < 90°$) waves, ω_e becomes negative. Because this may appear difficult to interpret, often (12.22) is given instead as

$$\omega_e = |\omega - kU \cos \mu|. \tag{12.23}$$

However, (12.23) is equivalent to (12.22) only if, for a ship overtaking the wave, in (12.21) and all other equations referring to the wave, k must be substituted by $-k$, and $\hat{\zeta}$ by its complex conjugate. Because a negative wave number appears also not intuitive, it is recommended to use only (12.22), not (12.23).

12.5 Long Encounter Periods and Surf-Riding

In this section, it is assumed that ζ_A is the amplitude of a regular wave. If $\hat{\mathbf{u}} = (\hat{u}_1, \hat{u}_2, \hat{u}_3)$ is the complex amplitude of the body translation, the ship speed component in the direction of wave propagation is oscillating between

$$v \cos \mu + \omega_e \zeta_A |\hat{u}_1 \cos \mu - \hat{u}_2 \sin \mu| \tag{12.24}$$

and

$$v \cos \mu - \omega_e \zeta_A |\hat{u}_1 \cos \mu - \hat{u}_2 \sin \mu|. \tag{12.25}$$

In following waves ($\cos \mu > 0$), the above speed interval may contain the phase speed ω/k of the wave. The ship will then be accelerated or decelerated so that its speed component in wave propagation direction is equal to the phase speed of the wave. It is then trapped by the wave, attaining a stationary floating condition near the wave trough. This condition is called surf-riding. Practically, due to nonlinear effects, surf-riding appears already if the wave phase speed is somewhat outside of (usually higher than) the above-given interval. The assumption of time-harmonic responses occurring with the encounter frequency given by (12.22), using the average forward speed U, is then violated, and the assumption of linear responses is grossly in error. Thus, predicting surf-riding requires nonlinear simulations of longitudinal motions.

Practically, this condition is dangerous because it may lead to broaching-to, i.e., a sudden course change of the ship to a direction more or less parallel to the wave crests accompanied by a severe list or even capsizing, like in a narrow turning circle at high speed.

12.6 Motion Restraints

The motion of a floating body may be restrained, e.g., by contact with one or several other fixed bodies or by cables or ropes. To describe such restraints, we assume that there are J (with $J \leq 6$) ship-fixed points with position vectors $\mathbf{x}_j$, $j = 1$ to J, at which the ship motion is suppressed in a given direction $\mathbf{n}_j$. Using linearization with respect to ship motions, for each restraint j we require

$$(\hat{\mathbf{t}} + \hat{\alpha} \times \mathbf{x}_j) \cdot \mathbf{n}_j = 0. \tag{12.26}$$

Here, $\hat{\mathbf{t}}$ is the complex amplitude of the ship's translation and $\hat{\alpha}$ that of the ship's rotation. One may specify also two restraints for the same point to model the case that the point is allowed to glide along a straight line or to fix a point in all three directions by giving 3 Eqs. (12.26) with three vectors $\mathbf{n}_j$ (not in a common plane) for the same point $\mathbf{x}_j$. However, elastic restraints or friction forces must be handled differently. Because of the linearization, we need not distinguish whether $\mathbf{x}_j$ and $\mathbf{n}_j$ are expressed in ship-fixed or inertial coordinates.

We want to determine ship motions and restraining forces acting at points $\mathbf{x}_j$ in direction of the unit vectors $\mathbf{n}_j$. To take account of these forces, the ship motion equation has to be complemented by the restraining forces:

$$\mathbf{A}\hat{\mathbf{u}} = \hat{\mathbf{F}}_e + \sum_{j=1}^{J} \hat{k}_j \mathbf{f}_j. \tag{12.27}$$

Here, $\mathbf{A}$ is the matrix $-\omega_e^2\mathbf{M} - \mathbf{B} + \mathbf{S}$ of (12.12); $\hat{\mathbf{u}} = (\hat{\mathbf{t}}, \hat{\alpha})^T$ is the six-component ship motion amplitude vector; $\hat{\mathbf{F}}_e$ is the six-component wave excitation amplitude vector; $\hat{k}_j$ is the jth scalar restraining force amplitude; and $\mathbf{f}_j$ characterizes of the jth restraint:

$$\mathbf{f}_j = \begin{pmatrix} \mathbf{n}_j \\ \mathbf{x}_j \times \mathbf{n}_j \end{pmatrix}. \tag{12.28}$$

With these symbols, (12.26) may be written as

$$\mathbf{f}_j^T \cdot \hat{\mathbf{u}} = 0. \tag{12.29}$$

From (12.27) follows:

$$\hat{\mathbf{u}} = \mathbf{A}^{-1} \left(\hat{\mathbf{F}}_e + \sum_{j=1}^{J} \hat{k}_j \mathbf{f}_j \right). \tag{12.30}$$

Multiplying this equation from the left by the J times 6 matrix

$$\begin{pmatrix} \mathbf{f}_1^T \\ \dots \\ \mathbf{f}_J^T \end{pmatrix} \tag{12.31}$$

and using (12.29) gives:

$$\begin{pmatrix} \mathbf{f}_1^T \\ \dots \\ \mathbf{f}_J^T \end{pmatrix} \mathbf{A}^{-1}(\mathbf{f}_1, \dots, \mathbf{f}_J) \begin{pmatrix} \hat{k}_1 \\ \dots \\ \hat{k}_J \end{pmatrix} + \begin{pmatrix} \mathbf{f}_1^T \\ \dots \\ \mathbf{f}_J^T \end{pmatrix} \mathbf{A}^{-1}\hat{\mathbf{F}}_e = 0. \tag{12.32}$$

This is a linear equation system for the J restraining force amplitudes $\hat{k}_j$. Once it is solved, the ship motions follow from (12.30).

If the floating body contains a suspended load, matrix $\mathbf{A}$ and $\mathbf{A}^{-1}$ have seven rows and columns (see (12.20). It is easily verified that, in this case, only the first six rows and columns of $\mathbf{A}^{-1}$ must be used in (12.32).

Reference

1. H. Söding, Limits of potential theory in rudder flow predictions. Ship Technol. Res. 141–155 (1998). Weinblum lecture

Chapter 13
Further Transfer Functions

Abstract This chapter explains how transfer functions (i.e., linear responses to regular waves) for hull pressure oscillations can be determined in strip methods more accurately than using the standard procedure, i.e., by eliminating second derivatives ('m terms') of the steady potential. For multi-body vessels (catamarans, etc.), a pressure formula is given that accounts also for the interaction of waves between the hulls. Furthermore, the chapter deals with computing relative motions between water (especially, the water surface) and ship-fixed points, including the disturbance of incoming waves by the ship; with forces and moments in cross sections of the hull girder; and with the wave- and motion-induced vertical water motion in a moonpool (a vertical channel in certain marine structures).

13.1 Hull Pressure

In three-dimensional panel methods, the complex amplitude of the pressure has been determined already (for computing the body motions) at the centers of body panels or as averages on body patches. However, because the motions were still unknown, they were determined

1. for unit motions in the 6 degrees-of-freedom: $\hat{p}_1$ to $\hat{p}_6$;
2. for the undisturbed wave (so-called Froude-Krylov pressure): $\hat{p}^w$;
3. for the diffraction (interaction between wave and non-moving body): $\hat{p}^d$.

The transfer function between pressure and wave is simply a superposition of these pressures, taking into account the transfer function of the six-component body motion $\hat{\mathbf{u}}$:

$$\hat{p} = \hat{p}^w + \hat{p}^d + (\hat{p}_1, \hat{p}_2, \hat{p}_3, \hat{p}_4, \hat{p}_5, \hat{p}_6) \cdot \hat{\mathbf{u}}. \tag{13.1}$$

In strip methods, the accurate determination of pressures is more difficult. Hachmann [1] derived a formula which appears to be better than previous attempts. It is based on the assumption that the steady potential due to forward speed follows the periodic body motions. This results in

B. O. el Moctar et al., *Numerical Methods for Seakeeping Problems*,
https://doi.org/10.1007/978-3-030-62561-0_13

$$\frac{\hat{p}(x,y,z)}{\rho} = \left(-i\omega_e + U\frac{\partial}{\partial l}\right)\left[(\hat{\varphi}_2, \hat{\varphi}_3, \hat{\varphi}_4)\frac{W}{i\omega_e}\hat{\mathbf{u}} + (\hat{\varphi}^w + \hat{\varphi}^d)e^{-ikx\cos\mu}\right] \tag{13.2}$$
$$+ (\phi_y^0, \phi_z^0, y\phi_z^0 - z\phi_y^0)W\hat{\mathbf{u}} + g(\hat{u}_3 + y\hat{u}_4 - x\hat{u}_5).$$

The symbols denote

ω, ω_e	incident wave frequency, encounter wave frequency
U	average ship speed ahead
$\partial/\partial l$	derivative along streamlines of the steady ship flow
$\hat{\varphi}_2, \hat{\varphi}_3, \hat{\varphi}_4$	2- d radiation potential transfer function for sway, heave, roll
W	matrix for transformation of ship motions (six components) into section velocities (three components; see (5.48)
$\hat{\mathbf{u}} = (\hat{u}_1, \ldots, \hat{u}_6)$	ship motion amplitudes
$\hat{\varphi}^w$	amplitude of incident wave potential $-ice^{-kz}e^{iky\sin\mu}$
$\hat{\varphi}^d$	amplitude of two-dimensional diffraction potential
k	wave number
μ	wave angle: 0° for following waves, 90° for waves from starboard
ϕ_0	potential of steady ship flow.

Assuming a slender body, the partial derivatives of ϕ_0 are approximated as

$$\phi_y = -\frac{dy}{dx}U, \qquad \phi_z = -\frac{dz}{dx}U, \tag{13.3}$$

where dy/dx and dz/dx denote the derivatives along the streamlines of the steady ship flow. The term between brackets in (13.2) is caused by the periodic potential and its interaction with the steady parallel flow; the second term represents the change of the steady disturbance potential at a point fixed in the inertial system; and the last term is the hydrostatic pressure oscillation.

The streamlines of the steady flow may be approximated as lines of constant relative girth length of the offset sections; however, this may be too inaccurate for sections in the range of a skeg and other narrow extensions. The two-dimensional potentials $\hat{\varphi}_2$, $\hat{\varphi}_3$, $\hat{\varphi}_4$, and $\hat{\varphi}^d$ are originally determined at the midpoints between section offset points; they are linearly interpolated and (at the waterline) extrapolated to equal-girth-length points.

In multi-hull vessels (catamarans, etc.), hull pressures are influenced also by the waves radiated and diffracted by one hull and acting on another hull; this is elaborated in Sect. 5.4. To take account of these waves, the term between brackets in (13.2) must be changed to

$$\left[(\hat{\varphi}_2, \hat{\varphi}_3, \hat{\varphi}_4)\frac{W}{i\omega_e} + (\hat{\varphi}^w_{+90} + \hat{\varphi}^d_{+90})\hat{\zeta}_{R,j_a,l}(x) + (\hat{\varphi}^w_{-90} + \hat{\varphi}^d_{-90})\hat{\zeta}_{R,j_a,r}(x)\right]\cdot\hat{\mathbf{u}}$$
$$+ (\hat{\varphi}^w_{+|\mu|} + \hat{\varphi}^d_{+|\mu|})\hat{\zeta}_{D,j_a,l}(x) + (\hat{\varphi}^w_{-|\mu|} + \hat{\varphi}^d_{-|\mu|})\hat{\zeta}_{D,j_a,r}(x). \tag{13.4}$$

Here, $\hat{\zeta}_R$ are complex amplitudes of radiation waves; six scalar amplitudes due to the six body motion components are combined to a vector. $\hat{\zeta}_D$ is the sum of the incident wave and diffracted waves. j_a is the index of the offset section at which the wave meets the pressure point originating at the left or right neighboring hull. Indices l and r designate left-running and right-running waves, respectively. These wave amplitudes have been determined previously to compute the body motions. $\hat{\varphi}^w i/\omega$ and $\hat{\varphi}^d i/\omega$ are the complex amplitudes of the incident (Airy) wave potential and the section diffraction potential, respectively, at the pressure point, both for wave amplitude 1. The lower indices +90, −90, $+|\boldsymbol{\mu}|$ and $-|\boldsymbol{\mu}|$ indicate the wave direction. Radiation waves propagate in direction ±90°; incident and diffracted waves, in direction $\pm\mu$. More details are given in Sect. 5.4.

13.2 Absolute and Relative Motions at Body-Fixed Points

The complex amplitude of the oscillatory shift (relative to the inertial coordinate system) of a body-fixed point at location $\mathbf{x}$ follows from the complex amplitudes of the body translation $\hat{\mathbf{t}}$ and rotation $\hat{\boldsymbol{\alpha}}$ as

$$\hat{\mathbf{v}} = \hat{\mathbf{t}} + \hat{\boldsymbol{\alpha}} \times \mathbf{x}. \tag{13.5}$$

The complex amplitudes of the velocity and acceleration of the body-fixed point follow simply by multiplying $\hat{\mathbf{v}}$ by $i\omega_e$ and by $-\omega_e^2$, respectively.

Often the complex amplitude of the relative motion $\hat{\mathbf{r}}$ between a body-fixed point on the hull and the fluid at the same point is of interest. If we omit the diffraction and radiation waves (their vertical component is, in this connection, often called 'pile-up'), the shift $\mathbf{v}_w$ of water particles by the wave from their mean position follows from the incident wave potential (all quantities being represented by their complex amplitude for wave amplitude 1):

$$\hat{\phi} = \frac{\omega}{ik} e^{\boldsymbol{\nu}\cdot\mathbf{x}}, \tag{13.6}$$

where $\boldsymbol{\nu}$ is given in (11.28). The velocity vector of fluid particles is

$$\nabla\hat{\phi} = \frac{\omega\boldsymbol{\nu}}{ik} e^{\boldsymbol{\nu}\cdot\mathbf{x}}, \tag{13.7}$$

and the shift $\hat{\mathbf{v}}_w$ follows from (13.7) by dividing the complex velocity amplitude by $i\omega$:

$$\hat{\mathbf{v}}_w = \frac{-\boldsymbol{\nu}}{k} e^{\boldsymbol{\nu}\cdot\mathbf{x}}. \tag{13.8}$$

The relative shift between a body-fixed point and a fluid particle follows from (13.5) and (13.8) as

$$\hat{\mathbf{r}} = \hat{\mathbf{v}} - \hat{\mathbf{v}}_w. \tag{13.9}$$

Often, however, the vertical relative motion $\hat{r}_3$ including pile-up is of interest, e.g., for estimating whether the maximum water level exceeds the height of an opening or a bulwark. $\hat{r}_3$ follows directly from the pressure at the waterline above or below the point of interest:

$$\hat{r}_3 = \frac{\hat{p}}{\rho g} \,. \tag{13.10}$$

This follows from the facts that the pressure (more exact: the difference to air pressure) is zero at the water surface and that the static vertical pressure gradient is ρg. The latter is sufficient to determine the first-order quantity $\hat{r}_3$.

Sometimes the relative motion velocity at a point P at some distance d from the hull is of interest, e.g., to determine the angle of attack of a roll-damping fin. If d is substantially larger than the panel size, the fluid velocity at P relative to the inertial coordinate system can be determined directly from (7.18) by taking analytical derivatives of the Green functions and the known source strengths. However, if d is in the size range of the panels or less, this would be too inaccurate. Instead, one should project the hull panel next to P normal to the hull to the hull distance d and determine the average velocity for this projected panel from the potential of the vertices of the projected panel (tangential components) and the flux through the panel (normal component), as explained in chapter 'Rankine source methods'.

13.3 Force and Moment at Transverse Cross Sections

For slender bodies like ships, it is often more convenient to base structural analyses not on hull pressure distributions and accelerations, but on sectional forces (including moments) in virtual transverse cross sections x_s = constant: normal force, horizontal and vertical shear forces, torsional moment, vertical bending moment (turning around the y axis), and horizontal bending moment (about z-axis).

At first, the case of loads linearized with respect to wave and ship motion amplitudes is discussed. The relation between forces and accelerations of the part of the ship in front of a cross section at longitudinal position x_s corresponds to the motion Eq. (12.12) for the total ship, except for the fact that the sectional force $\hat{\mathbf{F}}_s$ and moment $\hat{\mathbf{M}}_s$ have to be added:

$$[\mathbf{S}_s - \mathbf{B}_s - \omega_e^2 \mathbf{M}_s]\hat{u} - \hat{\mathbf{F}}_{es} = \begin{pmatrix} \hat{\mathbf{F}}_s \\ \hat{\mathbf{M}}_{s0} \end{pmatrix} . \tag{13.11}$$

Here, $\mathbf{S}_s$, $\mathbf{B}_s$, $\mathbf{M}_s$, $\hat{\mathbf{F}}_{es}$ are the restoring matrix, the complex added mass matrix times ω_e^2, the mass matrix, and the excitation, respectively, for the ship's part in front of x_s. (Here, the excitation force and moment $\hat{\mathbf{F}}_{es}$ is written as a single six-component vector, whereas the section force and moment are written separately as three-component vectors.) To determine these quantities by a patch method, contributions of body patches behind x_s are omitted, and for patches intersecting the area $x = x_s$, their contribution is reduced according to the ratio of patch area or moment in front of the cross section to total patch area or moment, respectively. When using a strip method to determine the restoring matrix $\mathbf{S}_s$ by using (5.40), all integrations over x are performed only from x_s to the forward end. Furthermore, in terms referring to the transom, the data of the cross section at x_s are used instead of the transom data. And for the mass data m, x_G, y_G, and z_G occurring in (5.40), the data for the part of the body in front of x_s have to be used. The center of gravity coordinates and the moments of inertia, however, refer as originally to the coordinate origin, not to $\mathbf{x}_s$.

Usually, the cross section moments are specified referring to a point $\mathbf{x}_s$ within the cross section, not referring to the global coordinate origin K. Sometimes the shear center of the section is used, but any other point may be selected as well. The conversion of the moment from the global coordinate origin to $\mathbf{x}_s$ is made as usual:

$$\hat{\mathbf{M}}_s = \hat{\mathbf{M}}_{s0} - \mathbf{x}_s \times \hat{\mathbf{F}}_s. \tag{13.12}$$

Loads are used for assessing stresses and deformations of the ship structure. Because the structure moves with the body, also the force $\hat{\mathbf{F}}_s$ and moment $\hat{\mathbf{M}}_s$ should be changed from the inertial to the ship-fixed coordinate system, using the inverse of the linearized $\mathbf{T}$ given in (2.47):

$$\mathbf{T}_{lin}^{-1} = \mathbf{T}_{lin}^{T} = \begin{pmatrix} 1 & \psi & -\theta \\ -\psi & 1 & \varphi \\ \theta & -\varphi & 1 \end{pmatrix}. \tag{13.13}$$

$\mathbf{T}_{lin}^{-1}$ is the sum of the unit matrix and elements proportional to the ship motions. If nonlinear contributions are omitted, only the product of $\mathbf{T}_{lin}^{-1}$ with loads of order zero, i.e., the static vertical force $F_z(x_s)$ and the vertical moment $M_y(x_s)$, must be considered. This results in a linear correction of the section force in ship-fixed directions of amplitude

$$\Delta\hat{\mathbf{F}}_s = (-\hat{\theta} F_z(x_s),\ \hat{\varphi} F_z(x_s),\ 0)^T, \tag{13.14}$$

and of the section moment amplitude

$$\Delta\hat{\mathbf{M}}_s = (\hat{\psi} M_y(x_s),\ 0,\ -\hat{\varphi} M_y(x_s))^T. \tag{13.15}$$

In case of nonlinear response prediction, instead of the complex amplitude, the time-dependent value of the section force is determined first in inertial coordinate directions, using a ship-fixed plane $\underline{x}_s = 0$ to decide which hull parts are in front of the section. The resulting section force is then changed to ship-fixed coordinate directions using the nonlinear matrix $\mathbf{T}^{-1}$ resulting from (2.33). Correspondingly, the section moment is first changed to the ship-fixed section reference point $\underline{\mathbf{x}}_s$ using (13.12), and then multiplied from the left by $\mathbf{T}^T$ resulting from (2.33).

13.4 Water Motion in a Moonpool

Some floating bodies have a vertical channel (moonpool) of constant cross section ending downward at the shell and being open at its upper end. It is used to perform operations in the water or on or under the ground below the hull. If the floating body has a substantial steady speed, oscillations of the water in the moonpool may occur even without waves. Furthermore, sloshing motions may occur within the moonpool. Here we deal only with another kind of water motion in the moonpool, namely, relative vertical motions of the water in the moonpool generated by incident waves and wave-induced motions of the floating body. In case of resonance, these motions may be large and may impede the intended operations; thus, it should be possible to predict them. The (usually small) influence of the water motion within the moonpool on the motions of the floating body may be easily accounted for if the relative water motion $z_M(t)$ is known.

The water motion relative to the inertial frame is driven by the oscillating pressure p_M on the shell in the neighborhood of the lower end of the moonpool. p_M is determined from the hull pressure distribution computed without the moonpool, interpolated at the location of the moonpool. May h designate the depth of the lower end of the moonpool below the undisturbed water surface. h is also the mean length of the water column within the moonpool; its actual length is $h - z_M$. z_B is the vertical motion of the body at the moonpool location.

The linearized motion equation for the water column of actual length $h - z_M$ within the moonpool is

$$p_M + \rho g h + \Delta p = \rho(g - \ddot{z}_B - \ddot{z}_M)(h - z_M). \tag{13.16}$$

On the right-hand side the vertical pressure gradient, depending on the 'effective acceleration' $g - \ddot{z}_B - \ddot{z}_M$ times the actual column length, gives the pressure difference between the lower and upper ends of the moonpool. On the left-hand side, in case of no flow through the moonpool, $p_M + \rho g h$ is also the pressure difference between the lower end of the moonpool and the atmosphere. In other cases (if $\dot{z}_M \neq 0$), we have to distinguish between inflow and outflow. If there is flow out of the moonpool, i.e., $\dot{z}_M > 0$, there is no substantial pressure difference Δp between the surrounding lower moonpool opening and the pressure within the moonpool near its lower end; thus, for outflow, the term Δp should be taken as zero.

In case of inflow, however, the flow relative to the shell is accelerated from zero to a speed v, giving rise to a pressure drop of $\frac{1}{2}\rho v^2$, and a contraction of the flow cross section to an area μA, which is smaller than the moonpool cross section A. For an approximately circular or quadratic cross section and a sharp edge between moonpool and hull, μ is about 0.63. Within some distance from the lower end of the moonpool there occurs mixing between the inner inflow and the outer stagnant water within the moonpool cross section. That causes a nearly uniform vertical velocity of $\mu v = -\dot{z}_M$. This mixing changes the momentum flux of the fluid in the moonpool: at the inlet, it is $\rho\mu A v^2 = \rho A/\mu \cdot \dot{z}_M^2$, while after mixing the momentum flux is $\rho A \dot{z}_M^2$.

That requires a pressure increase of $\rho\dot{z}_M^2(\frac{1}{\mu}-1)$. Together with the initial pressure drop of $\frac{1}{2}\rho v^2 = \frac{1}{2}\rho(\dot{z}_M/\mu)^2$, the total pressure change from outside to inside the moonpool is

$$\Delta p = \rho\dot{z}_m^2 \underbrace{\left(-\frac{1}{2\mu^2}+\frac{1}{\mu}-1\right)}_{-\beta}. \tag{13.17}$$

Δp does not contain the hydrostatic pressure change between the lower end of the moonpool and the cross section at which the flow speed is constant over the cross section, because that pressure drop is contained already in the g term in (13.16). The typical value $\mu = 0.63$ results in $\beta = 0.67$.

Equation (13.16) combined with (13.17) contains linear and quadratic terms. Quadratic terms will be neglected with one exception: the damping term Δp. Neglecting it would give infinite response in case of resonance because (13.16) does not contain other damping. Thus, also small contributions to damping can have a large effect on z_M. Linearization, except for Δp, gives the following equation:

$$\ddot{z}_M h + g z_M - \beta\dot{z}_M^2 = -\ddot{z}_B h - \frac{1}{\rho}p_M. \tag{13.18}$$

To determine transfer functions, the damping term must be linearized. Here, the so-called 'equivalent linearization' is applied. It substitutes the term $\dot{z}_M^2$ by $-\alpha\dot{z}_M$, where α is a real positive constant determined such that the energy dissipated by the original and the substituted terms within one encounter period T is the same. Expressing the energy as time integral over power and power as velocity times force, results in the following relation:

$$\int_0^{T/2} \dot{z}_M^3 dt = -\alpha \int_0^T \dot{z}_M^2 dt. \tag{13.19}$$

The upper integration limit $T/2$ in the left-hand integral is used because the damping term is non-zero only during inflow, not during outflow. To determine α, the time dependence of $\dot{z}_M$ must be used:

$$\dot{z}_M = \mathrm{Re}(i\omega_e\hat{z}_M e^{i\omega_e t}). \tag{13.20}$$

To have inflow during the first half of the encounter period requires that $\hat{z}_M$ is real and positive. If it were different, we have to use other integration limits on the left-hand side of (13.19). Evaluating the integrals after inserting (13.20) with real $\hat{z}_M$ into (13.19) gives

$$\alpha = \frac{4}{3\pi}\omega_e|\hat{z}_M|. \tag{13.21}$$

The result (13.21) allows us to write (13.18) as equation in complex amplitudes:

$$\left(-\omega_e^2 h + i\omega_e^2 \frac{4}{3\pi}\beta|\hat{z}_M| + g\right)\hat{z}_M = \omega_e^2 h\hat{z}_B - \frac{\hat{p}_M}{\rho}. \tag{13.22}$$

From this equation, the only unknown $\hat{z}_M$ can be determined. Strictly speaking, it is not a transfer function. If the wave amplitude is larger than 1, the damping (middle term between parentheses) increases with wave amplitude because $\hat{z}_M$ increases with wave amplitude.

Reference

1. D. Hachmann, Calculation of pressures on a ship's hull. Ship Technol. Res. **38**, 111–133 (1991)

Chapter 14
Drift Force and Added Resistance

Abstract This chapter deals with the stationary second-order force exerted on the body by regular waves. Of most interest is the longitudinal force component, the added resistance. The chapter gives a rigorous derivation of the second-order force, including effects which are usually neglected in the relevant literature. Results of the added resistance determined by a Rankine source method using these formulas are compared with results of model experiments for a sharp and a full ship.

Drift forces are time-averaged forces on a body due to waves. The negative longitudinal drift force is the wave-induced added resistance. For bodies without forward speed, drift forces act, roughly, in wave propagation direction. For bodies with steady speed ahead, drift forces tend to be directed backward. For surface vessels, normally only the longitudinal and transverse drift force and the yaw drift moment are relevant; the other drift components cause small changes of average draft, heel, and trim.

The leading order of drift forces is 2; that means, drift forces are, approximately, proportional to the square of the wave amplitude ζ_A. Sometimes the expression 'quadratic transfer functions' is used for second-order drift responses per wave amplitude squared. Here, it is shown how second-order drift force and moment can be determined from first-order responses.

Second-order wave responses consist of a stationary and an oscillatory contribution. The former constitutes the drift response; the latter varies periodically with twice the wave encounter frequency and is practically relevant only for vibrations of the elastic ship hull.

In the following, only first-order ship motions are taken into account. For drift forces, only the stationary part of the second-order motion is relevant. But because drift forces are interesting only for those degrees-of-freedom that lack a restoring force, second-order stationary ship motions have no effect on the interesting second-order drift force components.

Also, higher than first orders of the flow potential will be neglected. For the wave potential, the oscillatory part of the second-order contribution appears irrelevant, but the stationary part of the second-order contribution will generate a mean water flow that depends on the vertical coordinate. If the ship is assumed to follow this wave drift velocity (averaged over the vertical extent of the ship), there is no influence on

B. O. el Moctar et al., *Numerical Methods for Seakeeping Problems*,
https://doi.org/10.1007/978-3-030-62561-0_14

the drift force. The only simplification in the following elaboration of the second-order drift force is the neglect of the stationary part of the second-order disturbance potential. It is expected that this has only a small effect, which could also be quantified by further investigation.

Drift forces are sensitive to various approximations; thus, according to our experience, drift forces computed by strip methods are too inaccurate for practical applications.

Most papers on drift forces neglect many contributions that are included here; for instance:

- Interaction between the periodic velocity and the stationary disturbance flow due to ship speed
- Changes of the waterline due to steady ship waves and squat
- Shift of ship-fixed points in the inertial system due to squares of rotation angles
- Changes of normal vectors due to squares of rotation angles

The latter two contributions are, however, small.

Some other improvements have been investigated, but were found to be nearly negligible:

- To take account of the vertical pressure gradient in the stationary flow. In (14.39), instead of the gravity acceleration g, the quantity $g - a_z$ may be used, where a_z is the vertical fluid particle acceleration due to the stationary flow potential $-Ux + \phi_0$. This correction is not only small, but also sensitive to approximations of the steady ship waves, especially at the bow.
- The vertical drift force and the drift trim moment change the average floating position. Besides having other second-order effects on potential flow, this changes the wetted surface and thus the viscous resistance.
- The change of mean relative velocity $\mathbf{v}_r$ between fluid and hull due to the waves, which changes the friction resistance.

Besides drift forces, there exist also wave-induced drift moments. Practically interesting is the drift yaw moment and, possibly, the drift heel moment. Both are skipped in this book.

14.1 Preliminaries

The relation between the coordinates of a point expressed in the ship-fixed coordinate system, $\underline{\mathbf{x}}$, and the same point expressed in the inertial system, $\mathbf{x}$, is

$$\mathbf{x} = \underline{\mathbf{x}} + \mathbf{u} + \boldsymbol{\alpha} \times \underline{\mathbf{x}} + \mathbf{T}_2\underline{\mathbf{x}}. \tag{14.1}$$

Here and further in this section, terms of third and higher orders are omitted without notice. $\mathbf{u}$ is the translation (shift between ship-fixed and inertial coordinate origin), and $\boldsymbol{\alpha} = (\varphi, \theta, \psi)$ the rotation, i.e., the vector consisting of the components roll φ, pitch θ, and yaw ψ. $\mathbf{T}_2$ is the second-order contribution of matrix $\mathbf{T}$ defined in (2.33):

$$\mathbf{T}_2 = \begin{pmatrix} -\frac{1}{2}\theta^2 - \frac{1}{2}\psi^2 & \varphi\theta & \varphi\psi \\ 0 & -\frac{1}{2}\varphi^2 - \frac{1}{2}\psi^2 & \theta\psi \\ 0 & 0 & -\frac{1}{2}\varphi^2 - \frac{1}{2}\theta^2 \end{pmatrix}. \tag{14.2}$$

In the following, normal vectors on a body panel (directed into the body) are scaled so that their magnitude equals the panel area. Corresponding to (14.1), a panel normal vector expressed in the inertial frame follows from $\underline{\mathbf{n}}$ (expressed in the body-fixed coordinate system):

$$\mathbf{n} = \underbrace{\underline{\mathbf{n}}}_{\mathbf{n}_0} + \underbrace{\boldsymbol{\alpha} \times \underline{\mathbf{n}}}_{\mathbf{n}_1} + \underbrace{\mathbf{T}_2\underline{\mathbf{n}}}_{\mathbf{n}2}. \tag{14.3}$$

The product of two harmonically oscillating quantities $a = \mathrm{Re}(\hat{a}e^{i\omega_e t})$ and $b = \mathrm{Re}(\hat{b}e^{i\omega_e t})$ consists of a constant and a time-dependent part oscillating with twice the frequency of the factors:

$$ab = \mathrm{Re}(\hat{a}e^{i\omega_e t}) \cdot \mathrm{Re}(\hat{b}e^{i\omega_e t}) = \mathrm{Re}[\hat{a}e^{i\omega_e t} \cdot \mathrm{Re}(\hat{b}e^{i\omega_e t})] = \mathrm{Re}\left[\hat{a}e^{i\omega_e t} \cdot \frac{1}{2}(\hat{b}e^{i\omega_e t} + \hat{b}^* e^{-i\omega_e t})\right] = \frac{1}{2}\mathrm{Re}\left[\hat{a}\hat{b}e^{2i\omega_e t} + \hat{a}\hat{b}^*\right], \tag{14.4}$$

where $*$ indicates the complex conjugate. The oscillating part has a time average of zero; thus the average of the product (indicated here and in the following by over-lining) is

$$\overline{ab} = \frac{1}{2}\mathrm{Re}(\hat{a}\hat{b}^*). \tag{14.5}$$

Correspondingly, the following rule may be derived for the vector product of harmonically oscillating vectors:

$$\overline{\mathbf{a} \times \mathbf{b}} = \frac{1}{2}\mathrm{Re}(\hat{\mathbf{a}} \times \hat{\mathbf{b}}^*). \tag{14.6}$$

From that follows:

$$\overline{\dot{\boldsymbol{\alpha}} \times \boldsymbol{\alpha}} = \omega_e \mathrm{Re}(\hat{\boldsymbol{\alpha}}) \times \mathrm{Im}(\hat{\boldsymbol{\alpha}}). \tag{14.7}$$

In contrast to this, the time average of the product $\dot{a}a$ is zero, if a is a scalar, harmonically oscillating quantity.

The time average of another product of harmonically oscillating vectors $\mathbf{a}$, $\mathbf{b}$ is also required. For an arbitrary constant 3 by 3 matrix C, there is

$$\overline{\mathbf{a}C\mathbf{b}} = \frac{1}{2}\mathrm{Re}(\hat{\mathbf{a}}C\hat{\mathbf{b}}^*). \tag{14.8}$$

The shift of a ship-fixed point in the inertial system is, according to (14.1),

$$\mathbf{v}(\underline{\mathbf{x}}) = \mathbf{x} - \underline{\mathbf{x}} = \underbrace{\mathbf{u} + \boldsymbol{\alpha} \times \underline{\mathbf{x}}}_{\mathbf{v}_1} + \underbrace{\mathbf{T}_2 \underline{\mathbf{x}}}_{\mathbf{v}_2}. \tag{14.9}$$

$\mathbf{v}_1$ and $\mathbf{v}_2$ are the first- and second-order contributions at ship-fixed points, respectively. $\mathbf{v}$ can also be expressed as a function of $\mathbf{x}$ instead of $\underline{\mathbf{x}}$: Using the identity

$$-\boldsymbol{\alpha} \times (\boldsymbol{\alpha} \times \mathbf{x}) + \mathbf{T}_2 \mathbf{x} = -\mathbf{T}_2^T \mathbf{x} \tag{14.10}$$

gives

$$\mathbf{v}(\mathbf{x}) = \mathbf{u} + \boldsymbol{\alpha} \times (\mathbf{x} - \mathbf{u}) - \mathbf{T}_2^T \mathbf{x}. \tag{14.11}$$

If a quantity depends on both time and space, one has to distinguish partial time derivatives for fixed $\mathbf{x}$ from those for fixed $\underline{\mathbf{x}}$. In the following, $\dot{}$ indicates a time derivative for constant $\underline{\mathbf{x}}$, and $\breve{}$ that for constant $\mathbf{x}$. To illustrate this, and because the quantity is required later, the 'inertial' time derivative of $\mathbf{v}$ is considered. From (14.11) follows:

$$\breve{\mathbf{v}}(\mathbf{x}) = \dot{\mathbf{u}} + \dot{\boldsymbol{\alpha}} \times (\mathbf{x} - \mathbf{u}) - \boldsymbol{\alpha} \times \dot{\mathbf{u}} - \dot{\mathbf{T}}_2^T \mathbf{x}. \tag{14.12}$$

14.2 Drift Force Due to Pressure Acting on the Surface up to the Mean Waterline

The total pressure force $\mathbf{F}$ on the hull will be subdivided into contribution A due to the hull pressure up to the mean waterline in the ship-fixed coordinate system, and contribution B due to the pressure acting on the hull region between the actual and the mean waterline. Then, the pressure $p(\mathbf{x}, t)$ acting at $\mathbf{x}$ at time t generates the following force contribution A:

$$\mathbf{F}_A = \sum_{panels} p(\mathbf{x}, t)\mathbf{n}(t). \tag{14.13}$$

Pressure contributions which are time invariant, or vary as first-order or second-order quantities at ship-fixed points, will now be designated as p_0, p_1 and p_2, respectively. Thus,

$$\mathbf{F}_A = \sum_{panels} (p_0 + p_1 + p_2)(\mathbf{n}_0 + \mathbf{n}_1 + \mathbf{n}_2), \tag{14.14}$$

where the normal vectors are given by (14.3). Here, we are interested in the wave-induced drift force, which is the stationary part of the second-order force $\mathbf{F}_A^{(2)}$. It follows from (14.14) as

$$\mathbf{F}_A^{(2)} = \sum_{panels} (p_0 \mathbf{n}_2 + p_1 \mathbf{n}_1 + p_2 \mathbf{n}_0). \tag{14.15}$$

The pressure follows from Bernoulli's equation:

$$\frac{p(\mathbf{x},t)}{\rho} = \frac{1}{2}U^2 - \breve{\phi}(\mathbf{x},t) + gz - \frac{1}{2}|\nabla\phi(\mathbf{x},t)|^2. \tag{14.16}$$

The flow potential ϕ is superimposed according to the Hachmann method (7.41) (otherwise, the meaning of ϕ_1 would be different from that of Chap. 7):

$$\phi(\mathbf{x},t) = -Ux + \phi_0(\underline{\mathbf{x}}) + \underbrace{\mathrm{Re}(\hat{\phi}(\mathbf{x})e^{i\omega_e t})}_{\phi_1(\mathbf{x})}. \tag{14.17}$$

Potential ϕ_0 is the steady disturbance potential, a function that is independent of time in ship-fixed coordinates. However, we need to know its dependence on the inertial location $\mathbf{x}$. Thus, in (14.17) a Taylor expansion around $\mathbf{x}$ up to second order is made:

$$\phi(\mathbf{x},t) = -Ux + \phi_0(\mathbf{x}) - \mathbf{v}(\mathbf{x},t)\cdot\nabla\phi_0(\mathbf{x}) + \frac{1}{2}\mathbf{v}^T(\nabla\nabla\phi_0)\mathbf{v} + \phi_1(\mathbf{x}). \tag{14.18}$$

∇ refers to derivatives in the inertial coordinate directions. The partial time derivative of ϕ for fixed $\mathbf{x}$ follows as

$$\begin{aligned}\breve{\phi}(\mathbf{x},t) &= -\breve{\mathbf{v}}(\mathbf{x},t)\cdot\nabla\phi_0(\mathbf{x}) + \breve{\mathbf{v}}^T(\nabla\nabla\phi_0)\mathbf{v} + \breve{\phi}_1(\mathbf{x})\\ &= -\breve{\mathbf{v}}(\mathbf{x},t)\cdot\nabla\phi_0(\underline{\mathbf{x}}) + \breve{\phi}_1(\mathbf{x})\\ &= -\breve{\mathbf{v}}(\mathbf{x},t)\cdot(\underline{\nabla} + \boldsymbol{\alpha}\times\underline{\nabla})\phi_0(\underline{\mathbf{x}}) + \breve{\phi}_1(\mathbf{x}),\end{aligned} \tag{14.19}$$

where $\underline{\nabla}$ designates derivatives into the directions of the ship-fixed coordinates. Up to first order, both operators are related by

$$\nabla = \underline{\nabla} + \boldsymbol{\alpha}\times\nabla. \tag{14.20}$$

Because (14.15) requires seperating the pressure into orders zero, one, and two *at ship-fixed points*, $\mathbf{x}$ is changed to $\underline{\mathbf{x}}$ on the right-hand side of (14.19) using a Taylor expansion:

$$\breve{\phi}(\mathbf{x},t) = \breve{\phi}(\underline{\mathbf{x}},t) - (\nabla\breve{\mathbf{v}}\;\mathbf{v})\cdot(\underline{\nabla} + \boldsymbol{\alpha}\times\underline{\nabla})\phi_0(\underline{\mathbf{x}}) + \nabla\breve{\phi}_1\cdot\mathbf{v}, \tag{14.21}$$

where $\breve{\phi}(\underline{\mathbf{x}},t)$ is the expression (14.19) using $\underline{\mathbf{x}}$ instead of $\mathbf{x}$. The expression $\nabla\breve{\mathbf{v}}\;\mathbf{v}$ follows from (14.9):

$$\nabla\breve{\mathbf{v}}\;\mathbf{v} = \dot{\boldsymbol{\alpha}}\times\mathbf{v}_1. \tag{14.22}$$

To evaluate (14.16) we need also

$$|\nabla\phi(\mathbf{x},t)| = |\underline{\nabla}\phi(\mathbf{x},t)|. \tag{14.23}$$

The right-hand side is much easier to evaluate. Starting from (14.17) and changing, as above, from $\mathbf{x}$ to $\underline{\mathbf{x}}$ gives

$$\underline{\nabla}\phi(\mathbf{x},t) = \underline{\nabla}[(-U,0,0)\cdot(\underline{\mathbf{x}}+\mathbf{u}+\boldsymbol{\alpha}\times\underline{\mathbf{x}}+\mathbf{T}_2\underline{\mathbf{x}})+\phi_0(\underline{\mathbf{x}})+\phi_1(\underline{\mathbf{x}})+\mathbf{v}\cdot\nabla\phi_1] \quad (14.24)$$

$$= (-U,0,0)\cdot(\mathbf{E}+\boldsymbol{\alpha}\times\mathbf{E}+\mathbf{T}_2)+\underline{\nabla}\phi_0(\underline{\mathbf{x}})+\underline{\nabla}\phi_1(\underline{\mathbf{x}})+\nabla(\mathbf{v}\cdot\nabla\phi_1) \quad (14.25)$$

$$= \underbrace{(-U,0,0)+\underline{\nabla}\phi_0(\underline{\mathbf{x}})}_{\mathbf{w}_0}+\underbrace{(0,U\psi,-U\theta)+\nabla\phi_1(\underline{\mathbf{x}})}_{\mathbf{w}_1}$$

$$\underbrace{-\boldsymbol{\alpha}\times\nabla\phi_1+(-U,0,0)\mathbf{T}_2+\nabla\mathbf{v}\nabla\phi_1+\mathbf{v}^T\nabla\nabla\phi_1}_{\mathbf{w}_2}. \quad (14.26)$$

The vectors $\mathbf{w}_0$, $\mathbf{w}_1$, $\mathbf{w}_2$ are the contributions of zeroth, first, and second orders at ship-fixed positions, respectively. In the last line, the first and third terms cancel; thus,

$$\mathbf{w}_2 = (-U,0,0)\mathbf{T}_2+\mathbf{v}^T\nabla\nabla\phi_1. \quad (14.27)$$

Expressions written as vector×matrix are meant to designate

$$\mathbf{v}\times(\mathbf{m}_1,\mathbf{m}_2,\mathbf{m}_3) = (\mathbf{v}\times\mathbf{m}_1,\mathbf{v}\times\mathbf{m}_2,\mathbf{v}\times\mathbf{m}_3). \quad (14.28)$$

$\mathbf{E}$ is the 3 by 3 unit matrix. $\nabla\mathbf{v}$ is (up to first order) the matrix

$$\nabla\mathbf{v} = \boldsymbol{\alpha}\times\mathbf{E} = \begin{pmatrix} 0 & -\psi & \theta \\ \psi & 0 & -\varphi \\ -\theta & \varphi & 0 \end{pmatrix}. \quad (14.29)$$

Inserting (14.21) and (14.26) into (14.16) gives the pressure terms of different order:

$$\frac{p_0}{\rho} = (0,0,g)\cdot\underline{\mathbf{x}}-\frac{1}{2}|\underline{\nabla}\phi_0|^2-(-U,0,0)\cdot\underline{\nabla}\phi_0; \quad (14.30)$$

$$\frac{p_1}{\rho} = \check{\mathbf{v}}_1\underline{\nabla}\phi_0-\check{\phi}_1+(0,0,g)\cdot\mathbf{v}_1-\mathbf{w}_0\cdot\mathbf{w}_1; \quad (14.31)$$

$$\frac{p_2}{\rho} = \underline{\check{\mathbf{v}}_2\underline{\nabla}\phi_0}+\underline{(\boldsymbol{\alpha}\times\underline{\nabla}\phi_0)\times\check{\mathbf{v}}_1}+\underline{(\nabla\check{\mathbf{v}}_1\,\mathbf{v}_1)\underline{\nabla}\phi_0}-\nabla\check{\phi}_1\cdot\mathbf{v}_1+(0,0,g)\cdot\mathbf{v}_2-\frac{1}{2}|\mathbf{w}_1|^2-\mathbf{w}_0\cdot\mathbf{w}_2. \quad (14.32)$$

Here, quantities depending on location are all meant at $\underline{\mathbf{x}}$. The factor $\check{\mathbf{v}}_2$ in (14.32) is zero. The time averages of terms 2 and 3 on the right-hand side of (14.32) cancel. Because in (14.15) p_2 is multiplied by the constant factor $\mathbf{n}_0 = \underline{\mathbf{n}}$, the influence of these terms on the drift (i.e., time-averaged) force is zero. Thus, the underlined terms in (14.32) can all be omitted.

Inserting p_0, p_1 and p_2 into (14.15) gives

$$\frac{1}{\rho}\sum_{panels}\mathbf{F}_A^{(2)} = \left(g\underline{z}-\frac{1}{2}|\underline{\nabla}\phi_0|^2-(-U,0,0)\cdot\underline{\nabla}\phi_0\right)\mathbf{T}_2\underline{\mathbf{n}}+\left(\check{\mathbf{v}}_1\underline{\nabla}\phi_0-\check{\phi}_1+(0,0,g)\cdot\mathbf{v}_1-\mathbf{w}_0\cdot\mathbf{w}_1\right)\boldsymbol{\alpha}\times\underline{\mathbf{n}}$$

$$-\left(\nabla\breve{\phi}_1 \cdot \mathbf{v}_1 + (0,0,g)\cdot \mathbf{v}_2 + \frac{1}{2}|\mathbf{w}_1|^2 + \mathbf{w}_0 \cdot \mathbf{w}_2\right)\underline{\mathbf{n}}. \tag{14.33}$$

The time average of this expression, that is part A of the drift force, follows using the equations given in Sect. 14.1:

$$\frac{1}{\rho}\sum_{panels} \overline{\mathbf{F}_A^{(2)}} = -\left(\frac{1}{2}|\underline{\nabla}\phi_0|^2 + (-U,0,0)\cdot\underline{\nabla}\phi_0 - g\underline{z}\right)\bar{\mathbf{T}}_2\underline{\mathbf{n}}$$

$$+\frac{1}{2}\mathrm{Re}\left[\left(i\omega_e\hat{\mathbf{v}}_1\cdot\underline{\nabla}\phi_0 - i\omega_e\hat{\phi}_1 + (0,0,g)\cdot\hat{\mathbf{v}}_1 - \mathbf{w}_0\cdot\hat{\mathbf{w}}_1\right)\hat{\boldsymbol{\alpha}}^*\right]\times\underline{\mathbf{n}}$$

$$-\left(\frac{1}{2}\mathrm{Re}(i\omega_e\nabla\hat{\phi}_1\cdot\hat{\mathbf{v}}_1^*) - (0,0,g)\cdot\bar{\mathbf{v}}_2 + \frac{1}{4}|\hat{\mathbf{w}}_1|^2 + \mathbf{w}_0\cdot\bar{\mathbf{w}}_2\right)\underline{\mathbf{n}}. \tag{14.34}$$

$\bar{\mathbf{T}}_2$ follows from (14.2) as

$$\bar{\mathbf{T}}_2 = \begin{pmatrix} -\frac{1}{4}(|\hat{\theta}|^2+|\hat{\psi}|^2) & \frac{1}{2}\mathrm{Re}(\hat{\varphi}\hat{\theta}^*) & \frac{1}{2}\mathrm{Re}(\hat{\varphi}\hat{\psi}^*) \\ 0 & -\frac{1}{4}(|\hat{\varphi}|^2+|\hat{\psi}|^2) & \frac{1}{2}\mathrm{Re}(\hat{\theta}\hat{\psi}^*) \\ 0 & 0 & -\frac{1}{4}(|\hat{\varphi}|^2+|\hat{\theta}|^2) \end{pmatrix}, \tag{14.35}$$

and $\bar{\mathbf{w}}_2$ from (14.27):

$$\bar{\mathbf{w}}_2 = (-U,0,0)\bar{\mathbf{T}}_2 + \frac{1}{2}\mathrm{Re}(\nabla\nabla\hat{\phi}_1\hat{\mathbf{v}}_1^*). \tag{14.36}$$

14.3 Drift Force Due to Pressure Acting Between the Average and the Actual Waterline

For the drift force contribution $\mathbf{F}_B^{(2)}$ due to the variable hull submergence, the first-order pressure oscillations at ship-fixed points are relevant because the zero-order pressure p (the difference in the air pressure) is zero at the waterline, and the second-order pressure acting on the strip between actual and mean waterline results in a third-order force because the breadth of the strip is a first-order quantity.

If the first-order pressure at the mean waterline is $p_w > 0$, the pressure between actual and mean waterline changes approximately linearly from zero to p_w. On a length element $\boldsymbol{\Delta}\mathbf{s}$ along the waterline it produces a force element

$$d\mathbf{F}_B = \frac{1}{2}p_w\boldsymbol{\Delta}\mathbf{s}\times\boldsymbol{\Delta}\mathbf{t}. \tag{14.37}$$

Here, $\boldsymbol{\Delta}\mathbf{s} \times \boldsymbol{\Delta}\mathbf{t}$ is the area vector (pointing normal into the hull) between actual and mean waterline corresponding to s, which is directed forward on starboard and backward on port side. The vector $\boldsymbol{\Delta}\mathbf{t}$ is determined from the following conditions:

- The z component must be $p_w/\rho g$ to obtain the correct difference between actual and mean waterline.
- $\boldsymbol{\Delta}\mathbf{t}$ is tangential to the hull; thus, $\boldsymbol{\Delta}\mathbf{t} \cdot \underline{\mathbf{n}} = 0$.
- As the component of $\boldsymbol{\Delta}\mathbf{t}$ in direction $\boldsymbol{\Delta}\mathbf{s}$ has no influence on the result, we pose the condition $\boldsymbol{\Delta}\mathbf{s} \cdot \boldsymbol{\Delta}\mathbf{t} = 0$.

From these conditions follows:

$$\boldsymbol{\Delta}\mathbf{t} = \frac{p_w}{\rho g} \frac{\underline{\mathbf{n}} \times \boldsymbol{\Delta}\mathbf{s}}{(\underline{\mathbf{n}} \times \boldsymbol{\Delta}\mathbf{s})_z}. \tag{14.38}$$

This results in

$$\mathbf{F}_B = \sum_{WL-panels} \frac{p_w^2}{2\rho g} \frac{\boldsymbol{\Delta}\mathbf{s} \times (\underline{\mathbf{n}} \times \boldsymbol{\Delta}\mathbf{s})}{(\underline{\mathbf{n}} \times \boldsymbol{\Delta}\mathbf{s})_z}, \tag{14.39}$$

where the sum comprises all panels immediately below the waterline.

If $p_w < 0$, in the force contribution A the area between mean and actual waterline was included. Because the water pressure is lacking in this area, a correction must be applied. The force assumed in contribution A for this area must be subtracted. Because the pressure in this area is negative, the correction is, like for $p_w > 0$, a force directed into the body. Its size corresponds also to (14.39), which is, therefore, applicable both for positive and negative p_w.

The time average of $\mathbf{F}_B$, i.e., the drift force contribution B, follows from (14.39) by substituting p_w^2 by its average value $\frac{1}{2}|\hat{p}_w|^2$.

$\hat{p}_w$ can be determined only at the panel centers, not directly at the mean waterline. For the part of $\hat{p}_w$ caused by the wave and the diffraction potentials, $\hat{p}_w$ at the panel center should be multiplied by $e^{k\Delta z}$, where k is the wave number and Δz the vertical distance between panel center and waterline. This correction increases the drift force in short waves. For simplification, this correction may be applied also directly to $\hat{p}_w$ (comprising also other contributions) because in short waves contributions by ship motions are small, and in longer waves, where the ship moves substantially, the correction is small anyway.

14.4 Verification

The most important component of the drift force is the added resistance, the negative x component of the drift force $\mathbf{F}$. It is usually presented as the non-dimensional coefficient

$$C_{AR} = \frac{-F_x}{\rho g \zeta_A^2 B^2 / L_{pp}} \tag{14.40}$$

Main particulars of the WILS II containership

L_{pp}	321.0m	B	48.4m	$T_a = T_f$	15m
V	140200m^3	KG	21.30m	GM	2.0m
k_{xx}	19.07m	$k_{yy} = k_{zz}$	77.23m		

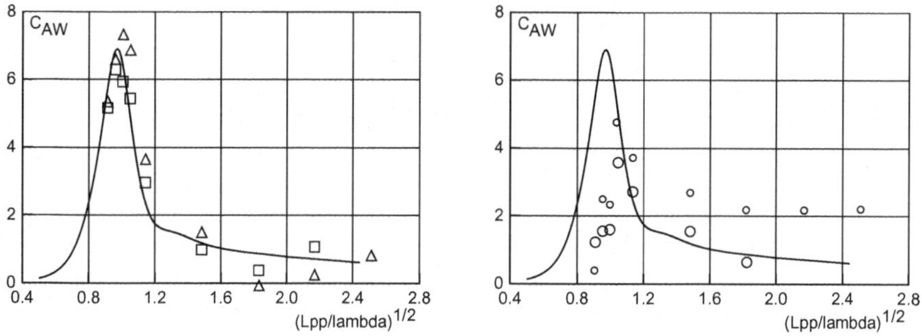

Fig. 14.1 Comparison of added resistance coefficient C_{AR} for the WILS II containership. Curves are the results of the formulae given above. Left: comparison with RANS calculations [1] using a coarse (triangles) and a fine (squares) grid. Right: comparison with model experiments [2] using two different evaluation methods

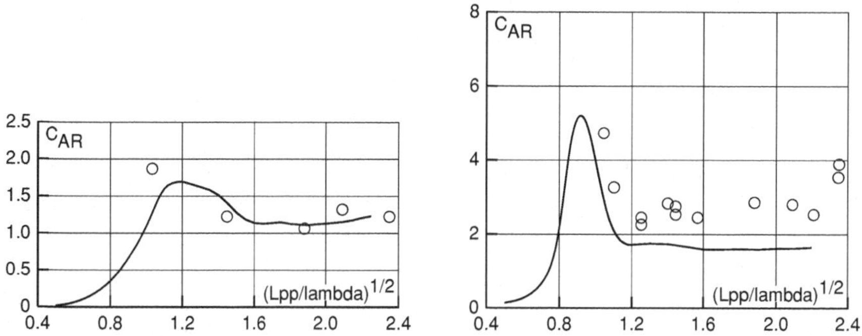

Fig. 14.2 Comparison of added resistance coefficient C_{AR} for the KVLCC2 tanker for $F_n = 0$ (left) and $F_n = 0.142$ (right). Curves are results of the formulae given above; markers from measurements by Bingje and Steen [3]

where B is the ship's waterline breadth, and ζ_A the wave amplitude.

Figures 14.1 and 14.2 show the added resistance coefficient in regular head waves for a containership and a tanker, respectively. The peak of C_{AR} in all curves corresponds to wavelengths causing large heave and pitch motions; it is important to estimate up to which wave conditions a ship can proceed against a heavy seaway, e.g., to avoid stranding. The short-wave region (larger abscissa values), on the other hand, is important for the average fuel consumption. Figure 14.2 for the tanker shows that the added resistance increases with forward speed, especially in the longer wavelength region. Only in the region of short wavelengths, added resistance is larger for full ships (tanker) than for slender (container) ships. Because the drift force is much smaller than the force amplitude, both measuring and computing it is difficult, especially in short waves; this results in large differences between the results of model experiments (circles), of CFD calculations (triangular and square symbols), and of potential calculations (curves).

14.5 Appendix: Determination of the Hesse Matrix of Potentials

To determine the first-order flow potential ϕ_1 around a ship, one needs to know the Hesse matrix $\nabla\nabla\phi_0$ of the steady disturbance potential ϕ_0 (due to the mean forward speed) on the wetted body surface if one does not use the Hachmann method (see Sect. 7.3.1). But even if this method is used, the Hesse matrix of ϕ_1 is required to determine the drift force and moment. Its contribution to added resistance is moderate. Panel methods, and the patch method as well, do not result, in the limit of vanishing panel size, in correct values for the Hesse matrices of the potential if one simply superimposes the second derivatives of the Green functions used to approximate first derivatives; and for the usual panel sizes, using second derivatives of the Green functions does not give reasonable approximations to the Hesse matrices. The reason is that, in case of usual panel methods, the first potential derivatives hold only at the collocation points in the panel centers; between these points, the potential is not sufficiently smooth.

Useful approximations for the second derivative are determined, instead, as finite differences between first derivatives sampled at the centers of the neighbor panels of the panel of interest O. For a triangular body panel having three neighbor panels, the midpoints of the neighbors are designated as A, B, and C (Fig. 14.3). If a panel O is located at the outer boundary of the panel mesh along a symmetry plane, the mirror image of O is used as one neighbor. If panel O is located at the waterline, Fig. 14.4 shows which panels are used as neighbors, because the panel centers A, B, C should approximate an equilateral triangle as far as possible.

The Hesse matrix of a panel O is defined as

$$\nabla\nabla\phi = \nabla\mathbf{u} = \begin{pmatrix} u_{1x} & u_{1y} & u_{1z} \\ u_{2x} & u_{2y} & u_{2z} \\ u_{3x} & u_{3y} & u_{3z} \end{pmatrix}, \tag{14.41}$$

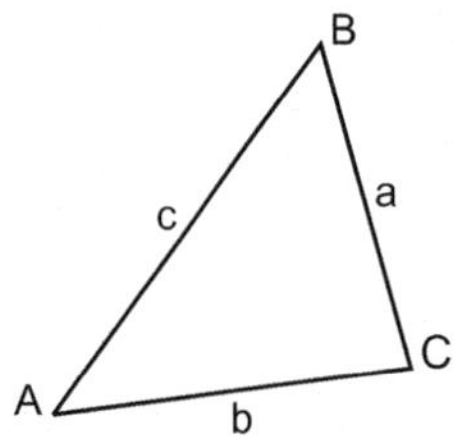

Fig. 14.3 Centers of direct neighbor panels A, B, C and meaning of vectors a, b, and c

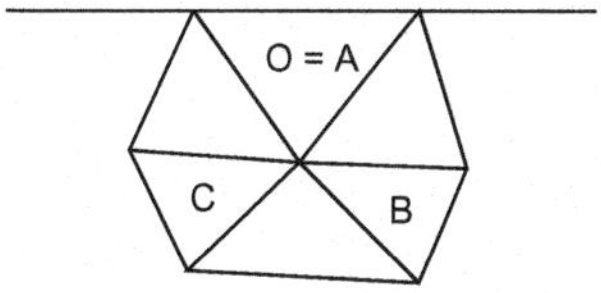

Fig. 14.4 Selection of neighbor panels A, B, C to the panel of interest O located at the waterline

where 1, 2, 3 indicate components of the velocity $\mathbf{u}$, and x, y, z partial derivatives in the coordinate directions. We define $\mathbf{a} = \mathbf{x}_C - \mathbf{x}_B$ and $\mathbf{b} = \mathbf{x}_A - \mathbf{x}_C$ (Fig. 14.3). The fluid velocities $\mathbf{u}_A$, $\mathbf{u}_B$, and $\mathbf{u}_C$ at the midpoints of the neighbor panels are supposed to be known. Then the following five relations are used to determine the elements of the Hesse matrix:

$$(\mathbf{a} \cdot \nabla)\mathbf{u} = \mathbf{u}_C - \mathbf{u}_B; \tag{14.42}$$

$$(\mathbf{b} \cdot \nabla)\mathbf{u} = \mathbf{u}_A - \mathbf{u}_C; \tag{14.43}$$

$$\mathbf{a} \cdot \text{rot}\mathbf{u} = 0; \tag{14.44}$$

$$\mathbf{b} \cdot \text{rot}\mathbf{u} = 0; \tag{14.45}$$

$$\text{div}\mathbf{u} = 0. \tag{14.46}$$

The first two of these equations (i.e., six scalar equations) are finite-difference equations for $\nabla\mathbf{u}$ in directions tangential to the hull surface. The following three scalar equations ensure the correct change of $\mathbf{u}$ in the direction normal to the hull. All together the nine scalar equations form a linear equation system for the nine elements of the matrix (14.41). Because it is a symmetrical matrix, the result can perhaps be improved by taking the average of the original solution and its transpose.

References

1. H. Söding, V. Shigunov, T.E. Schellin, O. el Moctar, A Rankine panel method for added resistance of ships in waves, in *Proceedings of the 31st ASME International Conference on Ocean, Offshore and Arctic Engineering* (2012)
2. S.Y. Hong, Wave induced loads on ships joint industry project - II. Technical report BSPIS503 A-2207-2 (confidential), MOERI, Daejeon (2010)
3. G. Bingje, S. Steen, Added resistance of a VLCC in short waves, in *Proceedings of 29th ASME International Conference on Ocean, Offshore and Arctic Engineering* (2010)

Chapter 15
Comparison Study

Abstract A more comprehensive comparison of responses to regular waves is given in this chapter for a 6500TEU containership. It shows the results of model experiments, of a linear strip method, a Green function method including nonlinear corrections, a linear and a fully nonlinear Rankine source method, and a RANS method. Compared are ship motions as well as linear and nonlinear section forces and moments in three transverse sections. The ship speed is varied between 0 and 15 knots. Only longitudinal (head and following) waves are covered; the maximum wave height (in full scale) is 10 m. Results show that motions can be well predicted by linear three-dimensional methods, but accurate section loads in steep waves require nonlinear computing methods. Apparently, both experiments and nonlinear computational methods have difficulties with loads in steep waves. Nonetheless, carefully computed results are of sufficient accuracy for practical application.

15.1 Description of Test Case

A 1/55-scale model of a 6500 TEU containership was tested by the Maritime and Ocean Engineering Research Institute (MOERI) in the joint industry project WILS [1]. Here, only results in head ($\mu = 180°$) and following regular waves ($\mu = 0°$) are shown: non-dimensional surge, heave, and pitch motion as well as vertical shear force F_z and vertical bending moment M_y in transverse sections at $\frac{1}{4}L_{pp}$, $\frac{1}{2}L_{pp}$, and $\frac{3}{4}L_{pp}$ ahead of the aft perpendicular. The model was tested in low-amplitude waves at full-scale forward speeds of 0, 5, 10, and 15 knots; additionally, runs for 5 knots speed in moderate and steep waves were performed to measure nonlinear effects on loads. Table 15.1 shows the full scale ship parameters.

B. O. el Moctar et al., *Numerical Methods for Seakeeping Problems*,
https://doi.org/10.1007/978-3-030-62561-0_15

Table 15.1 Main data of the container ship

L_{pp} 286.3 m	KG 17.816 m
B_{wl} 40.3 m	GM 1.140 m
T_m 13.05 m	k_{xx} 15.881 m
Trim −0.0308°	k_{yy} 68.880 m
Mass 95276.1 t	k_{zz} 68.552 m

15.2 Computational Methods

Five computational methods were applied to the above case:

- The strip method PDSTRIP [2] was applied, using 24 transverse sections spaced narrower at both ship ends than amidships; each half section was described by typically 20 (minimum 11, maximum 29) offset points.
- The Green function method GLPANEL [3] with nonlinear corrections used about 2000 panels.
- The linear Rankine source method GLRankine [4] used about 500 panels on one side of the wetted hull; on the free surface, depending on wavelength between 2300 and 3800 panels were used per symmetry half.
- The nonlinear Rankine panel method SIS used about 1900 triangular body panels on both ship sides up to the weather deck; on average, about 1200 of them were (at least partly) wetted simultaneously. The number of free-surface panels on both sides varied between ≈1100 in long waves and ≈5200 in the shortest wave (wavelength 96 m).
- The viscous field method OpenFOAM was used for CFD computations solving the RANS equations. Here, the $k - \omega SST$ turbulence model, PIMPLE for coupling pressure with velocity, and the VOF method were applied. The computational grid consisted of about 1.5 million cells.

For all methods, the discretization was verified as appropriate by computing nearly identical results for finer discretizations in test cases.

15.3 Results

Surge and heave amplitudes refer to the center of gravity G of the ship and are made non-dimensional by dividing them by wave amplitude A. The pitch motion is made non-dimensional by dividing it by the wave slope kA, where k is the wave number. Vertical shear force F_z and vertical bending moments M_y are made non-dimensional as indicated along the ordinate in the respective figures.

Figure 15.1 compares computed and measured ship motions in regular head waves of low amplitude at ship speeds from 0 to 15 knots. Except for the surge motion

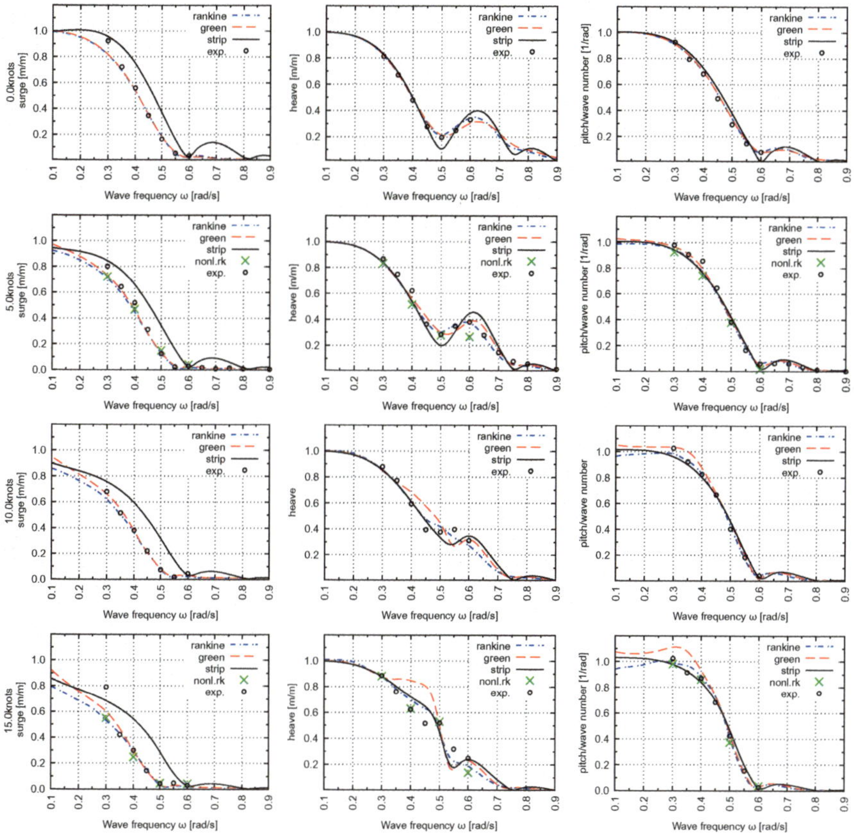

Fig. 15.1 Linear non-dimensional ship motions in head waves at various forward speeds

computed by the strip method, computed results are close to each other and to the experimental values. Regarding surge motion, good coincidence with experimental results was obtained for the linear and nonlinear Rankine source methods and the Green function method in spite of neglecting the change of propeller thrust with surge motion. The erroneous surge motion results of the strip method do not appear inherent to this method; instead, they can probably be removed by a minor improvement of the program used. The heave motion is sensitive to small changes of vertical forces at medium wave frequencies. The reason for this is a minimum of heave excitation (because forces on wave crest and in wave trough nearly cancel when integrated over ship length) at a wave frequency of about 0.5 rad/s and because of heave resonance (where mass and restoring force cancel) at about 0.63 rad/s.

Figure 15.2 shows non-dimensional motion amplitudes in following waves of low amplitude at a ship speed of 5 knots. Except for the surge motion computed by the strip method, the coincidence is good.

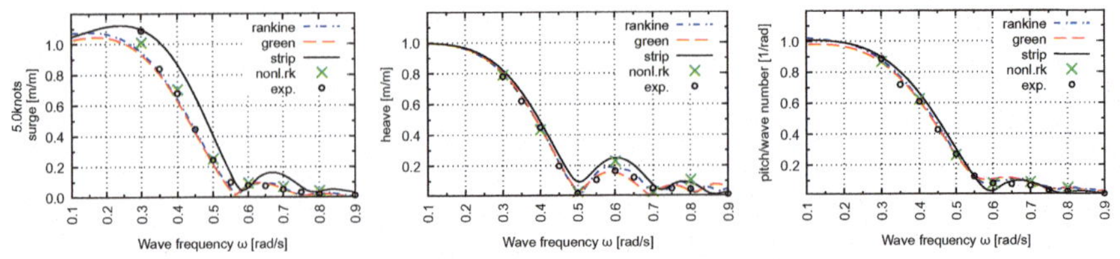

Fig. 15.2 Linear non-dimensional ship motions in following waves at 5 knots speed

Because loads in transverse ship sections result as the difference between large positive and negative contributions, they are more sensitive to small inaccuracies than ship motions. Figure 15.3 illustrates this for the ship in low-amplitude head waves at speeds from 0 to 15 knots (upper to lower diagrams) for three sections: The non-dimensional amplitude of vertical shear force is given at positions $\pm L_{pp}/4$ behind and in front of the midship section (left and right column, respectively), and of the vertical bending moment at midship section (middle column). The strip method is relatively inaccurate, as might be expected from its poor theoretical basis; but all other methods coincide reasonably with each other and with the measured results. The same is concluded from Fig. 15.4 for the ship in following waves.

The remaining figures show the same non-dimensional section loads, however, in steep waves, where nonlinear effects become substantial. No results for motions in steep waves are given because non-dimensional motion amplitudes are hardly influenced by nonlinear effects even in steep waves. The following figures give maximum and minimum loads of force and moment, which are exerted from the ship part behind the transverse section on the part in front of it. For the midship bending moment, that means this the bending moment is positive in hogging (on a wave crest) and negative in sagging (in wave trough) condition.

Here, postprocessing of the measured and the computed results is important. For the nonlinear Rankine source method, the time track of the load was approximated by a constant value, a linear trend (here rather unimportant), and a sinusoidal function. The approximation covered, typically, three encounter periods at the end of the simulation. Of the maximum and minimum values of this approximation, the load for the ship with forward speed, but without waves, was subtracted.

Figure 15.5 holds for 5 knots forward speed in head waves. Most interesting are the results for 10 m (uppermost row of diagrams) wave height at medium wave frequencies, where extreme loads occur. In sagging condition (negative M_y), large differences are found:

- The methods 'rankine' (program GLRankine) and 'green' (program GLPANEL) show large absolute bending moments; for 10 m wave height the nondimensional responses are about twice as large as the linear ones shown in Fig. 15.3. Both methods are essentially linear and use the same nonlinear correction explained in chapter 'Green function methods': they take account only of the difference between actual and average hull immersion.

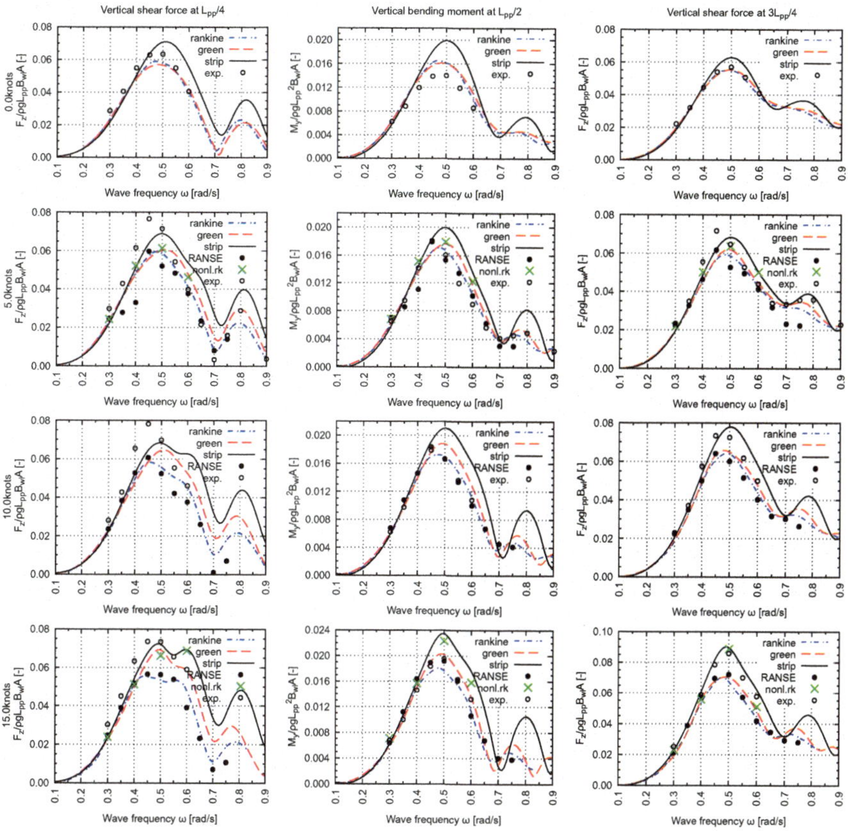

Fig. 15.3 Linear non-dimensional loads in head waves at various forward speeds

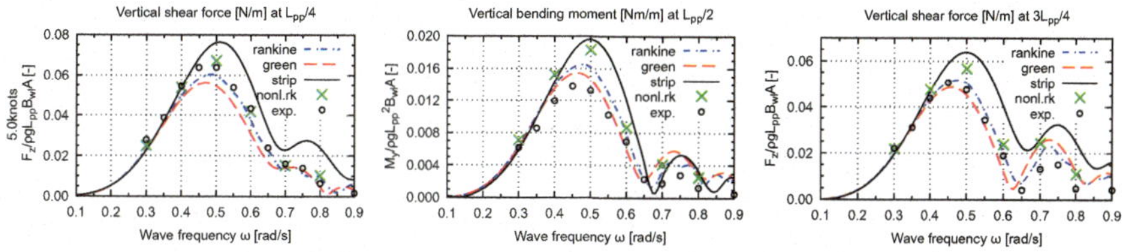

Fig. 15.4 Linear non-dimensional loads in following waves at 5 knots speed

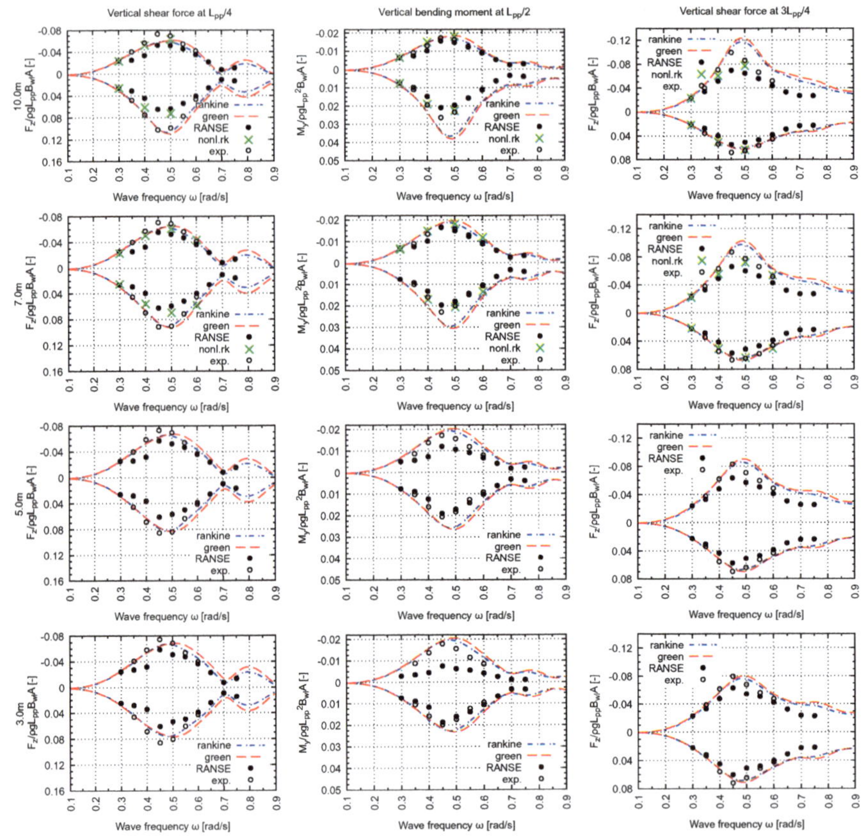

Fig. 15.5 Nonlinear non-dimensional loads in head waves of various heights at 5 knots speed

- For the same conditions, the fully nonlinear methods RANSE solver (OpenFOAM) and nonlinear Rankine source method (program SIS) show only moderately larger (≈25%) non-dimensional bending moments than in the linear case.

The experimental results for maximum sagging bending moments correspond more to those of the latter, fully nonlinear methods; however, the opposite is true for the vertical shear force at the section in the afterbody.

Corresponding results for following waves are shown in Fig. 15.6. Unfortunately, for this case, no results of the RANSE method are available. Also, the difference between sagging and hogging vertical bending moment computed by the linear methods with nonlinear corrections (rankine and green) is larger than that determined in experiments and the fully nonlinear computation.

One should be aware that the restriction of this chapter to head and following waves excludes the 'transverse' motions sway, roll, and yaw and the accompanied loads. Predictions of these are more difficult and inaccurate than predicting 'vertical' motions and loads. Main reasons for this are

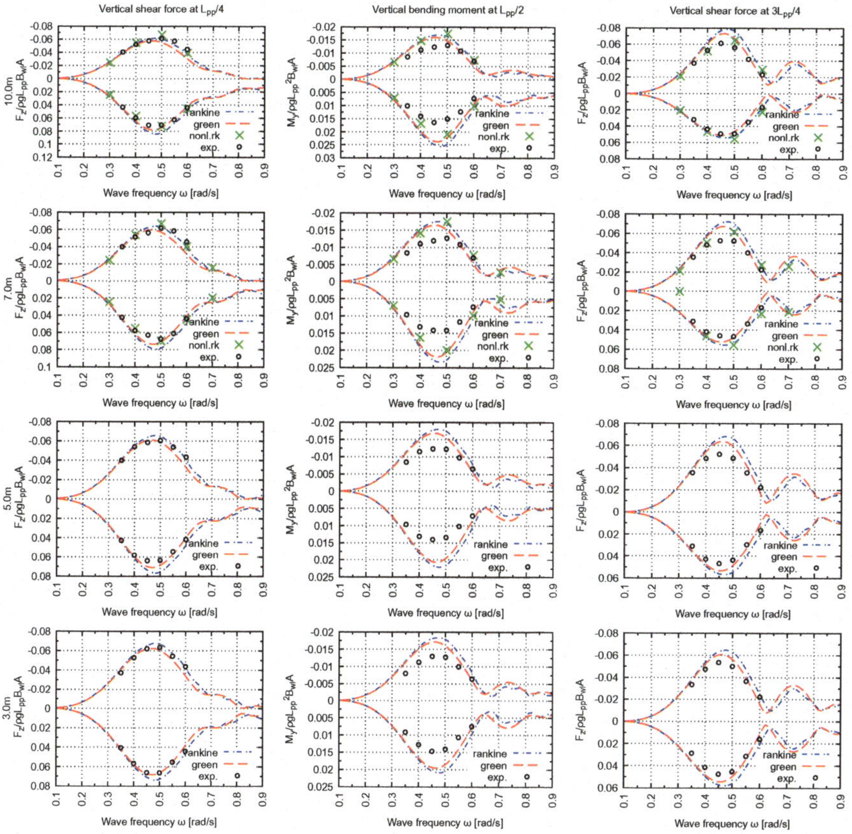

Fig. 15.6 Nonlinear non-dimensional loads in following waves of various heights at 5 knots speed

- the small roll restoring moment, which varies nonlinearly with heel and ship position in the wave, and
- viscous and transverse 'lift' forces on hull and appendages.

References

1. M. Song, K.H. Kim, Y. Kim, Analysis of linear and nonlinear structural loads on a 6500 TEU containership by a time-domain Rankine panel method, in *ISOPE-I-10-581* (2010), pp. 379–384
2. V. Bertram, B. Veelo, H. Söding, K. Graf, Development of a freely available strip method for seakeeping, in *COMPIT* (2006)
3. A.D. Papanikolaou, T.E. Schellin, A three-dimensional panel method for motions and loads of ships with forward speed. J. Ship Technol. Res. **39**, 147–156 (1992)
4. H. Söding, A. von Graefe, O. el Moctar, V. Shigunov, Rankine source method for seakeeping prediction, in *Proceedings of the ASME 2012 31st International Conference on Ocean, Offshore and Arctic Engineering, OMAE* (2012)

Chapter 16
Ships in Natural Seaways

Abstract This chapter starts with describing statistics of Gauß processes, which result from linear responses to a linearized stationary seaway. Then computing the distribution of nonlinear responses to one or several correlated Gauß processes is illustrated as an example. For more general nonlinear cases, simulation cannot be avoided, but this may require excessive time to directly count the rate with which seldomly occurring events like capsizing or extreme loads occur. A technique for reducing the effort of such computations by orders of magnitude is based on the fact that similar wave events occur in seaways of different significant height with rates differing by known factors. Finally, the superposition of probability distributions for stationary seaways to long-term distributions and the concept of design waves is described.

Statistical properties of seaways, for instance, the wave spectrum, vary continuously over time and space. The time and length scales of this variation are larger than a typical wavelength and period. Also, the transient behavior of ships responding to waves fades out within times substantially shorter than the time required for substantial changes of the wave spectrum. This fact allows predicting ship responses to the seaway in two separate steps:

1. Short-term investigation. The statistical parameters of the seaway are assumed constant over time and space; the seaway is then called stationary. The ship's responses are determined neglecting all transients. The probability distribution of responses is determined as if the seaway parameters remained the same for arbitrary long time.
2. Long-term investigation. The parameters of the seaway (e.g., significant height, mean period, main wave direction) and of the ship (e.g., speed, heading relative to the seaway, load case) are varied. The long-term probability distributions of ship responses are determined as a superposition of short-term distributions, taking into account the frequency distribution of the relevant wave and ship operating parameters.

Stationary seaways can be approximated as a superposition of regular waves of different lengths and directions. From measurements of the surface height, the amplitude and phase of each regular wave (for instance, at $t = \mathbf{x} = 0$) follow from a Fourier analysis. For response predictions, the seaway can be simulated from a

B. O. el Moctar et al., *Numerical Methods for Seakeeping Problems*,
https://doi.org/10.1007/978-3-030-62561-0_16

given wave spectrum $S(\omega, \mu)$, e.g., using the method of Ducrozet et al. (2016). This makes sense if also the ship responses are determined using all substantial nonlinear contributions. At first, however, a simpler method is described: Terms depending nonlinearly on wave amplitude are neglected both for the seaway and for the ship responses. Then the phase angles of the regular waves are uncorrelated to each other, and they are distributed with constant probability density in the range between zero and 2π. The same holds for the responses to the component waves, and each component of the response varies as a sinusoidal function (sine function with arbitrary factor, frequency, and phase angle) of time. Furthermore, the minimum difference between the frequency of the component waves is assumed tending to zero, and the number of component waves to infinity. Thus, also the amplitude of each component tends to zero. Then the response is called an ergodic Gauß process. Ergodic means that averages over time (for the seaway also over space within the region for which the seaway is stationary) are the same as averages over phase angles. In the following, the seaway (and then also linear responses to the seaway) are presumed to be ergodic Gauß processes.

16.1 Statistics of Linear Responses in a Stationary Seaway

The seaway is approximated from a finite number J times L of elementary regular waves:

$$z_s(\mathbf{x}, t) = \sum_{j=1}^{J} \sum_{l=1}^{L} \sqrt{2S_\zeta(\omega_{jl}, \mu_{jl})\Delta\omega_j \Delta\mu_l} \, \mathrm{Re}(e^{i(\omega_{jl}t + \boldsymbol{\nu}_{jl}\cdot\mathbf{x} + \epsilon_{jl})}). \tag{16.1}$$

Here, z_s is the vertical coordinate of the water surface at position $\mathbf{x} = (x, y, 0)$ and time t. The ranges of wave frequency and wave direction are subdivided into J and L, respectively, intervals of (possibly different) breadths $\Delta\omega_j$ and $\Delta\mu_l$. The expression under the square root is the amplitude of one elementary wave; it follows from (4.74) of chapter Water Waves by substituting the integrals by sums using the rectangle integration rule. ν is the 'complex vector wave number' defined in (4.63). ϵ is a random phase angle. ω_{jl} and μ_{jl} are wave frequency and wave direction, respectively. Details for choosing ω_{jl}, μ_{jl} and ϵ_{jl} are given in Sect. 4.2.5.

A response r depending linearly on wave height (e.g., motion, acceleration, stress, bending moment) is characterized by its response amplitude operator $\hat{Y}_r(\omega, \mu)$, i.e., the ratio of the complex amplitude of the response to that of the regular wave causing that response. The time history of r in a natural seaway described by (16.1) is obtained by superimposing the time histories of all regular waves contributing to the seaway:

$$r(t) = \sum_{j=1}^{J} \sum_{l=1}^{L} \sqrt{2S_\zeta(\omega_{jl}, \mu_{jl})\Delta\omega_j \Delta\mu_l} \cdot \text{Re}\left(\hat{Y}_r(\omega_{jl}, \mu_{jl})e^{i(\omega_{jl}t+\boldsymbol{\nu}_{jl}\cdot\mathbf{x}+\epsilon_{jl})}\right). \quad (16.2)$$

By decomposing $\hat{Y}_r$ into modulus Y_r and argument ϵ_r,

$$\hat{Y}_r(\omega, \mu) = Y_r(\omega, \mu)e^{i\epsilon_r(\omega,\mu)},$$

we obtain the response r(t) in the form

$$r(t) = \sum_{j=1}^{J} \sum_{l=1}^{L} \sqrt{2Y_r^2(\omega_{jl}, \mu_{jl})S_\zeta(\omega_{jl}, \mu_{jl})\Delta\omega_j \Delta\mu_l} \text{Re}\left(e^{i(\omega_{jl}t+\epsilon_{jl}+\epsilon_r(\omega_{jl},\mu_{jl}))}\right). \quad (16.3)$$

Comparing this with (16.1) shows that the response spectrum of r is

$$S_r(\omega, \mu) = Y_r^2(\omega, \mu)S_\zeta(\omega, \mu). \quad (16.4)$$

The phase angles ϵ_r are, contrary to ϵ_{jl}, not random quantities. Because they are added to the random quantities ϵ_{jl}, they have no influence on the statistical properties of $r(t)$, but they are important for the correlation between different responses or for that between surface height z_s (at a reference location) and response r.

In the following, some statistical properties of the ergodic Gauß processes $r(t)$ are stated:

- The average of $r(t)$ is zero. This follows from (16.3): Each term of the double-sum assumes positive and corresponding negative values with even probability due to the constant distribution of random phase angles ϵ_{jl}.
- The (probability) distribution of $f(r)$ is the normal (Gauß) distribution (Fig. 16.2):

$$f(r) = \frac{1}{\sqrt{2\pi}\sigma_r} e^{-\frac{r^2}{2\sigma_r^2}}, \quad (16.5)$$

  where σ_r is the standard deviation (the square root of the variance) of r. This follows from the central limit theorem, which states that the sum of n random quantities tends to be normally distributed for $n \to \infty$. Loosely stated, the theorem holds under the condition that the random quantities are not or only weakly correlated, and that the sum is not dominated by a few terms only.

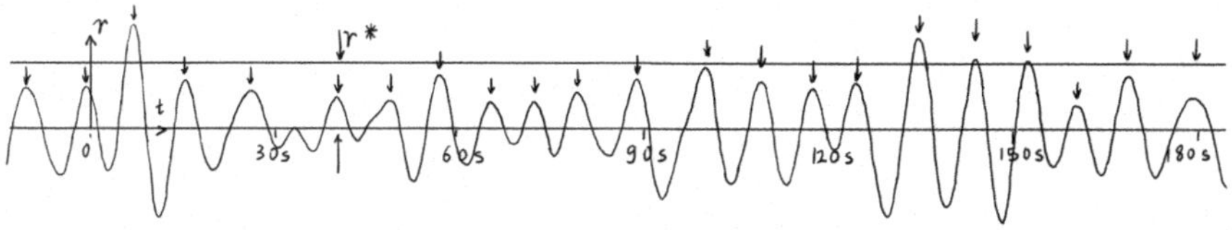

Fig. 16.1 Response r(t), threshold value r^*, and amplitudes (arrows)

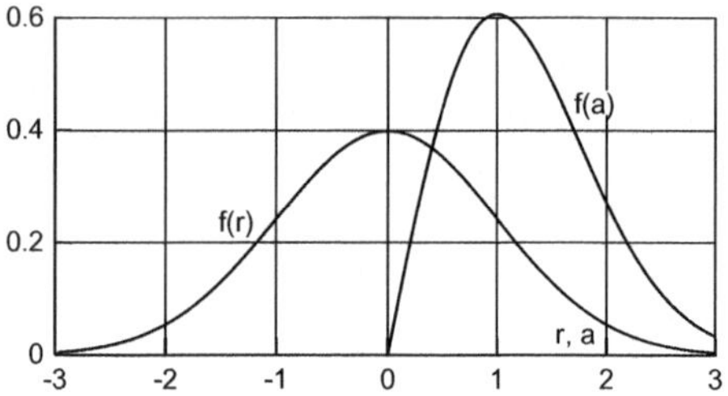

Fig. 16.2 Normal distribution $f(r)$ of a response r, and Rayleigh distribution $f(a)$ of the amplitudes a of r, for the case $\sigma_r = 1$

- The variance σ_r follows from the spectrum of r as

$$\sigma_r^2 = \int_0^\infty \int_0^{2\pi} S_r(\omega, \mu)\, d\mu d\omega = \int_0^\infty \int_0^{2\pi} Y_r^2(\omega, \mu)\, S_r(\omega, \mu)\, d\mu d\omega. \quad (16.6)$$

This follows from the fact that the variance of a sum of uncorrelated random quantities is the sum of the variances of the random quantities. If r_{jl} is one element of the double sum in (16.3), its variance is defined as

$$\sigma_{rl}^2 = \overline{(r_{jl} - \overline{r_{jl}})^2}, \quad (16.7)$$

where over-lining designates the average. As stated before, $\overline{r_{jl}} = 0$. Thus, inserting r_{jl} from (16.3) gives

$$\sigma_{rl}^2 = \overline{2Y_r^2(\omega_{jl}, \mu_{jl}) S_\zeta(\omega_{jl}, \mu_{jl}) \Delta\omega_j \Delta\mu_l \cos^2(\omega_{jl} t + \epsilon_{jl} + \epsilon_r(\omega_{jl}, \mu_{jl}))} \quad (16.8)$$

$$= 2Y_r^2(\omega_{jl}, \mu_{jl}) S_\zeta(\omega_{jl}, \mu_{jl}) \Delta\omega_j \Delta\mu_l \underbrace{\overline{\cos^2(\omega_{jl} t + \epsilon_{jl} + \epsilon_r(\omega_{jl}, \mu_{jl}))}}_{\frac{1}{2}} \quad (16.9)$$

$$= Y_r^2(\omega_{jl}, \mu_{jl}) S_\zeta(\omega_{jl}, \mu_{jl}) \Delta\omega_j \Delta\mu_l. \quad (16.10)$$

This is the contribution of the element jl to the integral (16.6).

- The probability $P(r > r^*)$ that, at a fixed time instant, r exceeds a threshold value r^* (Fig. 16.1) is

$$P(r > r^*) = \int_{r^*}^\infty f(r) dr = \frac{1}{\sqrt{2\pi}\sigma_r} \int_{r^*}^\infty e^{-\frac{r^2}{2\sigma_r^2}} dr$$
$$= \frac{1}{\sqrt{2\pi}} \int_{r^*/\sigma_r}^\infty e^{-t^2/2} dt = 1 - \Phi(r^*/\sigma_r), \quad (16.11)$$

where the substitution $t = r/\sigma_r$ is used, and

$$\Phi(x) = \frac{1}{\sqrt{2\pi}} \int_{-\infty}^x e^{-t^2/2} dt \quad (16.12)$$

is the cumulative probability function of the normal distribution, for which programs and tables are readily available. If r is, for instance, the relative motion between water surface and ship at a position x, and r^* is the freeboard (height of weather deck above average waterline), then $P(r > r^*)$ can be interpreted as the fraction of time during which the water surface exceeds the deck level at x. Table 16.1 illustrates the exceedance probability.

- The positive maximum values of a response (indicated by arrows in the figure above; negative maxima may occur and are excluded) are called amplitudes. Their distribution depends on the 'breadth' of the response spectrum, which is quantified by the nondimensional parameter ϵ between 0 and 1:

$$\epsilon = \sqrt{1 - \frac{m_2^2}{m_0 m_4}}, \tag{16.13}$$

where the so-called spectral moments m_i, $i = 0, 1, 2, \ldots$, are defined as

$$m_i = \int_0^\infty \int_0^{2\pi} \omega^i S(\omega, \mu) d\mu d\omega. \tag{16.14}$$

Equation (16.6) shows that m_0 is the variance σ_r^2 of the response. Typical seaway spectra have maximum breadth $\epsilon = 1$ because m_4 is infinite due to the decrease of S for high frequencies proportional to $1/\omega^5$. Depending on case, $\epsilon = 1$ may be avoided by introducing a 'cut-off frequency'. For responses, the response amplitude operators of which tend to zero for high frequency, ϵ is < 1.
The fraction of amplitudes a exceeding a limit r^* can be well approximated as

$$P(a > r^*) \approx \underbrace{\sqrt{1 - \epsilon^2}}_{\approx 1 \text{ for small } \epsilon} \exp\left(\frac{-r^{*2}}{2m_0}\right) \tag{16.15}$$

if $r^* > 2\sqrt{m_0}$ and $\epsilon < 0.95$, i.e., in nearly all cases of practical interest. For smaller $\epsilon < 0.25$ and again $r^* > 2\sqrt{m_0}$, the factor $\sqrt{1 - \epsilon^2}$ can be omitted. In naval architecture, the formula is nearly always used without this factor. The distribution of a is then called Rayleigh distribution. Table 16.1 gives values of $P(a > r^*)$ for small spectrum breadth ϵ. Figure 16.2 shows the density function of the Rayleigh distribution

$$f(a) = -\frac{dP(a > r^*)}{dr^*}|_{a=r^*} = \frac{a}{m_0} \exp\left(\frac{-a^2}{2m_0}\right). \tag{16.16}$$

- The rate (average number of occurrences per time) of zero up-crossings of $r(t)$ is

$$f_0 = \frac{1}{2\pi}\sqrt{\frac{m_2}{m_0}}. \tag{16.17}$$

- The rate of amplitudes exceeding r^* is, approximately,

Table 16.1 Probability that a response r or an amplitude a exceeds a threshold r^*

r^*/σ_r	1.0	1.5	2.0	2.5	3.0
$P(r > r^*)$	0.159	0.067	0.023	0.006	0.0014
$P(a > r^*)$	0.61	0.32	0.14	0.044	0.011

$$f_a(r^*) \approx f_0 \exp\left(\frac{-r^{*2}}{2m_0}\right). \tag{16.18}$$

Here, f_0 is an approximation for the rate of amplitudes of any size, and $\exp(-r^{*2}/2m_0)$ approximates the probability that an amplitude exceeds r^*. For larger values of ϵ and r^*, the errors in both approximations nearly cancel; thus, (16.18) appears applicable if $\epsilon < 0.95$ and $r^* > 2\sqrt{m_0}$, i.e., in nearly all cases of practical interest.—Often instead of f_a the inverse rate $1/f_a$ is considered and called 'average time to failure', if exceeding r^* is valuated as a failure (e.g., of a structure).

The following questions can be addressed using (16.18) and some of the previous relations:

1. What is the average number $z(r^*)$ of response amplitudes exceeding a threshold value r^* during a time interval of length T?

$$z\left(r^*\right) = Tf_a\left(r^*\right) = Tf_0 \exp\left(\frac{-r^{*2}}{2m_0}\right). \tag{16.19}$$

2. Which threshold r^* is exceeded on average by exactly one amplitude during T? Loosely formulated: Which maximum value must be expected during T? From (16.18) and $z(r^*) = 1$ follows

$$r^* = \sqrt{2m_0 \ln(Tf_0)}. \tag{16.20}$$

3. What is the probability $P(r^*)$ that a threshold value r^* is not exceeded during time T? If n is the number of amplitudes during T, the probability that an arbitrary amplitude among these amplitudes is less than r^* is

$$P_1 = \frac{n - z(r^*)}{n}. \tag{16.21}$$

The probability that all n amplitudes are $<r^*$ is equal to the probability that the first, the second, ..., the nth amplitude are all $<r^*$. Except for lightly damped responses like rolling motions and hull vibrations, successive response amplitudes can be assumed to be (nearly) uncorrelated. Then the product rule of probabilities of uncorrelated events gives

$$P\left(r^*\right) = P_1^n \underset{(16.21)}{\uparrow=} \left(1 - \frac{z\left(r^*\right)}{n}\right)^n \underset{\text{for } z(r^*)\ll n}{\uparrow=} \left(e^{-z(r^*)/n}\right)^n = e^{-z(r^*)}, \tag{16.22}$$

where $z(r^*)$ follows from (16.19). For example, the threshold r^* according to (16.20), which is exceeded once on the average during T, is exceeded (also during T) with probability $e^{-1} \approx 0.37$.

4. What is the probability $\bar{P}\,(r^*)$ that a threshold r^* is exceeded during T? Solution:

$$\bar{P}(r^*) = 1 - P(r^*) = 1 - e^{-z(r^*)} \tag{16.23}$$

If the average number of exceedances $z(r^*)$ is much less than 1, this can be simplified to

$$\bar{P}(r^*) \approx 1 - (1 - z(r^*)) = z(r*) = T\, f_a(r^*). \tag{16.24}$$

5. Which threshold r^* is exceeded with a given small probability $\bar{P}\,(r^*)$ during time T? This is a frequent question in dimensioning, where $\bar{P}\,(r^*)$ is the tolerated 'failure level'. From (16.24) and (16.18) follows

$$r^* = \sqrt{2m_0 \ln(T f_0/\bar{P}(r^*))}. \tag{16.25}$$

Table 16.2 lists the probability $\bar{P}\,(r^*)$ that the response $r(t)$ exceeds the threshold r^* at least once during time T, where σ_r is the standard deviation of $r(t)$, and f_0 is the rate of zero up-crossings of $r(t)$. The table was computed using the relation following from (16.19), (16.22) and (16.23), avoiding the simplifications for special cases:

$$\bar{P}(r^*) = 1 - \left[1 - \exp\left(-\frac{r^{*2}}{2m_0}\right)\right]^{T f_0}.$$

The table shows that, for long registration time T, large values of the response may be exceeded; an upper limit does not exist.

Table 16.2 Probability $\bar{P}(r^*)$ that a response $r(t)$ exceeds a threshold r^* during T

$T f_0$	$r^*/\sigma_r =$							
	1.0	1.5	2.0	2.5	3.0	3.5	4.0	4.5
1	0.607	0.325	0.135	0.044	0.011	0.0022	0.00034	0.00004
10	0.99991	0.980	0.766	0.362	0.106	0.022	0.0034	0.00040
100	≈1	≈1	≈1	0.989	0.673	0.197	0.033	0.0040
1000	≈1	≈1	≈1	≈1	0.99999	0.888	0.285	0.039
10000	≈1	≈1	≈1	≈1	≈1	≈1	0.965	0.330

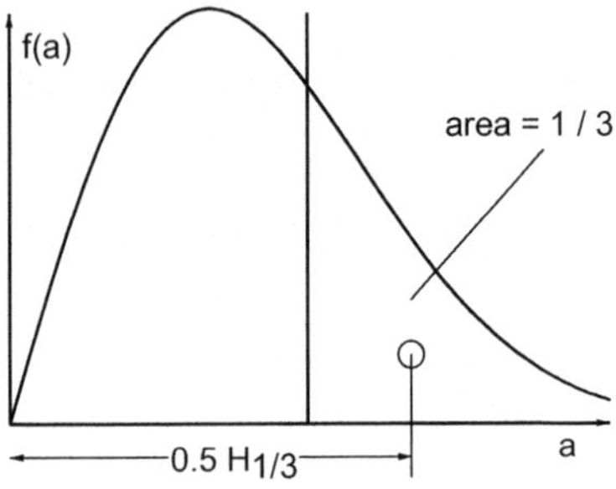

Fig. 16.3 Rayleigh distribution of amplitudes a of a seaway, and definition of H_s

6. What is the average height of the subset of the 1/3 largest wave heights? This quantity H_s is called significant wave height. It is used to characterize the 'strength' of a seaway. Up to now, we considered not wave heights, but amplitudes of a seaway. A wave height might be defined as the difference between a maximum and the following minimum of the vertical coordinate of the water surface, or a range-pair counting (explained in Chap. 10) of positive and negative extremes of the surface height. Here, we approximate a wave height simply as twice the amplitude a. Then the definition of H_s together with (16.30) gives

$$\begin{aligned} H_s &= 2\,\frac{\int_{r^*}^{\infty} af(a)da}{\int_{r^*}^{\infty} f(a)da} = \frac{2}{1/3}\int_{r^*}^{\infty}\frac{a^2}{m_0}\exp\left(\frac{-a^2}{2m_0}\right)da \\ &= 6\sqrt{m_0}\int_{r^*/\sqrt{m_0}}^{\infty} b^2\exp(-b^2/2)db, \end{aligned} \tag{16.26}$$

where r^* follows from $P(a > r^*) = 1/3$, using (16.15), as

$$r^* = \sqrt{2\ln(3)m_0} \approx 1.48\sqrt{m_0}. \tag{16.27}$$

Evaluating numerically the right-most integral in (16.26) gives 0.668; thus, we obtain

$$H_s = 4.0\sqrt{m_0}. \tag{16.28}$$

This is illustrated in Fig. 16.3.

16.2 Statistics of Nonlinear Responses in a Stationary Seaway

In this section, we assume that the seaway is approximated as a Gauß process; only the responses of interest are nonlinear.

16.2.1 Nonlinear Function of a Linear Response

If a nonlinear response $s(t)$ depends on a linear response $r(t)$, i.e., if

$$s(t) = S(r(t)), \tag{16.29}$$

where S is a monotonically increasing function, then the probability that an amplitude a_s of s exceeds a threshold s^* is equal to the probability that an amplitude a_r of r exceeds r^*,

$$P(s > s^*) = P(r > r^*) = \exp\left(\frac{-r^{*2}}{2m_{0r}}\right), \tag{16.30}$$

if

$$s^* = S(r^*). \tag{16.31}$$

Also the rate of amplitudes of s exceeding s^* is the same as that given in (16.18) for the linear response r if r^* is determined from s^* using the inverse function $S^{-1}(s^*)$.

16.2.2 Function of Several Linear Responses

If a response $s(t)$ is a (linear or nonlinear) monotonic function S of several linear responses (each with zero mean),

$$s(t) = S(\mathbf{r}(t)) \tag{16.32}$$

with

$$\mathbf{r} = (r_1, r_2, \ldots, r_n(t)), \tag{16.33}$$

we use the fact that the distribution of $\mathbf{r}$ is the multidimensional normal distribution. Its density is

$$f(\mathbf{r}) = \frac{1}{\sqrt{(2\pi)^n \det(\mathbf{\Sigma})}} \exp\left(-\frac{1}{2}\mathbf{r}^T \mathbf{\Sigma} \mathbf{r}\right), \tag{16.34}$$

where $\mathbf{\Sigma}$ is the co-variance matrix of $\mathbf{r}$:

$$\mathbf{\Sigma} = \begin{pmatrix} \mathrm{Var}(r_1) & \mathrm{Cov}(r_1, r_2) & \ldots & \mathrm{Cov}(r_1, r_n) \\ \mathrm{Cov}(r_1, r_2) & \mathrm{Var}(r_2) & \ldots & \mathrm{Cov}(r_2, r_n) \\ \ldots & \ldots & \ldots & \ldots \\ \mathrm{Cov}(r_1, r_n) & \mathrm{Cov}(r_2, r_n) & \ldots & \mathrm{Var}(r_n) \end{pmatrix}. \tag{16.35}$$

In (16.34) $\det(\mathbf{\Sigma})$ is the determinant of $\mathbf{\Sigma}$. As before, the variance $\mathrm{Var}(r_i)$ is $\sqrt{m_{0,i}}$. The co-variance between responses r_i and r_j follows from the two complex transfer functions $\hat{Y}_i$ and $\hat{Y}_j$ as

$$\text{Cov}(r_i, r_j) = \int_0^\infty \int_0^{2\pi} \text{Re}(\hat{Y}_i \hat{Y}_j^*) S_\varsigma(\omega, \mu) d\mu d\omega, \tag{16.36}$$

where the asterisk designates the complex conjugate. Equation (16.34) simplifies to a product of one-dimensional normal distributions if the co-variance matrix $\mathbf{\Sigma}$ is a diagonal matrix, i.e., if the responses r_i are mutually uncorrelated.

An application example is the wave-induced moment acting on a bilge keel and turning around the weld between bilge keel and hull (more exactly, hull plate doubling). This moment can be expressed depending on two roughly linear transfer functions: the acceleration r_1 and the velocity r_2 of the relative flow between water and hull, both computed for the hull without bilge keel at the position of the bilge keel (distance from the hull about 1/2 bilge keel height), taking the component transverse to the bilge keel and parallel to the hull. The bilge keel moment depends approximately linearly on the acceleration r_1 because it is an added-mass effect, whereas the resistance effect gives another contribution proportional to $r_2|r_2|$.

Generally, we want to determine the probability $P(a_s > s^*)$ that an amplitude a_s of the random process $s(t) = S(\mathbf{r}(t))$ exceeds a threshold s^*. To illustrate the proposed procedure, a simple example is used. Determine $P(a_s > s^* = 5)$ for the case

$$s = 2r_1 + \frac{1}{2} r_2 |r_2|. \tag{16.37}$$

1. Determine the co-variance matrix $\mathbf{\Sigma}$ according to (16.35), using the relations (16.36) and

$$\text{Var}(r_i) = \text{Cov}(r_i, r_i) = m_{0i}. \tag{16.38}$$

For the example, let us assume that the co-variance matrix is

$$\mathbf{\Sigma} = \begin{pmatrix} 1 & -0.5 \\ -0.5 & 2 \end{pmatrix}. \tag{16.39}$$

(Actually, $\text{Cov}(r_1, r_2) = 0$ if $r_1 = \dot{r}_2$.)

2. Generate new linear responses $\bar{\mathbf{r}} = (\bar{r}_1, \ldots, \bar{r}_n)$ which are mutually independent:

$$\bar{r}_i = \mathbf{e}_i \cdot \mathbf{r} \tag{16.40}$$

for $i = 1$ to n, where $\mathbf{e}_i$ are the eigenvectors of the co-variance matrix.

For the example: The eigenvectors (their magnitude is arbitrary) of $\mathbf{\Sigma}$ are

$$\mathbf{e}_1 = (1, -(1+\sqrt{2})); \quad \mathbf{e}_2 = (1, \sqrt{2} - 1). \tag{16.41}$$

The independent linear responses follow from (16.40) as

$$\bar{r}_1 = r_1 - (1+\sqrt{2})r_2; \quad \bar{r}_2 = r_1 + (\sqrt{2}-1)r_2. \tag{16.42}$$

3. Determine the variance of $\bar{r}_i$ for $i = 1$ to n using the formula (for real constants α and β)

$$\text{Var}(\alpha r_1 + \beta r_2) = \alpha^2 \text{Var}(r_1) + \beta^2 \text{Var}(r_2) + 2\alpha\beta \text{Cov}(r_1, r_2). \tag{16.43}$$

For example:

$$\begin{aligned}\text{Var}(\bar{r}_1) &= \text{Var}(r_1 - (1+\sqrt{2})r_2)\\ &= \underbrace{\text{Var}(r_1)}_{1} + (1+\sqrt{2})^2 \underbrace{\text{Var}(r_2)}_{2} - 2(1+\sqrt{2}) \underbrace{\text{Cov}(r_1, r_2)}_{-0.5} = 15.07,\end{aligned} \tag{16.44}$$

and correspondingly $\text{Var}(\bar{r}_2) = 0.929$.

4. Determine normed linear mutually independent responses r_i' (normed meaning $\text{Var}(r_i') = 1$) for $i = 1$ to n:

$$r_i' = \bar{r}_i / \sqrt{\text{Var}(\bar{r}_i)}. \tag{16.45}$$

For example:

$$\begin{pmatrix} r_1' \\ r_2' \end{pmatrix} = \begin{pmatrix} \bar{r}_1/\sqrt{15.07} \\ \bar{r}_2/\sqrt{0.929} \end{pmatrix} = \begin{pmatrix} 0.258 & -0.622 \\ 1.038 & 0.430 \end{pmatrix} \begin{pmatrix} r_1 \\ r_2 \end{pmatrix}. \tag{16.46}$$

5. Determine $\mathbf{r}$ as function of $\mathbf{r}'$.
For example:

$$\begin{pmatrix} r_1 \\ r_2 \end{pmatrix} = \begin{pmatrix} 0.258 & -0.622 \\ 1.038 & 0.430 \end{pmatrix}^{-1} \begin{pmatrix} r_1' \\ r_2' \end{pmatrix} = \begin{pmatrix} 0.568 & 0.822 \\ -1.372 & 0.341 \end{pmatrix} \begin{pmatrix} r_1' \\ r_2' \end{pmatrix}. \tag{16.47}$$

6. Determine s as a function of $\mathbf{r}'$.
For example: From (16.37) and (16.47) follows

$$s = 2(0.568 r_1' + 0.822 r_2') + \frac{1}{2}(-1.372 r_1' + 0.341 r_2') \mid -1.372 r_1' + 0.341 r_2' |. \tag{16.48}$$

Figure 16.4 shows the threshold $s^* = 5$ in the (r_1', r_2')-plane derived from (16.48), together with isolines of the density of the two-dimensional normal distribution for unit variance of both random variables (circles). The density above the threshold line could be integrated numerically to determine the probability of s exceeding $s^* = 5$ at an arbitrary time instant. However, in the typical case of seldom exceeded threshold values, the density decreases rapidly with distance. Then, only minor errors occur if the curved integration limit is substituted by the tangent line to it at the point (dot

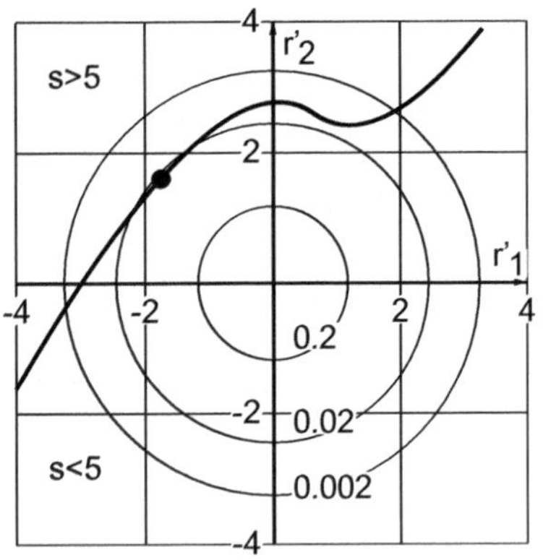

Fig. 16.4 Contours of constant density (numbers at circles) of the two-dimensional normal distribution (thin lines) and limit between $s > 5$ and $s < 5$ (thick line)

in Fig. 16.4) where the limit curve is closest to the coordinate origin. This point, here $r_1' = -1.75$, $r_2' = 1.60$, is the most probable combination of r_1' and r_2' when $s = 5$. The circular isolines of the probability density show that the distribution of the combined response depends only on the distance $\sqrt{r_1'^2 + r_2'^2}$ from the origin, not on the actual linear combination of r_1' and r_2'. Thus, the relations given before for a single random variable hold also here if, instead of the value of the single variable, the distance $\sqrt{r_1'^2 + r_2'^2}$ or, in the general case, $|\mathbf{r}'|^2$ is used. In the example, the probability of an amplitude a_s of the combined response exceeding $s^* = 5$ follows from (16.30):

$$P(a_s > 5) = \exp\left(-\frac{1.75^2 + 1.60^2}{2}\right) = 0.060. \tag{16.49}$$

For other quantities of interest, we need the average zero-up frequency f_0 of the combined response. It follows from (16.17) and (16.14) using for the spectrum of the combined response

$$S_{r'}(\omega, \mu) = |\hat{Y}_1'(\omega, \mu) + \hat{Y}_2'(\omega, \mu)|^2 S_\zeta(\omega, \mu), \tag{16.50}$$

where $\hat{Y}_1'$ and $\hat{Y}_2'$ are the complex transfer functions of r_1' and r_2', respectively. They follow from the transfer functions $\hat{Y}_1$ and $\hat{Y}_2$ of r_1 and r_2, using (16.46), as

$$\begin{pmatrix} \hat{Y}_1' \\ \hat{Y}_2' \end{pmatrix} = \begin{pmatrix} 0.258 & -0.622 \\ 1.038 & 0.430 \end{pmatrix} \begin{pmatrix} \hat{Y}_1 \\ \hat{Y}_2 \end{pmatrix}. \tag{16.51}$$

16.2.3 Other Nonlinear Responses

An example of more complicated relations between waves and response are roll motions of ships. In long-crested head or following regular waves, the parametric roll may occur if the wave amplitude exceeds a threshold. In a natural seaway of any direction, the roll response is a superposition of parametric (due to time-variable restoring) and synchronous (due to exciting moments) motion. Also the latter is a highly nonlinear response, not alone because the righting moment is a nonlinear, not even a monotonous function of heel.

For such nonlinear responses, simulation (computing the time history of the response in one or several example seaways) is the only practical way to estimate the probability $P(a_r > r^*)$ of response amplitudes exceeding a threshold r^* during a given time, or the rate $f_a(r^*)$ with which response amplitudes exceed r^*. The latter is approximated simply as

$$f_a(r^*) = n_E/T, \tag{16.52}$$

where n_E is the number of cases that a response amplitude exceeded r^* during the simulation time T.

Typically, large thresholds r^* are interesting, which are seldom exceeded. Then long simulation times T are required to find at least a few exceedance cases to apply (16.52). The computational effort can be reduced very much by using an idea of Tonguc and Söding [1]. The simulation is performed in two or several very steep seaways where exceedances occur with a high rate. The rates found are then extrapolated to those in the actually interesting, lower seaway. The spectra of the different seaways must be the same except for a scale factor, and also the ship's operating conditions must be the same. The extrapolation of the exceedance rate over significant wave height H_s is based on the fact that a large threshold is exceeded when unusually high waves are encountered by the ship. The same wave event can occur in seaways of different significant height H_s, however with a different rate. The extrapolation of exceedance rate over H_s uses knowledge of the rate with which similar large wave events occur in seaways of different H_s.

For instance, we may assume that the nonlinear response amplitude $a_r(t)$ exceeds the threshold r^* with a probability P if n successive wave amplitudes exceed a threshold ζ^*. Then the rate $f_a(r^*)$ follows from (16.30), when substituting m_0 by $H_s/16$ according to (16.28), as

$$f_a(r^*) = P f_0 \left[\exp\left(-\frac{8\zeta^{*2}}{H_s^2}\right)\right]^n. \tag{16.53}$$

(Here it was assumed that successive amplitudes are statistically independent.) Taking the logarithm gives

$$\ln(f_a) = \ln(P f_0) - n\frac{8\zeta^{*2}}{H_s^2} = A + \frac{B}{H_s^2}. \tag{16.54}$$

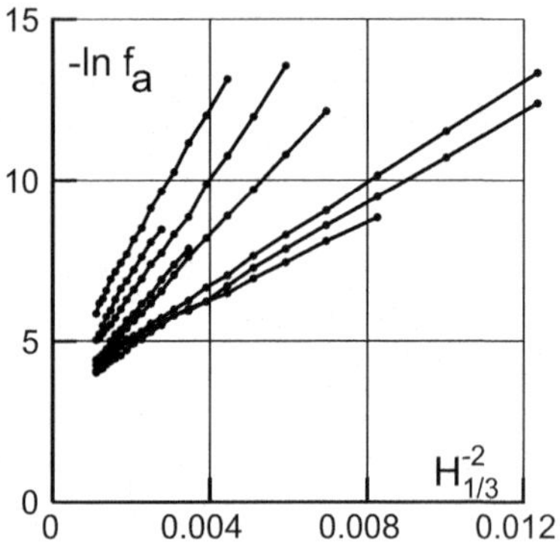

Fig. 16.5 Dependence of $-\ln(f_a)$ on $1/H_s^2$. Here f_a is the rate (in 1/s) of roll amplitudes exceeding a safety limit: $\varphi = 40°$ and transverse acceleration in ship-fixed y direction exceeding $g/2$. Each curve is for a different combination of Froude number, seaway direction, and characteristic period. Results computed by Shigunov

This shows that, if the assumed wave scenario for exceeding r^*, i.e., the parameters P, f_0, n, and ζ^*, are held constant while varying H_s, then the logarithm of the exceedance rate $\ln(f_a)$ is a linear function of $1/H_s^2$. Figure 16.5 shows this relation for the case of roll motion amplitudes exceeding a limit for one ship at various speeds and wave conditions (direction, period, height). The wave conditions comprise those leading to intense parametric roll, resonant rolling in beam sea, and capsizing due to negative righting arms on the wave crest (called pure loss of stability). The curves show that the linear relationship of (16.54) is correct except for the left-most part of the curves corresponding to very high rates of exceedance.

Actual values of A and B in (16.54) follow from two or more simulations, using a least-squares fit if more than two f_a values are available. Each simulation should use new sets of random phases, frequencies, and wave propagation angles. Equation (16.52) is used to derive an f_a value from the simulations. For Fig. 16.5 200 exceedances were used to derive a single f_a value. This requires a fast simulation method; for the curves of Fig. 16.5 the method Rolls [2] for simulating nonlinear roll motions was used. However, much less exceedances appear sufficient, especially if, after finding fewer exceedances, it becomes obvious that a set of conditions is either far below or far above the tolerable failure rate. Presently, the discussed IMO rules require at least three simulations per case within a range of H_s values spanning at least a range of 2 m; each simulation should cover at least 20 natural roll periods. Shigunov [3] recommends:

- to extend each simulation not too long (3 h) to avoid repetition effects of the seaway;
- to omit the beginning of each simulation to avoid transient motion effects; and
- to start a new simulation (new random seaway parameters) after each exceedance.

The method is based on the seaway being approximated as a Gauß process. Thus, the simulations in exaggerated seaways may not be performed using nonlinear natural seaways. It is expected that, for most responses of practical interest, nonlinear effects of the amplitude response operators are more important than those of the seaway.

16.3 Long-Term Distribution of Responses

The short-term rate $f_a(r^*)$ of response amplitudes exceeding a limit r^* depends on the seaway and on the ship's operating conditions. The seaway may be characterized by the parameters significant height H_s, zero-up-crossing period T_m (the inverse of the previouslyused zero-up-crossing rate f_0), and heading μ relative to the ship's longitudinal axis. Other parameters of the wave spectrum like the breadth parameter β of the angular spreading and the peak enhancement factor Γ are usually held constant for simplification. Swells, and swell contributions to the wave spectrum, are usually neglected because they are important only in low and moderate seaways and, thus, have an influence on the exceedance rate of small and moderate limits r^*, whereas in most cases only large, seldomly exceeded limits r^* are interesting. (An exception is the fatigue load of ship structures.) Of the ship's operating conditions, speed v has a strong influence on the exceedance rate $f_a(r^*)$, but also the ship's draft (especially the forward draft d_f in case of slamming) has influence on the exceedance rate.

In case of variable seaway and operating conditions, one wants to know the so-called long-term exceedance rate $f_a^L(r^*)$ (or its inverse $1/f_a^L(r^*)$, the average time between successive exceedances) of the threshold r^*. It results from the short-term rate $f_a(r^*; H_s, T_m, \mu, v, d_f, \ldots)$ as

$$f_a^L(r^*) = \int \ldots \int f_a(r^*; H_s, T_m, \mu, v, d_f, \ldots)\, p(H_s, T_m, \mu, v, d_f, \ldots)\, dH_s\, dT_m\, d\mu\, dv\, dd_f, \ldots, \tag{16.55}$$

where $p(H_s, \ldots)$ is the combined probability density of the variables $H_s, \ldots$ To determine the rate with which the sum of a still-water load l_s and a wave load l_w exceeds a limit r^*, the above formula is slightly changed:

$$f_a^L(r^*) = \int \ldots \int f_a(r^* - l_s; H_s, T_m, \mu, v, d_f, \ldots)\, p(H_s, T_m, \mu, v, d_f, \ldots)\, dH_s\, dT_m\, d\mu\, dv\, dd_f, \ldots \tag{16.56}$$

Obviously, for practical application, the formula must be strongly simplified with respect to the combined probability density p. Usually, the multiple integration is substituted by a multiple sum of 'cases', i.e., combinations of the integration variables, estimating or ruling a probability of each case to occur. For the speed v, often only one or two values are selected: zero to cover the case of machine failure and (if higher speed is expected to give higher responses) the maximum speed the ship can attain, or is estimated to be commanded, taking account of forced and intentional speed reductions in a heavy seaway. For the wave angle μ, in most cases a uniform distribution between 0 and 2π is assumed, numerically approximated by seaway directions every 20 or 30°. For the combined probability density of H_s and T_m times $\Delta H_s\ \Delta T_m$ (the height and period range comprised to a single case), so-called scatter

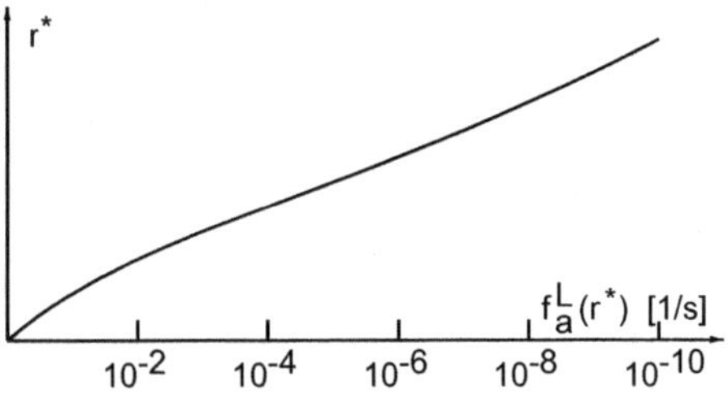

Fig. 16.6 Example for the relation between threshold value r^* and the long-term rate $f_a^L(r^*)$ that a response exceeds r^*

tables are used; an example is given in the chapter Water Waves as Table 3.1. Tables used for ships in worldwide service may be generated as a suitable mean over shipping routes; however, often a table for the North Atlantic is used, combined with a rate reduction factor because most time the ship is in less adverse ocean areas or a harbor. Further rate reductions are recommended to account for weather routing, harbor times, more favorable loading cases, etc.

Of the different load cases, sometimes only that with the maximum still-water bending moment is taken into account if one is interested in bending moment; or that having minimum metacentric height, if large roll angles may cause problems; or that with maximum metacentric height, if large transverse accelerations in beam sea may become dangerous.

Figure 16.6 shows the typical relationship between the long-term exceedance rate f_a^L and the threshold value r^*. A rate of 10^{-9} Hz corresponds to a 'return period' (i.e., average time between successive exceedances) $= 1/f_a^L = 10^9\,\text{s} = 31.6$ years, which roughly corresponds to the expected lifetime of a ship. However, whether an accepted exceedance rate is adequate does not, or at most slightly, depend on whether a ship is substituted by another one after 1 or 100 years.

The above formulas for $f_a^L(r^*)$ are applicable both to linearly and nonlinearly determined short-term exceedance rates f_a. In case of rates f_a which have been determined by a strongly simplified (e.g., a linear) method A, one may correct the design value r^* for this exceedance rate using the 'design wave concept'. That means that one uses method A to determine the amplitude of a regular wave that produces a response amplitude of size r^*. Wavelength and direction are those where the magnitude of the transfer function Y of the response r is maximum. The ship is then simulated in this 'design wave' using a more accurate method B, giving another response amplitude r^{**}. It is then assumed that the exceedance rate of r^* using method A is the same as that which would have been found for design value r^{**} using method B. This concept is often used for extreme wave loads. To our knowledge, a validation of this intuitively developed concept is still lacking.

References

1. E. Tonguc, H. Söding, Computing capsizing frequencies of ships in a seaway, in *3rd Conference on Stability of Ships and Ocean Vehicles (STAB'86)* (1986)
2. H. Söding, Method ROLLS for simulating roll motions of ships. Ship Technol. Res. **60/2**, 70–85 (2013)
3. V. Shigunov, Direct counting method and its validation, in *International Ship Stability Workshop Helsinki* (2019)

Chapter 17
Miscellaneous Topics

Abstract In this chapter 'Miscellaneous topics', the following is shortly dealt with

- A frequently-used method for simulating roll motions of ships, which treats surge, roll, and the coupling of roll with the other four kinds of motion nonlinearly, while these other motions themselves are linearized, using pre-calculated transfer functions. The method uses the extremely fast method by Grim to determine the heel restoring moment in waves from previously computed results.
- The above method is extended to damaged ships by simulating the in- and outflow of water through openings and the water motion within ship compartments and on deck, using special methods for (a) extremely shallow and (b) deeper flooded compartments.
- Nonlinear motions of planing boats in waves are simulated by extensions of the classical Wagner method developed already about 1930 for starting and landing sea planes.
- Mathematical quantities ('perturbators') and their operations are constructed to simplify both the derivation and the programming of second-order formulas, e.g., drift forces and springing excitation.
- For passenger transport, often the occurrence of sea sickness must be reduced as much as possible. An ISO norm for estimating the frequency of occurrence is sketched and evaluated for normal catamarans and those with longitudinally staggered hulls ('weinblums'). The weinblums have not only less resistance, but are also less prone to generate sea sickness.

17.1 Simulating Nonlinear Roll Motions

On November 11, 1981, the containership E.L.M.A. Tres capsized in a hurricane, causing the death of 23 crew members. To investigate the case, the program ROLLS (roll simulation; [1, 2]) was developed, which—with various extensions and modifications—is being used until today for predicting roll motions of ships in a natural seaway. The method simulates the nonlinear surge and roll motion, including the coupling with the other four rigid-body motions. The sway, heave, pitch, and yaw motions are treated linearly, using transfer functions derived from a strip method.

B. O. el Moctar et al., *Numerical Methods for Seakeeping Problems*,
https://doi.org/10.1007/978-3-030-62561-0_17

Also, the wave-induced roll exciting moment is linearized and determined using a transfer function. This concept leads to the following roll motion equation:

$$M_x = [\Theta_{xx} - \Theta_{xz}(\psi \sin\varphi + \theta\cos\varphi)]\ddot{\varphi} + \Theta_{xz}[(\ddot{\theta} + \dot{\varphi}^2\theta)\sin\varphi - (\ddot{\psi} + \dot{\varphi}^2\psi)\cos\varphi]. \quad (17.1)$$

Here, M_x is the x component of the moment around the center of gravity G of ship mass m, and Θ_x and Θ_{xz} are the mass moments of inertia around G.

In case of severe rolling, the largest contribution to M_x is the restoring moment, caused mainly by the hydrostatic pressure increase with z. Under static conditions, this moment divided by gm is the righting arm $h(\varphi)$. In a seaway, instead of the gravity acceleration g, the expression $g - \ddot{z}$ should be used to determine the righting moment, where z is the heave motion at G. Further contributions to M_x stem from the wind, the roll damping, the roll exciting moment caused by the waves, and the moment due to sway and yaw motions of the ship. The latter two contributions are determined from transfer functions computed using a strip method, assuming a linear dependence on wave height.

Most important for ship rolling is the change of $h(\varphi)$ with the wave contour along with the ship; this change is responsible for the so-called parametric roll excitation. To approximate it, an elegant concept of Grim [3] is used. The wave contour along the ship length is approximated, using the least-squares method, as

$$\zeta(x,t) = a(t) + b(t)x + c(t)\sin(kx) \quad (17.2)$$

with a fixed wave number k. It is easy to determine the transfer functions of a, b, and c; thus, these parameters can be simulated along with the linearized ship motions. To accelerate the simulation, righting arm curves for the ship under the wave according to (17.2) are computed before the simulation and stored depending on draft, trim, and wave amplitude c, assuming a static pressure increase below the wave contour ζ. During the simulation, h is interpolated from the stored data. This makes ROLLS extremely fast compared to other methods of comparable accuracy, allowing numerous and long-during simulations.

To validate the method, parametric roll motions in head regular waves were measured in model experiments and computed using ROLLS and another simulation method SIMBEL for a containership ($L_{pp} = 317$ m, $GM = 1.26$ m; see Fig. 17.1). Predominantly, cases near the most intense 'parametric resonance' were tested, i.e., cases, where the wave encounter frequency is two times the roll eigenfrequency. For this frequency ratio, the deceleration of roll velocity $\dot{\varphi}$ during the increase of φ from zero to φ_{max} may be small because the ship is near a wave crest, where righting arms are small; during the return to $\varphi = 0$ the negative acceleration of roll velocity may be large because then the ship is near a wave trough; thus, the ship returns to $\varphi = 0$ with a higher negative roll speed than it started therewith positive $\dot{\varphi}$. Thus, roll amplitudes increase, until either quadratic roll damping becomes dominant, or the change of roll eigenfrequency with roll amplitude changes the frequency ratio substantially from 2.

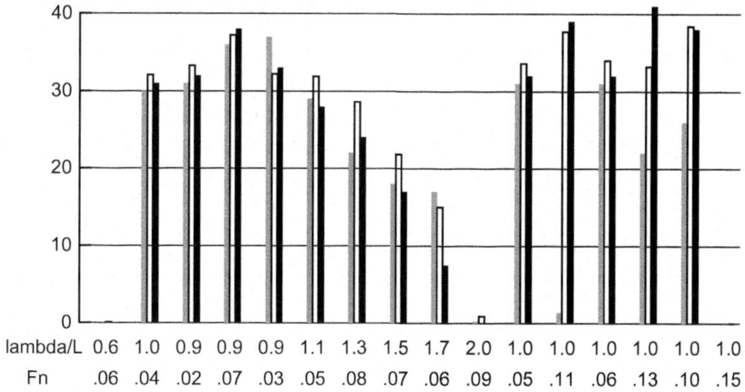

Fig. 17.1 Maximum heel angle in regular head waves ($\mu = 175$–$178°$) for various wave lengths λ and Froude numbers F_n. Gray: model experiments, white program SIMBEL, black program rolls [2]

In the experiments and in simulations of Fig. 17.1, slight deviations from $\mu = 180°$ were used to accelerate the beginning of parametric roll motions. In the case $\lambda = L_{pp}$, $F_n = 0.11$, both computations showed intense rolling, while it was lacking in the experiment. In all other cases, both computations and the experiment coincided in showing the presence or absence of parametric rolling, and even roll amplitudes were found similar in most cases in spite of their high sensitivity to estimated parameters like nonlinear roll damping. Also, other validations came to the conclusion that the method is well suited to predict wave-induced roll motions of ships.

17.2 Simulating Damaged Ships in a Seaway

In case of collision, grounding, structural failure, or unclosed openings, water may flow into ship compartments and/or on deck. Hydrostatic computations are routinely applied to determine the equilibrium floating position if it exists. Then it should be estimated whether this condition is safe in an assumed seaway, or whether, to which extent, and how fast progressive flooding and/or capsizing will occur.

To tackle such problems, the above-described program ROLLS was extended to ships containing compartments partly filled with water [4, 5]. The compartments may contain openings between each other and to outer space. The water flow through the openings is determined using empirical relations of hydraulic engineering, based on the difference of water surface height between inside and outside the openings.

To model the water flow in each compartment, two cases are distinguished. Deep-water compartments have the lowest sloshing eigenfrequency, which is substantially higher than the eigenfrequency of ship rolling and the peak encounter frequency of the waves. In these compartments, the water surface is assumed plane, but it may be inclined longitudinally and transversely. The motion of the water in such compartments is determined like that of a mass point moving on the surface on which the center or gravity of fluid mass is located for arbitrary surface inclinations. The fluid motion is excited by the tank accelerations and rotations. The main effect of such compartments is to reduce the righting moments in the case of heel.

Shallow water compartments, on the other hand, are those in which the smallest sloshing eigenfrequency is lower because of a small filling depth. They do not only reduce the righting moments, but they may also increase roll damping substantially (like roll-damping tanks). The fluid flow in these compartments is determined using a cartesian two-dimensional grid of cells extended in the ship-fixed horizontal plane. Fluid velocities are assumed parallel to the ship's decks and independent of vertical coordinates. The two velocity components and the filling height in each cell follow from mass and momentum conservation equations.

To solve this shallow-water equation system is not straightforward if the maximum flow velocity exceeds the 'surge velocity' $v_s = \sqrt{gh}$, where g is gravity acceleration and h the local fluid depth. Usually, v_s is exceeded locally, e.g., because h approaches zero in some cells if parts of the compartment bottom are dry. The flow has then similarity to supersonic flows in that jumps of surface height occur, accompanied by a jump in velocity from $>v_s$ to $<v_s$. In [4], the Glimm method is used. It uses analytical solutions of the 'dam break problem' (the flow occurring after removing a barrier between stagnant fluids of different surface height) applied at each cell

Table 17.1 Minimum metacentric height GM_D [m] to avoid capsizing [5]

Draft of intact ship [m]	$H_s = 4$ m, $T_p = 8$ s				$H_s = 4$ m, $T_p = 12$ s	
	Damage case A		Damage case B		Damage case A	
	Test	Simulation	Test	Simulation	Test	Simulation
	No subdivision on vehicle deck; no initial heel					
5.9	2.54	2.60	3.28	3.25	2.32	2.45
6.2	3.55	3.65	4.09	4.14	–	3.33
6.4	–	–	4.30	4.50	–	–
	No subdivision on vehicle deck; initial heel $-1°$					
5.9	3.20	2.95	–	–	2.76	2.80
	With longitudinal bulkheads on vehicle deck					
5.9	1.53	1.67	1.90	1.69	–	1.62
6.2	1.82	2.22	2.36	1.90	–	2.17
6.4	–	–	2.36	2.00	–	–
	With 2 transverse bulkheads on vehicle deck					
5.9	1.38	1.34	1.78	1.91	–	1.29
6.2	1.51	1.57	2.18	2.28	–	1.57
6.4	–	–	2.24	2.46	–	–

boundary, in combination with a random sampling of these solutions for constructing an approximation for the next time step.

For validation, in [5], results of model experiments and simulations are compared. As an example, Table 17.1 shows metacentric height GM_D in damaged condition required to survive two damage cases of a roro ship in two stationary seaways for different initial drafts. Regarding the complexity of the problem for experiments and simulations and the sensitivity of the results (see the case of initial heel $-1°$), the similarity of measured and computed results compared favorably.

17.3 Simulating Planing Boats

In planing boats which typically, operate at Froude numbers F_n exceeding unity, other fluid force components are dominant than in the low-speed cases treated above, the combined longitudinal and downward flow of the water along the inclined hull bottom. Whereas field methods solving the Euler or RANS equations can be applied to cases of large and small F_n, potential flow methods applied to both cases are different. Because spray is generated along the waterline. In low-speed ships, it is simply a nuisance to be skipped, whereas in high-speed craft the momentum of the fluid in the spray corresponds to the major part of the vertical force.

Here, the 'modified Wagner method' [6] will be sketched. Wagner [7] proposed the following formula for the fluid force per length f (positive downward) on one half of a symmetric prismatic body with horizontal axis moving downward with velocity V into the water:

$$f = -\frac{d}{dt}\left(\rho\frac{\pi}{4}c^2 V\right). \tag{17.3}$$

Figure 17.2 shows the definition of waterline half breadth c. The factor $\rho\frac{\pi}{4}c^2$ approximates the added mass per length of the half cross section up to the spray root. Important is the pile-up of the water surface beside the body; according to the Wagner theory, it increases the force on a circular cross section by the factor 2, and for a wedge (triangle) by the factor $\pi^2/4 \approx 2.5$.

For application to planing boats, the formula is modified:

- For larger ratios T/c, the actual force is smaller.

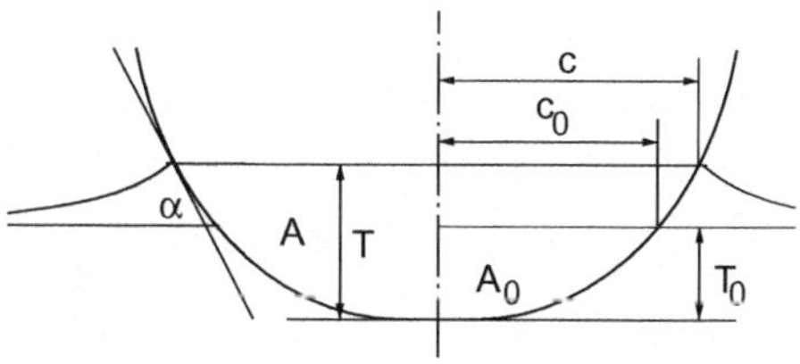

Fig. 17.2 Cross section definitions for Wagner formula (17.3). T_0 is section draft below the undisturbed water surface

- The pile-up is approximated for arbitrary section shapes characterized by area A, draft T, and c.
- For application to craft, where cross sections and V vary with longitudinal coordinate x, the time derivative d/dt is substituted by $\partial/\partial t - u\partial/\partial x$, where u is forward speed.
- In sections with immersed knuckles at which the flow separates, a modification depending on angle α below the knuckle is made.
- The formula must be changed in case of negative V.
- The hydrostatic force $-\rho g A_0$ is added.
- Ventilation (air drawing from the surface in case of negative pressure) is roughly accounted for.

For oblique waves and horizontal motions, both half sections are treated separately according to this scheme, and approximations of horizontal force and heel moment are added.

A small model $L_{oa} = 0.625$ m of a planing boat was towed in regular head waves [8] at the unusually high Froude number $F_n = 3.63$. The model was free only with respect to heave (at its center of gravity) and pitch. Corresponding simulations [6] were made using the above-described method as well as two different field (RANSE) methods. For lower wave amplitudes, the model moved with wave encounter frequency ω_e, but in higher waves with $\frac{1}{2}\omega_e$. In the highest waves tested, the model jumped out of the water, passed the next wave still in air, and entered the water at the following wave. (In still higher waves or for higher speed, the responses would become periodic with smaller ratios ω/ω_e, and finally would become chaotic.)

For small wave amplitudes, measured and all three simulated motions coincided reasonably, but for larger wave amplitude severe differences were found; Fig. 17.3 shows an example. Reasons for the extremely different responses in higher waves are:

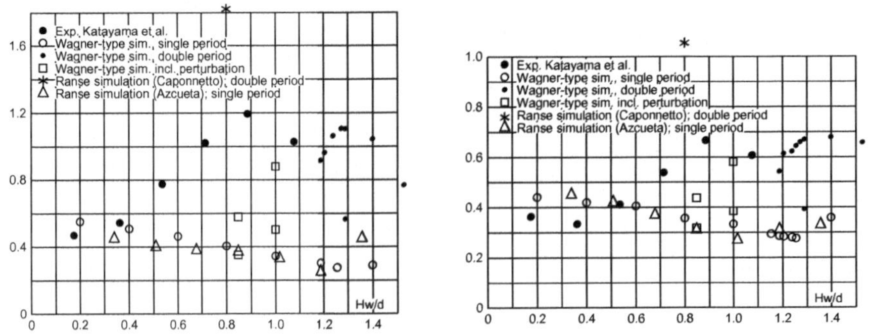

Fig. 17.3 Non-dimensional heave (left) and pitch (right) double amplitude in regular head waves of length = 2.49 times boat length 0.625 m for various wave heights H_w. d = draft at rest = 0.059 m. Froude number $F_n = 3.63$ from [6]; test results from [8]

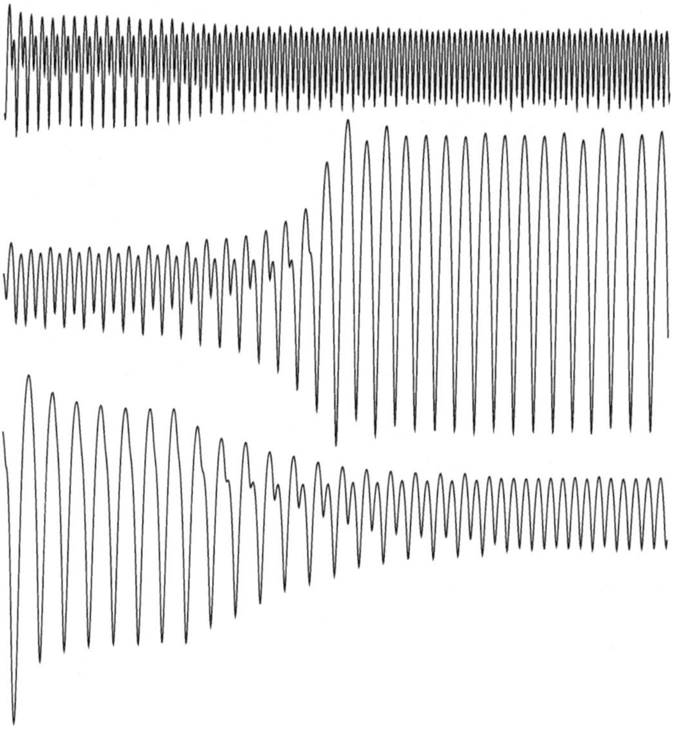

Fig. 17.4 Simulated heave motions in waves as in Fig. 17.3; $H_w = 1.4d$ (top), $1.28d$ (middle) and $1.2d$ (bottom curve), where d = draft at rest (5.9 cm). Curves have different time scales

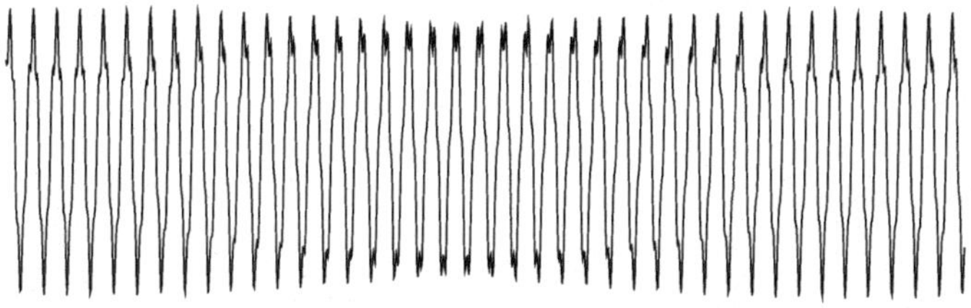

Fig. 17.5 Wave track with perturbation

- Often the final periodic response was attained only after very long simulation times, which could not be realized in the experiments and sometimes not even in RANS calculations (Fig. 17.4).
- Small to moderate disturbances of the exciting wave (Fig. 17.5), which are difficult to avoid in experiments, may cause totally different responses. This is shown by circles and squares in Fig. 17.3 and is accompanied by a change between response frequency ω_e and $\frac{1}{2}\omega_e$.

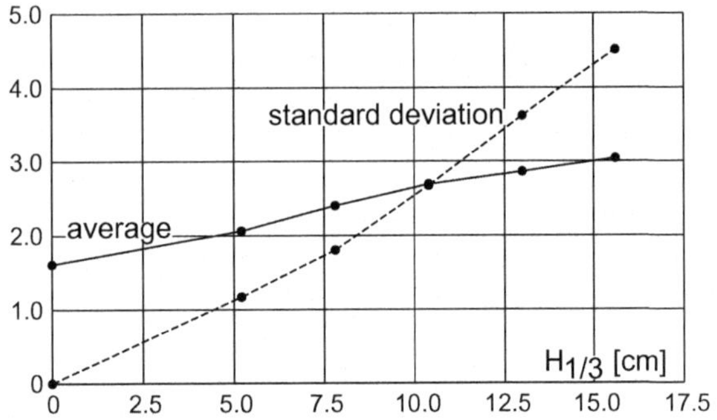

Fig. 17.6 Average and standard deviation of trim angle (in degree) depending on significant wave height $H_{1/3}$

- In some cases, the final response depended on the procedure to start the simulation (note the two square symbols for $H_w/d = 0.85$ and 1.0) This start procedure differed between experiments and the three simulations. (In simulations using the modified Wagner method, in most cases, the model was dropped with full speed from above the water.)

It may be concluded that, if nonlinear effects are strong, validation using periodic waves, or more generally periodic excitation, may be inconclusive because periodic excitation leads to various peculiar effects, which depend discontinuously on parameters of the process and which are known also from many other nonlinear systems. Stochastic excitation, on the other hand, does not only correspond more closely to reality; averages and variances generated by stochastic excitation lack also such discontinuous dependencies. This is demonstrated for the present case by Fig. 17.6. However, due to the higher simulation duration required for conclusive simulations in irregular waves, no results from other methods are available.

17.4 Perturbators

In previous chapters concerned with linear wave responses, we used complex amplitudes instead of harmonic time functions because this is simpler, and complex amplitudes allow not only generating the time functions but also determine the probability distribution of responses in a natural seaway. Perturbators extend that concept to second-order (and, if required, to higher order) responses [9, 10]. But, whereas in first-order responses the simplification by using complex amplitudes is marginal, second-order perturbators cause substantial simplifications in deriving and programming formulas for boundary conditions, flow potentials, and all kinds of wave responses up to second order. Chapter 14, which deals with drift forces, is an example where second-order contributions were discussed. Using perturbators would have simplified that chapter considerably.

17.4.1 One-Frequency Second-Order Perturbators

Consider a real, scalar quantity $a(t)$ oscillating harmonically over time t with angular frequency ω:

$$a(t) = a_0 + \mathrm{Re}(a_1 e^{i\omega t}), \tag{17.4}$$

where a_0 is a real constant, and a_1 a complex constant called complex amplitude of a. a_0 and a_1 are considered as quantities of zeroth and first orders, respectively.

A corresponding time dependence is assumed to hold for another quantity b:

$$b(t) = b_0 + \mathrm{Re}(b_1 e^{i\omega t}). \tag{17.5}$$

Here, a and b are supposed to oscillate with the same frequency ω; for a moving ship, ω is the encounter frequency with a regular wave.

The product ab of functions (17.4) and (17.5) can be written in the form

$$c(t) = a(t)b(t) = \underbrace{a_0 b_0}_{c_0} + \mathrm{Re}[\underbrace{(a_0 b_1 + a_1 b_0)}_{c_1} e^{i\omega t}] + \underbrace{\frac{1}{2}\mathrm{Re}(a_1 b_1^*)}_{c_2} + \mathrm{Re}\left(\underbrace{\frac{1}{2} a_1 b_1}_{c3} e^{2i\omega t}\right), \tag{17.6}$$

where the asterisk designates the complex conjugate. c_0 and c_2 are real, c_1 and c_3 complex. The right-hand side shows the general form of time functions excited by regular waves in second-order approximations: They contain constant terms of zeroth (c_0) and second (c_2) orders, a first-order term oscillating with frequency ω, and a second-order term oscillating with frequency 2ω. For this case, the perturbator of $c(t)$ is defined as the ordered set of c_0, c_1, c_2 and c_3:

$$\mathbf{c} =< c_0, c_1, c_2, c_3 > . \tag{17.7}$$

The time function can be constructed from the perturbator (compare (17.6)) as

$$c(t) = c_0 + \mathrm{Re}(c_1 e^{i\omega t}) + c_2 + \mathrm{Re}(c_3 e^{2i\omega t}). \tag{17.8}$$

Addition, subtraction, multiplication, and division are defined for perturbators such that they correspond to these operations applied to the associated time functions. Thus, addition and subtraction are defined elementwise like in vectors. Multiplication of two perturbators $\mathbf{a} =< a_0, a_1, a_2, a_3 >$ and $\mathbf{b} =< b_0, b_1, b_2, b_3 >$ is defined (compare (17.6)) as

$$\mathbf{c} = \mathbf{ab} = \left\langle \underbrace{a_0 b_0}_{c_0}, \underbrace{a_0 b_1 + a_1 b_0}_{c_1}, \underbrace{a_0 b_2 + a_2 b_0 + \frac{1}{2}\mathrm{Re}(a_1 b_1^*)}_{c_2}, \underbrace{a_0 b_3 + a_3 b_0 + \frac{1}{2} a_1 b_1}_{c_3} \right\rangle . \tag{17.9}$$

The usual rules for addition and multiplication of numbers (existence of inverse, unit element, associative, commutative, distributive law) hold also for (scalar) perturbators. Further, multiplication with constants and an operation which corresponds to taking the time derivative of the associated time functions are defined. The latter operation is designated by $^\bullet$, its definition is

$$\dot{c} =< 0, i\omega c_1, 0, 2i\omega c_3 > . \tag{17.10}$$

In a program, these operations must be defined only once for application to perturbators of this type.

Instead of scalars c_0 to c_3, also vectors and matrices can be used. Depending on the type of elements, instead of products of numbers, also scalar products of vectors, vector products of three-dimensional vectors, and matrix products of perturbators are defined exactly corresponding to (17.9). This allows to write equations for time functions directly as corresponding equations for perturbators.

For example, in [9] the following free-surface boundary condition is to be satisfied by a suitable set of source strength q_j:

$$\sum_j \left[-G(\mathbf{x}^I - \mathbf{x}_j)\ddot{q}_j + gG_z(\mathbf{x}^I - \mathbf{x}_J)q_j\right] =$$

$$-g\alpha^I \phi_{dy} - \frac{\partial^2}{\partial t^2}(\beta\nabla\phi_d) + g\beta\nabla\phi_{dz} + \frac{1}{g}\frac{\partial}{\partial t}(\dot{\phi}\dot{\phi}_z) + \nabla\phi\nabla\dot{\phi} + \frac{\partial}{\partial y}(\dot{\phi}\phi_y). \tag{17.11}$$

Here, $G_z(\mathbf{x}^I - \mathbf{x}_J)$ are time-independent Green functions, g is gravity acceleration, α^I is a time-dependent scalar, β is a time-dependent vector, and ϕ is the scalar time-dependent potential. In the corresponding perturbator equation, all time-dependent quantities have been substituted by perturbators (printed boldface):

$$\sum_j \left[-G(\mathbf{x}^I - \mathbf{x}_j)\ddot{\mathbf{q}}_j + gG_z(\mathbf{x}^I - \mathbf{x}_J)\mathbf{q}_j\right] =$$

$$-g\mathbf{a}^I\mathbf{\Phi}_{dy} - (\mathbf{b}\nabla\mathbf{\Phi}_d)^{\bullet\bullet} + g\mathbf{b}\nabla\mathbf{\Phi}_{dz} + \frac{1}{g}(\dot{\mathbf{\Phi}}\dot{\mathbf{\Phi}}_z)^\bullet + \nabla\mathbf{\Phi}\nabla\dot{\mathbf{\Phi}} + \frac{\partial}{\partial y}(\dot{\mathbf{\Phi}}\mathbf{\Phi}_y). \tag{17.12}$$

Solving the equation system for the perturbators $\mathbf{q}_j$ of source strengths is done by successively solving real and complex systems for the elements q_0 to q_3; however, one or two of these systems may be skipped if not all elements of $\mathbf{q}$ are of interest. Establishing the coefficient matrices and the right-hand sides of the four systems together by using the arithmetic of perturbators is simpler than transforming the condition to real and complex variables. The same holds for the perturbators $\mathbf{a}^\mathbf{I}$ and $\mathbf{b}$ occurring in (17.12), and for the following computations of forces and motions.

The right-hand side of (17.12) is of second order; thus, it is zero in the systems for q_0 and q_1, and it requires only quantities up to first order for the remaining two equation systems; therefore, no iteration is required to solve (17.12). The results given in Figs. 5.2, 5.3 and 5.5 were computed using this concept.

17.4.2 Two-Frequency Second-Order Perturbators

Computing responses up to second order in irregular waves, we need a more general type of perturbators, which takes into account interactions between pairs of two regular waves having encounter frequencies ω_1 and ω_2. If we include all second-order responses of a pair of regular waves, the time function of the response is

$$a(t) = a_0 + \mathrm{Re}(a_1 e^{i\omega_1 t}) + \mathrm{Re}(a_2 e^{i\omega_2 t}) + \mathrm{Re}(a_3 e^{i(\omega_1-\omega_2)t}) + \mathrm{Re}(a_4 e^{i(\omega_1+\omega_2)t}) + \mathrm{Re}(a_5 e^{2i\omega_1 t}) + \mathrm{Re}(a_6 e^{2i\omega_2 t}). \tag{17.13}$$

In most cases, one is interested only in some of these terms. For example, for springing excitation one is interested in the term oscillating with the sum frequency $\omega_1 + \omega_2$, whereas for low-frequency oscillations of a moored ship, only oscillations with the difference frequency $\omega_1 - \omega_2$ are of interest. In such cases, second-order terms corresponding to oscillations in which one is not interested should be omitted from the definition of perturbators to reduce the program complexity and the computing time. In the most general case, on the other hand, the second-order perturbator is

$$\mathbf{a} =< a_0, a_1, a_2, a_3, a_4, a_5, a_6 > \tag{17.14}$$

with a real element a_0 and complex elements a_1 to a_6. Note that here only one element a_0 corresponding to the time-independent term in (17.13) is used; it contains contributions of order 0 and 2.

For this type of perturbator, the product rule is, as before, defined to correspond to the product of two time functions $a(t)$ and $b(t)$. This results in

$$\mathbf{ab} =< a_0 b_0 + \frac{1}{2}\mathrm{Re}(a_1 b_1^* + a_2 b_2^*), \quad a_0 b_1 + a_1 b_0, \quad a_0 b_2 + a_2 b_0,$$

$$a_0 b_3 + a_3 b_0 + \frac{1}{2}(a_1 b_2^* + b_1 a_2^*), \quad a_0 b_4 + a_4 b_0 + \frac{1}{2}(a_1 b_2 + b_1 a_2), \quad a_0 b_5 + b_0 a_5, \quad a_0 b_6 + b_0 a_6 > . \tag{17.15}$$

The dot operator, which corresponds to a partial time derivative of the corresponding time function, is defined here (compare (17.13)) as

$$\dot{\mathbf{a}} =< 0, \quad i\omega_1 a_1, \quad i\omega_2 a_2, \quad i(\omega_1 - \omega_2)a_3, \quad i(\omega_1 + \omega_2)a_4, \quad 2i\omega_1 a_5, \quad 2i\omega_2 a_6 > . \tag{17.16}$$

17.5 Seakeeping of Catamarans and Weinblums

For medium and high ship speeds, catamarans (twin-hull ships) and trimarans (three hulls) are often preferred to single hulls because of lower resistance and possibly better seakeeping behavior. Special types of twin hulls are

- SWATH (small waterplane area twin hulls), in which the small waterline area reduces the wave excitation force and moment and, thus, ship motions except in very long waves; and
- Weinblums [11], in which both hulls are staggered (see Fig. 17.8). This leads to partially canceling of stationary transverse waves generated by both hulls in case of forward speed, thus reducing ship wave resistance. (Weinblums, which are not yet realized, are called with reference to the ship hydrodynamicist Georg Weinblum, 1897–1974.)

Starting from an existing ferry design (Fig. 17.7, Table 17.2), various alternative hull arrangements (Fig. 17.8) were investigated. The trimarans are unusual in that they

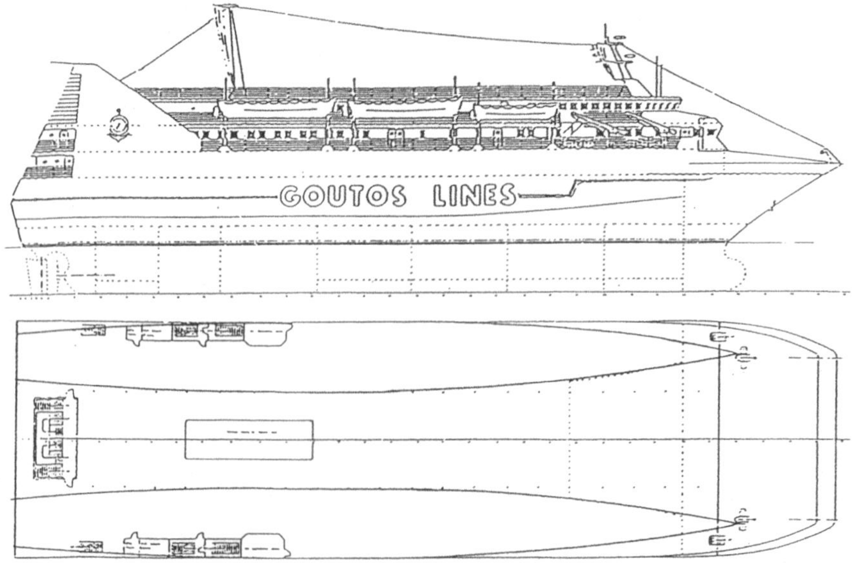

Fig. 17.7 Catamaran ferry used as reference. *Source*: A. Papanikolaou

Table 17.2 Main particulars of the ferry used as reference

Length over all	77.85 m	Length b.perp.	65.00 m	Hull centerline spacing	15.00 m
Design draft	5.10 m	Deadweight	458 t	Design speed	20 kn

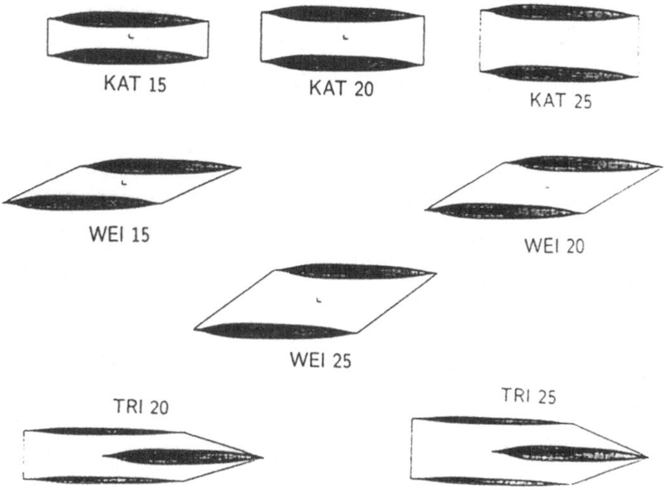

Fig. 17.8 Investigated hull arrangements

use long, narrow, backward staggered side hulls to reduce the trimaran resistance like in weinblums. Figure 17.9 shows that the principle of staggered hulls can result in a drastic resistance decrease compared to symmetrical catamarans. The resistance was computed using the stationary Rankine source method described in Sect. 7.2; some of the results were validated by model experiments.

Naturally, whether the unsymmetrical hull arrangement in Weinblums is acceptable depends on their seakeeping performance. It was tested using the strip method,

Fig. 17.9 Total resistance depending on speed

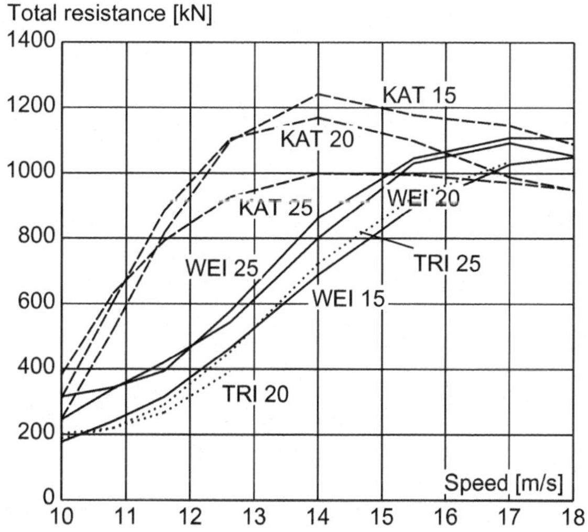

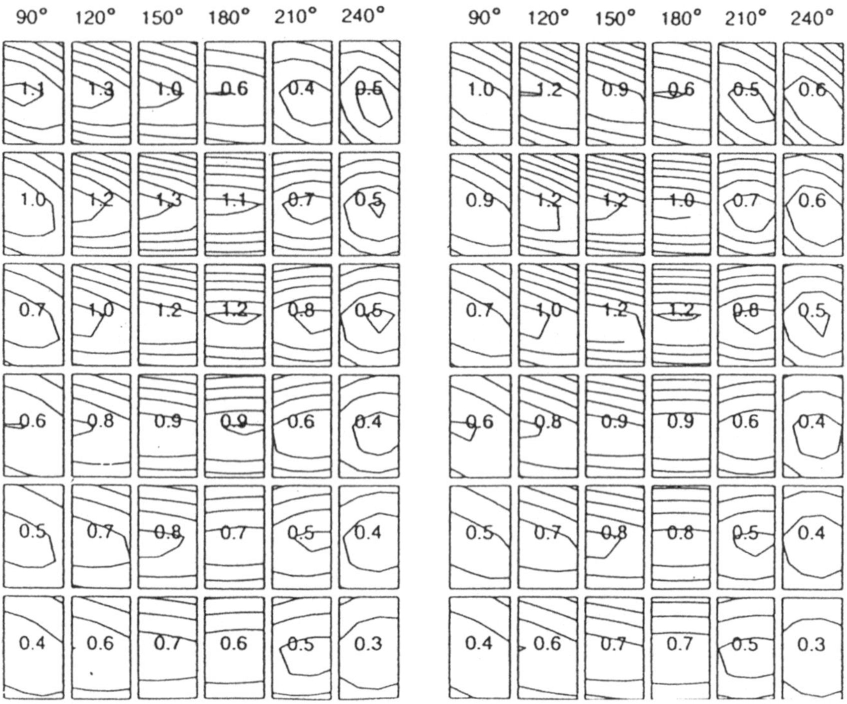

Fig. 17.10 Weighted root mean square vertical acceleration at different ship positions in seaways of $H_{1/3} = 1$ m for seaways from different directions (given on top of each column) for peak periods $T_p = 5$ s (top row), 6 s, ..., 10 s (bottom row); left for WEI 20, right for WEI 25

including hull interaction, as described in Sect. 5.4. It turned out that the Weinblum WEI 15 (see Fig. 17.8), which has a transverse hull distance of 15 m like the reference ship, is rolling too much. For the larger hull distances 20 and 25 m, however, rolling is not much larger than for the reference catamaran.

For passenger comfort, more important than rolling is the low-frequency vertical acceleration because it is the main cause of sea sickness. The norm ISO 2631 states that, approximately, 10% of all passengers will suffer from severe motion sickness (vomiting) if the root mean square vertical acceleration exceeds 0.5 m/s^2 for 2 h. For different exposure time T the acceleration limit changes proportional to $T^{-\frac{1}{2}}$. Only frequencies $\leq$0.315 Hz must be taken into account directly; accelerations of higher frequency must be reduced by 5 dB (corresponding to a factor of 0.31) per octave when computing the weighted root mean square acceleration A.

A was determined in natural seaways of JONSWAP spectra with peak enhancement factor 3.3 for different wave angles and peak periods. Significant wave height was 1 m; thus actual A values must be multiplied by the actual significant wave height. As A depends on the position on deck, A was computed at different positions within a rectangle of 96 m length and 21 m breadth (same for all hull spacings)

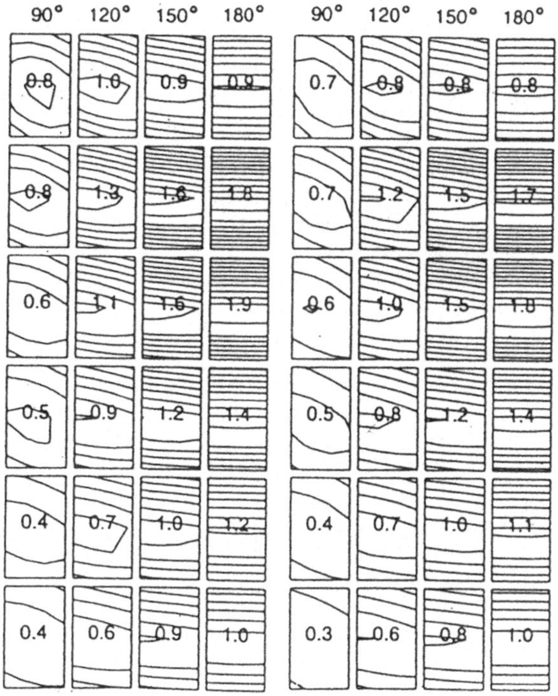

Fig. 17.11 Like Fig. 17.10, however, for KAT 20 (left) and KAT 25 (right)

centered at the midship section and midship plane. In this rectangle, A is given by isolines corresponding to the values 0.3, 0.6, 0.9, …, m/s^2. In Figs. 17.10 and 17.11, for each wave case this rectangle is plotted, with length and breadth scaled differently and forward direction plotted upward, together with the isolines of A. The minimum A value, which occurs normally a little behind midship section near midship plane, is also indicated.

The results show that the Weinblums are substantially better than the catamarans also with respect to the occurrence of sea sickness. Further results given in [11] show that the trimarans are, on average, somewhat more seakindly than the Weinblums in this respect.

References

1. P. Kröger, Roll simulations of ships in a seaway (in German). Ship Technol. Res. **33**(4), 187–216 (1986)
2. H. Söding, V. Shigunov, T. Zorn, P. Soukup, Method rolls for simulating roll motions of ships. Ship Technol. Res. **60**(4), 70–84 (2013)
3. O. Grim, Contribution to the problem of ship safety in a seaway (in German). Schiff und Hafen 490 (1961)

4. F. Petey, Numerical calculation of forces and moments due to fluid motion in tanks and damaged compartments, in *3rd Conference on Stability of Ships and Ocean Vehicles (STAB'86)* (1986)
5. B.-C. Chang, On the survivability of damaged Ro-Ro vessels using a simulation method. Technical report Nr. 597, University of Technology Hamburg (1999)
6. M. Caponnetto, H. Söding, R. Azcueta, Motion simulations for planing boats in waves. Ship Technol. Res. **50**(4) (2003)
7. H. Wagner, About slamming and planing on a fluid surface (in German). ZAMM **12** (1932)
8. T. Katayama, T. Hinami, Y. Ikeda, Longitudinal motion of a super high-speed planing craft in regular head waves, in *Proceedings of 4th Osaka Colloquium on Seakeeping Performance of Ships* (2000)
9. H. Söding, A method to simplify perturbation analyses of periodical flows. Eng. Marit. Environ. (2014)
10. H. Söding, Second-order seakeeping analyses using perturbators. Ship Technol. Res. **61**(1), 4–15 (2014)
11. H. Söding, Seakeeping of multihulls, in *Proceedings of High-Performance Marine Vehicles, Stellenbosch, South Africa* (1999)

Printed by Printforce, the Netherlands